The
PRINCETON
FIELD GUIDE *to*
PTEROSAURS

The PRINCETON FIELD GUIDE *to* PTEROSAURS

GREGORY S. PAUL

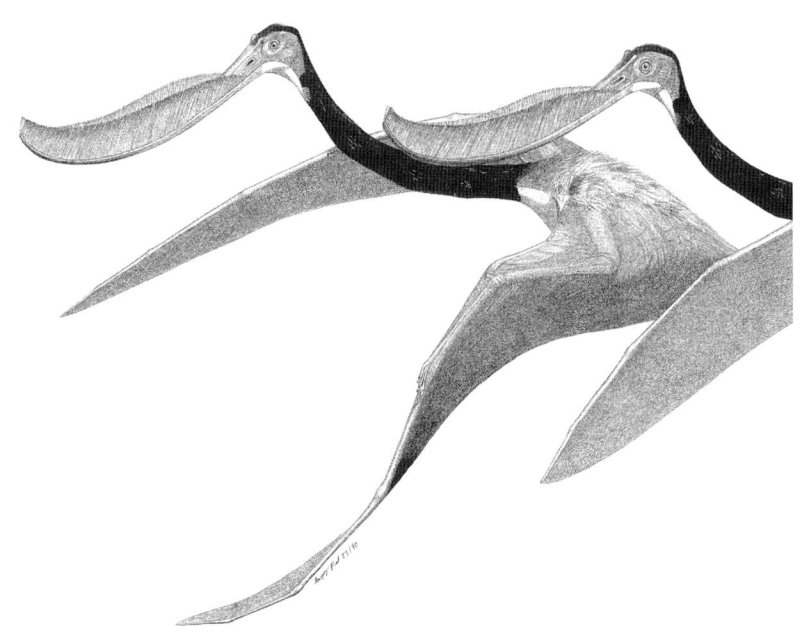

Princeton University Press

Princeton and Oxford

Published by Princeton University Press
41 William Street, Princeton, New Jersey 08540
6 Oxford Street, Woodstock, Oxfordshire OX20 1TR

press.princeton.edu

Library of Congress Cataloging-in-Publication Data

Names: Paul, Gregory S., author.
Title: The princeton field guide to pterosaurs / Gregory S. Paul.
Description: Princeton : Princeton University Press, [2022] | Series: Princeton field guides | Includes bibliographical references and index.
Identifiers: LCCN 2021015050 (print) | LCCN 2021015051 (ebook) | ISBN 9780691180175 (hardback) | ISBN 9780691232218 (ebook)
Subjects: LCSH: Pterosauria—Identification.
Classification: LCC QE862.P7 P38 2022 (print) | LCC QE862.P7 (ebook) | DDC 567.918–dc23
LC record available at https://lccn.loc.gov/2021015050
LC ebook record available at https://lccn.loc.gov/2021015051

British Library Cataloging-in-Publication Data is available

Editorial: Robert Kirk and Abigail Johnson
Production Editorial: Kathleen Cioffi
Jacket Design: Layla Mac Rory
Production: Steven Sears
Publicity: Matthew Taylor and Caitlyn Robson
Copyeditor: Laurel Anderton

This book has been composed in Goudy Old Style (introduction), ITC Galliard Pro (species section and headings), and Optima (labels)

Printed on acid-free paper. ∞

Typeset and designed by D & N Publishing, Wiltshire, UK
Printed in Italy

10 9 8 7 6 5 4 3 2 1

CONTENTS

PREFACE

If I were, at about age twenty as a budding paleoresearcher and artist, handed a copy of this book by a mysterious time traveler, I would have been startled as well as delighted. The pages would have revealed a world of new pterosaurs and ideas that I had only a hint of, if any such ideas existed at all. My head would have spun at the revelation that *Pteranodon* was not, as had been calculated, the largest flying creature possible, but that *Quetzalcoatlus* was even more colossal, with the wingspan of a small human-carrying glider, and a weight exceeding that of land-bound ostriches manyfold. Visually pleasing would have been the array of pterosaurs bearing fantastical head crests, including *Tupuxuara*, *Thalassodromeus*, *Shenzhoupterus*, *Tapejara*, *Sinopterus*, *Tupandactylus*, and most especially the superelongated, psychedelic prongs of *Nyctosaurus*, which had long been thought to be dully crestless. Equally surreal would have been the hyperelongated, superslender head of *Moganopterus*, which looks like it is out of an avant-garde Looney Tunes cartoon. Then there would have been the weirdly twisted beak of *Cycnorhamphus*. Plus the novel formations, at least to my eyes and ears, of Dolomia di Forni, Nugget Sandstone, Zorzino, Tiaojishan, Yixian, Jiufotang, Tangshang, Toolebuc, Bissekty, Densus Ciula, Csebanya, Jagu, Santana, Crato, and Javelina. The sheer number of new pterosaurs would tell that an explosion in pterosaur discoveries and research, well beyond anything that had previously occurred, and often based on new high technologies, marked the end of the twentieth century going into the twenty-first.

Confirmed would be the long-held speculation that pterosaurs—often called pterodactyls by the public (but those are just a portion of the group)—were not reptiles limited to gliding and weakly powered flight, but highly energetic fliers like their distant archosaur relatives the birds, as well as mammalian bats. And, like bats, pterosaurs were covered with fuzz. How pterosaurs got around when not in the air has long been contentious—trackways have shown that they often walked on all fours like bats, but the situation is otherwise complicated. Pterosaurs are often seen as living in warm Mesozoic climes, but some at least visited chilly polar regions and high altitudes.

Remaining frustratingly unresolved are the origins of pterosaurs. Unlike with birds, whose descent from advanced dinosaurs has become well understood in many respects, but like with bats, the early evolution of which remains little documented in the fossil record, we do not yet have a detailed understanding of which archosaur group pterosaurs evolved from, or how.

Producing this volume has been satisfying in that it has given me reason to illustrate the skeletons of almost all pterosaur species for which sufficiently complete material is available. These have been used to construct the most extensive library of side-view life studies of these archosaurs of the air to date in print. An advantage of producing large sets of rigorous profile-skeletals is that the restorations can reveal information that would not otherwise arise. One pertinent example, first appearing here, is the discovery of the differing head/body proportions of *Pteranodon*, one species of which was startlingly big beaked. The overall result is a work that covers what is already well over two centuries of scientific investigation into the group of tetrapods that inhabited the skies for up to 170 million years. Enjoy the travel back in time.

Acknowledgments

Many thanks to those who have provided the assistance over the years that has made this book possible, including Peter Wellnhofer, Wann Langston, Paul MacCready, Alec Brooks, Asier Larramendi, Kenneth Carpenter, Michael Brett-Surman, Frank Boothman, Brian Andres, Kevin Padian, David Peters, Michael Habib, Sandra Chapman, David Martill, Robert Telleria, and others. I would also like to thank all those who worked on this book for Princeton University Press: Robert Kirk, Abigail Johnson, Kathleen Cioffi, Layla Mac Rory, Steven Sears, and Namrita and David Price-Goodfellow.

INTRODUCTION

Anhanguera

HISTORY OF DISCOVERY AND RESEARCH

Pterosaur remains have been found by humans for millennia and may have helped form the basis for belief in mythical beasts, including winged dragons. In the prescientific West, the claim in the Genesis creation story that the planet and all life were formed just two or three thousand years before the great Egyptian pyramids were built hindered the scientific study of fossils. The discovery and subsequent detailed illustration of the virtually complete and articulated skull and skeleton of the small *Pterodactylus* found in fine-grained Lagerstätte sediments of Bavaria in the late 1700s began modern pterosaur paleozoology decades before the discovery of teeth and a few bones led to the recognition of land-bound dinosaurs in the 1820s. The strange pterodactyl skeleton posed a major problem for early science because it was still thought that major extinctions had not occurred. The creature had a hyperelongated single finger, which was radically different from the feathery airfoils of birds and also quite distinctive from the multifingered wings of bats. Nor did the very long, spike-toothed jaws belong to either group.

In the early 1800s, various researchers believed that the peculiar fossil, as well as more remains found in the Bavarian quarries that would later produce the feather-winged dinobird *Archaeopteryx*, belonged to flippered swimmers, or to mammalian bats, perhaps of a marsupial nature. However, in the first decade of the 1800s, the famed comparative anatomist Georges Cuvier realized that pterosaurs were a distinctive group that used the single elongated finger to support a flight membrane. The new German remains further showed that there were two major subgroups of pterosaurs—the long-tailed rhamphorhynchoids epitomized by *Rhamphorhynchus*, and the short-tailed pterodactyloids epitomized by *Pterodactylus*. English sediments began to produce pterosaurs, some found by the famed early field paleozoologist Mary Anning, and others described by Richard Owen, who decided that the ancient fliers were low-energy reptiles. In the later 1800s, Harvey Seeley became the leading pterosaur expert. He concluded that because pterosaurs had fully developed wings like birds and bats, they should have had similarly high metabolic rates and were therefore not classic reptiles. This led to a bitter dispute between Owen and Seeley. At about the same time, it was realized that fossil thecodonts, as well as crocodilians, dinosaurs, birds, and pterosaurs, are all archosaurs.

The pterosaurs coming out of Europe were generally small forms, often very well preserved, from the Jurassic period of the middle Mesozoic, but some less complete English fossils showed that Early Cretaceous pterodactyloids were of considerable size, competitive with and even exceeding the size of the largest living birds. What caused some head scratching was that these remains were too fragmentary to reveal the identity of what seemed like bizarre rounded snouts and other oddities. Pterosaurian gigantism was confirmed when the focus of pterosaur paleozoology shifted to the newly open plains of North America. In the deposits of the Late Cretaceous seaway that had then covered the area were discovered, first by Othniel Marsh and soon after by his equally notorious competitor Edward Cope, the often largely complete skulls and skeletons of the classic, largest known pterosaur—and future star of feature films—*Pteranodon*. With wings spanning 6.4 m (21 ft), it was a monster of the skies. At the turn of the next century, the much less enormous *Nyctosaurus* was found in the same beds, but unfortunately, those specimens did not include its amazing head crest.

After the loss of Seeley early in the 1900s, the new century saw an extended slowdown in the paleozoology of pterosaurs. Good specimens continued to emerge from Europe, especially from the Bavarian quarries, but nothing particularly novel. The same was true for the American plains, where *Pteranodon* specimens with broader, more erect crests than the better-known, more backward-bladed structures showed up. The onset of the age of aviation did inspire a pre–Great War study on the aerodynamics of the big pterosaurs, which were not all that much smaller than the first airplanes. And in the 1920s, Tilly Edinger did innovative work on the avian-like structure of pterosaur brains. In the 1950s a fossil trackway was attributed to a pterosaur, but this assignment was disputed and little was done with it. Overall, remarkably little effort was expended on better understanding of pterosaur biology and flight, and a similar disturbing lack of vigorous science afflicted the study of dinosaurs during the same period. Like dinosaurs, pterosaurs were seen as an extinct group, good for getting crowds through museum doors, but not of great biological importance. And their origins remained vexingly obscure. The two superwars and grim economic troubles did not help matters.

Again in parallel to dinosaur science, pterosaur science started to revive in the 1960s and 1970s. In China, despite extreme political strife, new pterosaurs such as *Dsungaripterus* were described. Meanwhile, in Argentina the remarkable *Pterodaustro*, which sported long, slender, reverse flamingo-like filter-feeding jaws, was revealed. Of much greater paleobiological importance was the publication of the central Asian *Sordes*. The specimen was a typical small pterosaur, but what was remarkable was that it included an extensive covering of fur-like fibrous insulation, solid evidence for an elevated metabolic rate. That boosted the hypothesis that pterosaurs were energetic at the same time that this same idea was being applied to dinosaurs. Peter Wellnhofer conducted an extensive examination of the anatomy and biology of pterosaurs, German and otherwise. During this preliminary pterosaur renaissance, attention was again directed toward giant pterosaur flight, with the conclusion that pterosaurs were superlight aircraft that wafted around in the light airs of the warm and calm Mesozoic.

The stunning event that truly ignited the second golden age of pterosaur paleozoology was the announcement in 1975 of an extreme pterosaur, *Quetzalcoatlus*, from just before the end of the Mesozoic in Texas. Although the early wingspan estimates were too high by as much as a factor of two, it was soon settled that these ultrafliers were borne by wings spanning 10 to 12 m (33

9

to 40 ft), the same as World War II fighters and small gliders. This far exceeded even the extinct condor-like *Teratornis* and albatross-like pelagornithid superbirds whose wings spanned up to 7 m (23 ft), much less the modern condors and albatrosses at 3 to 3.7 m (10–12 ft).

Quetzalcoatlus and other similarly gargantuan azhdarchid pterosaurs posed a big problem. With fully developed wings, they obviously could fly, but how? One idea was that because pterosaurs were archosaurian ultralights, they simply did not weigh that much, about the same as a human being. That was the premise accepted by the QN Project of the mid-1980s, an effort to replicate the superpterosaurs via a full-size flying robot. It was led by the famed unconventional aerodynamicist Paul MacCready, who had designed the first successful human-powered flying machines, which were superultralights. Brought into the project to produce the paleozoological design of the biomachine, I soon realized that a creature with wings as long as a 70-passenger school bus could not have the mass of just one human. There would not be enough muscle to operate the hindlegs, which would be as tall as a human, much less the tremendous wings. My calculation that the biggest pterosaurs must have weighed much more than the biggest living ground birds (Paul 1991, 2002) was initially controversial, but it has since become widely accepted (Habib 2008, 2010; Witton 2013; Venditti et al. 2020; Larramendi et al. 2021).Another interesting aspect of azhdarchids is where they lived. Most pterosaurs are found in lagoonal or lake deposits, which, along with their anatomy, indicates that they were water pterosaurs, like waterbirds. And the big pteranodonts are found only in marine sediments, indicating that, like albatrosses, they strictly avoided large land areas. But the azhdarchids are largely found in land deposits or nearshore sediments, meaning they usually stayed over or near land.

The rapid rise in popular and scientific interest in dinosaurs spilled over to pterosaurs, as discoveries of the bat-winged archosaurs exploded around the world. About a third of valid species have been named since the turn of this century. It helped that many of the new forms sported incredible head crests, in some cases made at least partly of soft tissues that look like they belonged on cartoon or alien creatures. Even boring, medium-sized *Nyctosaurus* came in for a major remake when its complete and amazing crests were finally found (Bennett 2003). And the old classic *Pteranodon* underwent a notable revision with the realization that it had a major overbite, with the lower beak markedly shorter than the upper (Bennett 2001). Modern discoveries have been made worldwide, including at high latitudes. The most important centers of discovery have been in South America, which has produced many of the Cretaceous crested examples, some of which were giants exceeded only by azhdarchids. This helped solve a lot of the questions about the fragmentary finds of similar-aged European pterosaurs, as well as those in Asia. In the latter, the Middle Jurassic to Early Cretaceous Jehol lake-bed deposits in northeastern China have been especially productive, not only of pterodactyloids but also of the early birds that were then beginning to challenge pterosaurs as vertebrate rulers of the skies.

More pterosaur footprints confirmed that they often walked on all fours. One trackway seems to even show a pterosaur landing. As for taking off, recent analysis has indicated that they could take off by pushing off with their powerful arms. Pterosaur eggs and even nesting colonies have been found, and the age at which pterosaur young left the nest and how fast they grew are becoming better understood. The peculiar pterosaur respiratory complex that combines aspects of those of crocodilians and birds is being detailed. Although finds in Italy have produced the earliest pterosaurs from the Triassic, what has not yet shown up is fossils of protopterosaurs, because suitable fine-grained lake or lagoon bottom deposits from that earlier geological time are not known.

The evolution of human understanding of pterosaurs has not undergone as dramatic a transformation as has our view of dinosaurs over the last quarter millennium. This is true partly because the very first described pterosaur fossil was nearly complete and articulated, and within a few decades all acknowledged it was not a bat or a bird or another dinosaur. And while many contended for a long time that pterosaurs were low-energy reptilian gliders, others have long argued that they were high-energy powered fliers. The latter view has now nearly entirely won out. The basic paleobiology of pterosaurs is unlikely to undergo a dramatic change in the future. Even so, the research and discovery is nowhere near its end. To date, about ten dozen valid pterosaur species in around a hundred genera have been discovered and named. This probably represents at most a quarter, and perhaps a much smaller fraction, of the species that have been preserved in sediments that can be accessed. And, as astonishingly strange as many of the pterosaurs uncovered so far have been, there are equally odd species waiting to be unearthed. New fossils are needed to better determine how big pterosaurs got. Reams of work based on as-yet-undeveloped technologies and techniques are required to further detail both pterosaur biology and the world they lived in.

WHAT IS A PTEROSAUR?

Pterosaurs are often called flying reptiles, but they were not particularly closely related or anatomically similar to what are popularly considered modern reptiles, and in technical terms birds are also flying reptiles. Nor were pterosaurs at all close to being flying dinosaurs, because it is birds that are literally aerial dinosaurs directly descended from theropod dinosaurs, just as bats are flying mammals.

To understand what a pterosaur actually is, we must first start higher in the scheme of animal classification. The Tetrapoda are the vertebrates adapted for life on land—amphibians,

Comparative wingspan and weights of man-made aircraft, pterosaurs, and wandering albatross

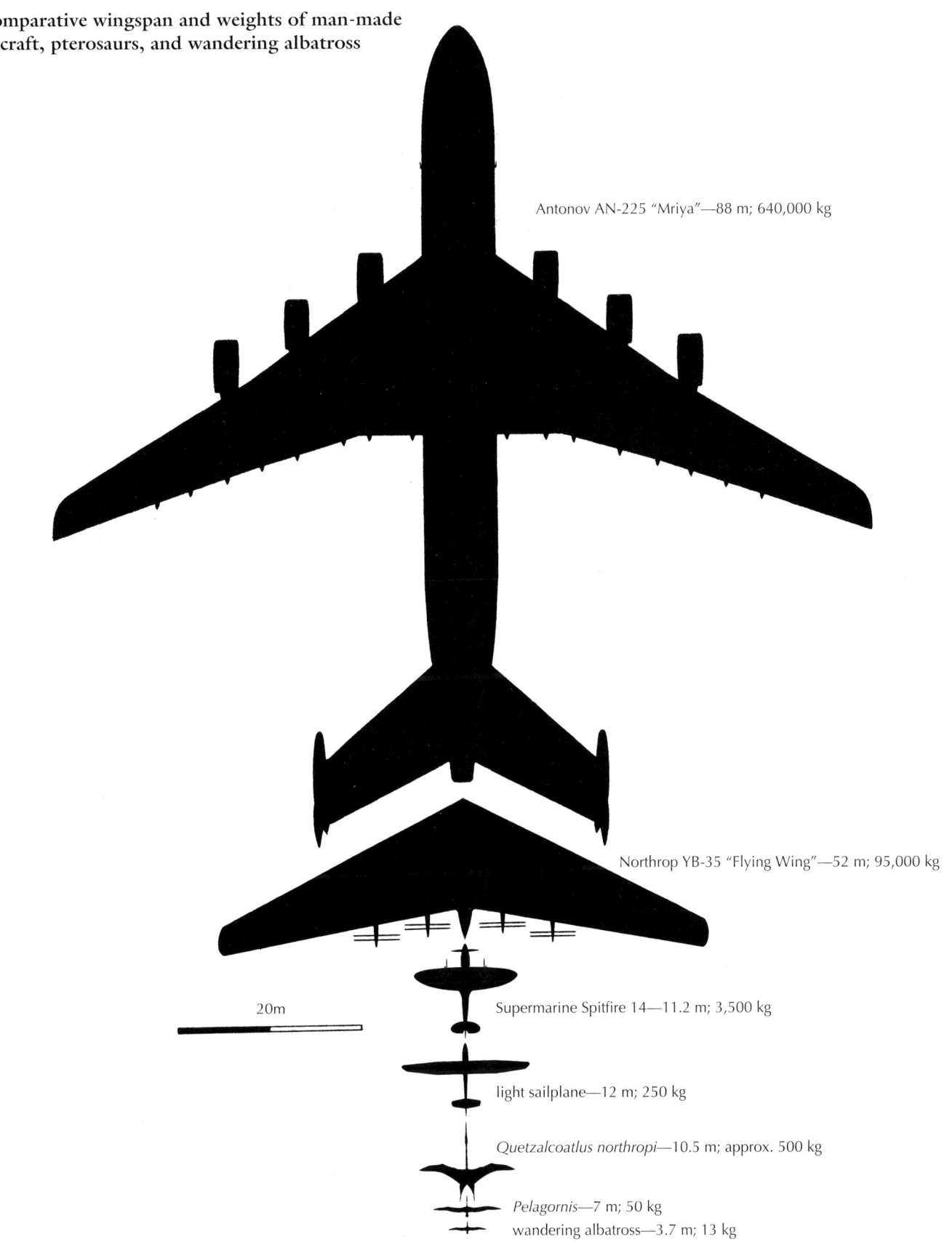

Antonov AN-225 "Mriya"—88 m; 640,000 kg

Northrop YB-35 "Flying Wing"—52 m; 95,000 kg

20m

Supermarine Spitfire 14—11.2 m; 3,500 kg

light sailplane—12 m; 250 kg

Quetzalcoatlus northropi—10.5 m; approx. 500 kg

Pelagornis—7 m; 50 kg

wandering albatross—3.7 m; 13 kg

reptiles, mammals, birds, and the like. Amniota comprises those tetrapod groups that reproduce by laying shelled eggs, with the proviso that some have switched to live birth. Very early in their evolution, amniotes split into two great groups. One is the Synapsida, which includes the archaic pelycosaurs, the more advanced therapsids, and mammals, which are the only surviving synapsids. The other is the Diapsida. Surviving diapsids include the lizard-like tuataras, actual lizards and snakes, crocodilians, and birds. The Archosauria is the largest and most successful group of diapsids and includes crocodilians and dinosaurs, of which birds are the one surviving group.

Archosaurs also include the basal forms informally known as thecodonts because of their socketed teeth, themselves a diverse group of terrestrial and aquatic forms that included the ancestors of crocodilians and the flying pterosaurs, which are not intimate relatives of dinosaurs and birds.

The great majority of researchers agree that the pterosaurs were archosaurs that were not thecodonts or dinosaurs, avian or otherwise, and probably evolved from some form of basal archosaur, probably in the Early to Middle Triassic near the beginning of the Mesozoic. That is where our certain knowledge of pterosaur relationships pretty much ends. The kind of fine-grained Lagerstätte deposits that would preserve the delicate little protopterosaurs have yet to be found before the Late Triassic, when the first pterosaurs show up. By then, even the early pterosaurs were so profoundly pterosaurian that their ancestry is obscured. This is very similar to the situation with bats, but quite unlike that with birds, in which a fairly extensive series of fine-grained sediments from the Middle Jurassic to the Early Cretaceous has told us a lot about how they evolved from flightless dinosaurs. The origins of dinosaurs themselves are also fairly well known thanks to a number of Triassic protodinosaur fossils—although they were not big beasts, protodinosaur remains are large and robust enough to have been preserved in coarser-grained sediments.

Although there is little in the way of fossil evidence, there has been considerable speculation as to which archosaurs the pterosaurs are most closely related to. Most researchers consider them fairly close relatives of the dinosaurs, or more broadly the dinosauromorphs, which include the protodinosaurs. In part this is because the latter were small, lightly built archosaurs

with long, slender limbs and modest-sized, light skulls at the end of medium-length necks. Most critically, the dinosaur group was distinctive in having a simple, fore-and-aft hinged ankle joint, unlike the more complex ankles sporting a large backward projection, like those of mammals, characteristic of many thecodonts and retained by crocodilians. Pterosaurs had a simple dinosaur-type ankle. Recent work suggests that the lagerpetids, which were semibipeds known only from partial remains, shared significant features of the skull and skeleton with pterosaurs that suggest that they were, among other things, developing sensory and locomotory adaptations retained by pterosaurs (Ezcurra et al. 2020). That the arms of lagerpetids were not strongly reduced is compatible with them being close pterosaur relatives if not ancestors, although their hands do not appear to have been elongated. A key feature of dinosaurs not seen in pterosaurs is a cylindrical hip joint with a large internal opening that prevents the femur from sprawling out to the sides—you can check it out the next time you have a chicken or turkey thigh. Lagerpetids did not have this internal opening, and while their limbs could be directed in a vertical fore-and-aft working plane, they could also be directed more out to the sides, as is true of many mammals, humans included. If lagerpetids are protopterosaurs, then they were pterosauromorphs rather than dinosauromorphs, as has been thought. If lagerpetids, pterosaurs, protodinosaurs, and dinosaurs are a distinct group separate from other dinosaurs, then the collective is known as the Ornithodira. Some of the lagerpetids lived earlier than the first know pterosaurs, so it is possible that the former group included the direct ancestors of the latter. Alternatively, the two clades may have been sister groups that shared a common ancestor. That the lagerpetid foot was more asymmetrical toward the outside yet had a very reduced outer toe, as well as the absence of evidence for well-developed climbing adaptations, tends to favor the latter relationship. Because lagerpetids are fairly widely distributed above and below the equator, they cannot tell us much about where pterosaurs originated geographically even if they are intimate pterosaur relations.

Much attention has been directed toward the Late Triassic *Scleromochlus* as the closest known pterosaur relative. But aside from being present rather late, *Scleromochlus* is poorly preserved, it lacks the simple hinged ankle of pterosaurs and dinosaurs,

A basal archosaur,
Euparkeria

new analysis indicates that it was a sprawl-legged hopper like frogs (Bennett 2020), and its quite short arms are not what is expected in an archosaur close to the origins of very long-armed winged fliers. This archosaur has been the focus of so much attention more because so little else has been on hand than because it is a well-suited candidate for a close relative of pterosaurs.

It can be difficult to tell whether some early members of the dinosaur group are protodinosaurs or dinosaurs, or whether some early members of the bird complex are protobirds or birds. That is because in both cases we have a good series of transitional forms, so there is a blurring gradation rather than the sharply defined separation that results from a large gap in the fossil record. Because there are no known, partly flight-capable protopterosaurs, this problem does not apply to known pterosaurs, which are all very distinctive from other vertebrates. Because all known pterosaurs are fliers, they all share a basically similar body form—that is, like flying birds and all bats—meaning that pterosaurs are not as extremely diverse as dinosaurs as a whole, much less mammals, which include everything from marine swimmers to fliers. It is possible but not at all certain that some pterosaurs swam underwater, but none were highly adapted to do so.

The most obvious universal pterosaur feature is the extreme elongation of an outer finger into a large, clawless, stiff spur that supports the leading edge of the outer main wing membrane. The great finger articulates with the rest of the hand via a large pulley joint. The other three fingers form a short, clawed, grasping hand, although this was lost in the most extreme soaring pterodactyloids. The chest features a large sternal plate similar to that of birds, except that it lacks the deep bony keel found in most flying avians. There is no wishbone furcula, and the shoulder girdle is highly modified. The body is short, and in the pelvis the front part of the ilium is long and slender, while below the vertical pubis is short, with a prepubic process projecting forward toward the sternum. A cup-shaped hip joint may have allowed the femur to swing well out to the sides to help support flight membranes during flight, as in bats, while also allowing the femur to project strongly downward when moving on the ground. The foot was plantigrade so that the heel contacted the ground, as in humans and bears. This is unlike the feet of digitigrade dinosaurs and birds, as well as cats and dogs, in which the ankle is well off the ground. That pterosaur eggs were soft shelled, like those of most reptiles and perhaps some other archosaurs, is another feature that distinguishes them from most or all dinosaurs, including birds whose calcified eggshells are quite hard.

Even though flier groups like pterosaurs, birds, and bats always share a basic body plan within each assembly, considerable variation also exists—so sparrows are quite different from albatrosses, which differ greatly from eagles, crows, and flamingos. Among pterosaurs, most had short or moderately long necks, but in some the neck was more elongated than in giraffes. Heads tended to be long, low, subtriangular, narrow, and lightly built, but in some the skull was deep and/or broad, and a few pterosaur skulls were robustly built. As for the crests, they varied from possibly absent to enormous, with major variations in shape, and they were not always made of only bone.

Possible pterosaur relative

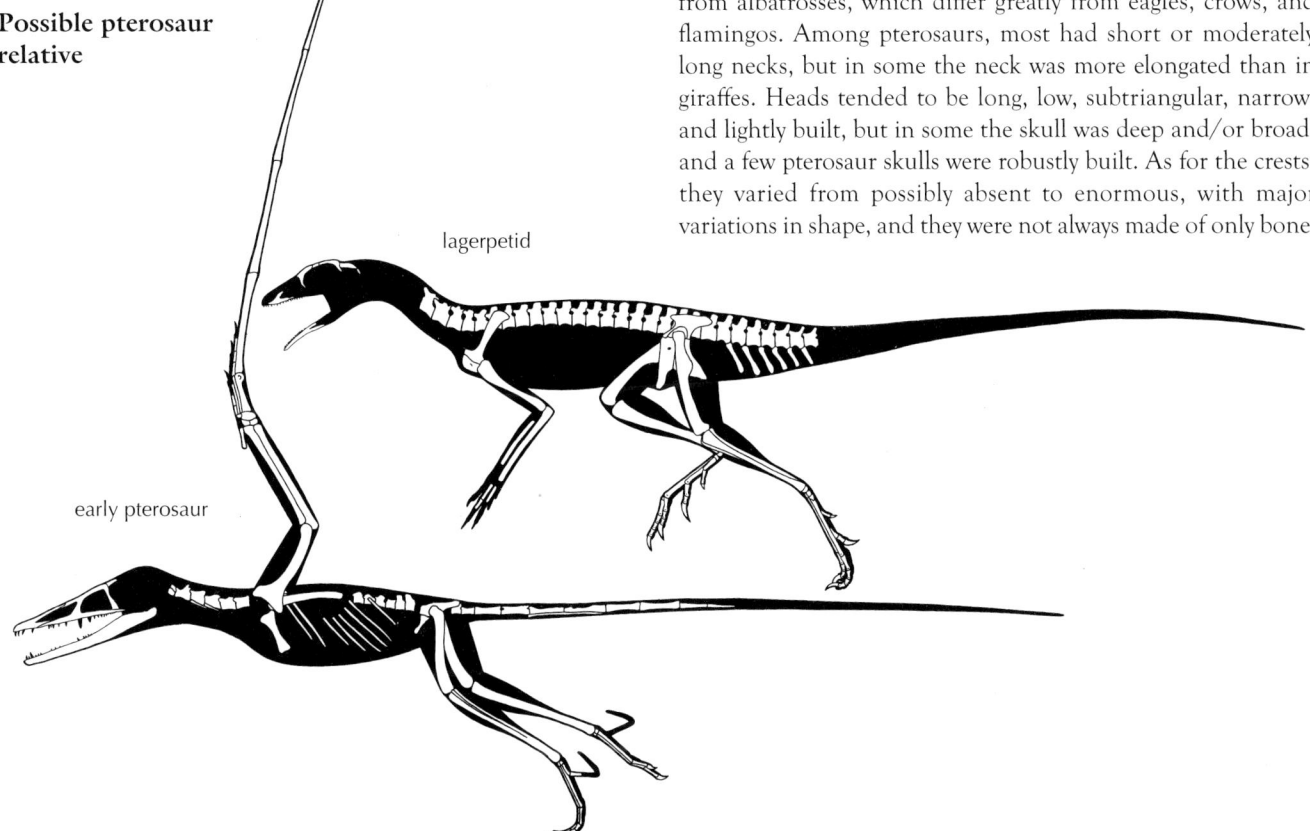

lagerpetid

early pterosaur

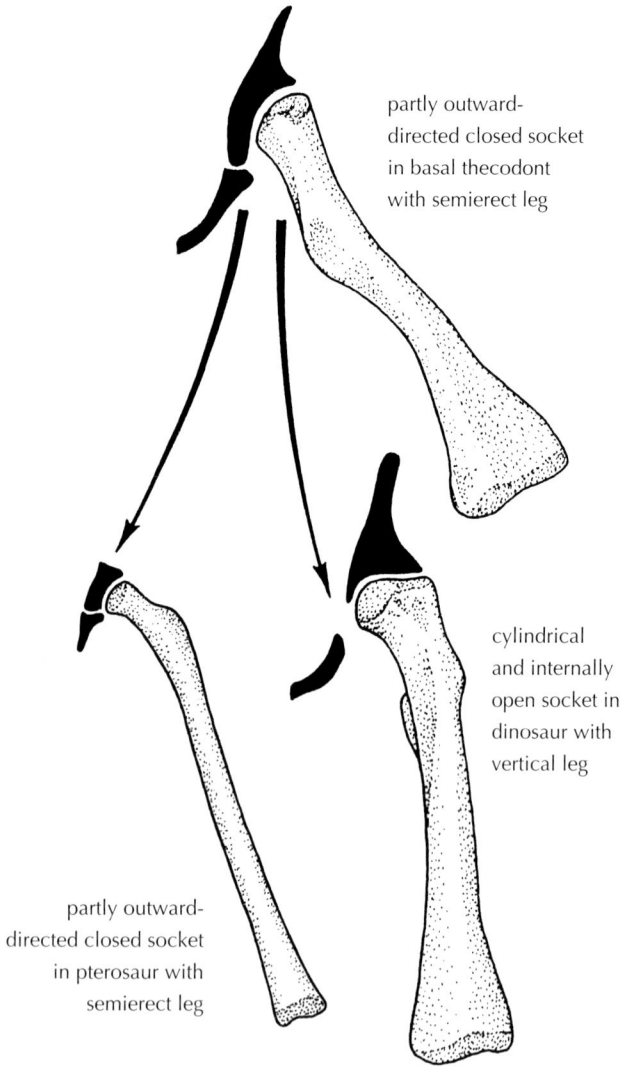

partly outward-
directed closed socket
in basal thecodont
with semierect leg

cylindrical
and internally
open socket in
dinosaur with
vertical leg

partly outward-
directed closed socket
in pterosaur with
semierect leg

**Hip socket articulation
in archosaurs**

Pterosaurs were monophyletic in that they shared a common ancestor that made them distinct from all other archosaurs, much as all mammals share a single common ancestor that renders them distinct from all other synapsids. The same is true of dinosaurs. There does not seem to be a great division within the Pterosauria in which two or more distinct groups split apart soon after the initial appearance of the group, as occurred in mammals, dinosaurs, and birds. Over time there seems to have been a general mainline evolution among pterosaurs, with modest offshoots forming branches from a central trunk (Witton 2013; Baron 2020). Within this single-trunked tree, pterosaurs come in two basic flavors: toward the bottom are the basal forms with long tails, and further along the tree are the more derived examples with short tails. In one sense this shift is similar to that of birds, which started out with and then lost long tails. But in birds the change occurred early, so long-tailed birds are quite a small minority of short duration. In pterosaurs the change occurred

relatively late, and the species ratio is about four to one, with short tails being much more numerous species-wise. Most of the short-tailed pterosaurs are the pterodactyloids. Those with long tails were tagged rhamphorhynchoids. Because the latter were not a clade that left no descendants, many pterosaur researchers have dropped the term in favor of "nonpterodactyloids." But calling such a large portion of the pterosaurs by a name based on what the long-tailed pterosaurs were not is like calling all apes except the human clade nonhuman apes. And there is a basic problem with giving those groups of organisms that happened to be among the latest to evolve a formal title that applies only to them, while those that diverged earlier are left without their own title. So this work retains the use of "rhamphorhynchoids" to include the species that were not pterodactyloids.

All rhamphorhynchoids had teeth. They were also usually characterized by a short inner arm/hand, mainly because the metacarpals between the wrist and finger base were not strongly elongated. Basal pterodactyloids retained their teeth, while more derived groups often lost them. Pterodactyloids had more elongated inner wings, the metacarpals in particular being longer, sometimes extremely so. Both groups had crests, which were somewhat more strongly expressed in the pterodactyloids.

Unlike birds, which have lost flight fairly frequently, and like bats, which apparently never have, no known pterosaur was flightless, or particularly close to it—the outer wing from the finger base outward was always at least about as long as the middle wing from the elbow to the finger base. It is possible, however, that pterosaurs dwelling on small islands became largely or entirely nonaerial and were not preserved as fossils; on the other hand, no island bat is known to have become flightless, while birds have often done so.

Overall, pterosaurs have been less variable than birds, even when only flying birds are considered. The pterosaurs, some with very long tails and others with quite short tails, some with teeth and others without, some with long, narrow heads and others with short crania, showed more differentiation than the always toothy, short-headed, short-necked, short-tailed bats.

Nearly all pterosaurs were predators or one kind of another, downing invertebrates and vertebrates, generally small, both terrestrial and aquatic. Only one pterosaur group shows evidence of being predominantly herbivorous, but it is possible that many were omnivorous.

Pterosaurs seem strange, but that is just because we are mammals biased toward assuming the modern fauna is familiar and normal, and past forms are exotic and alien. Consider that elephants are bizarre creatures with their combination of big brains, massive limbs, oversized ears, a pair of teeth turned into tusks, and noses elongated into hose-like trunks. And if animals could think about it, they might find humans bizarre. Nor were pterosaurs part of an evolutionary progression that was necessary to set the stage for mammals culminating in humans. What pterosaurs do show is a parallel world, one in which the familiar birds were at first absent, and then for a time less numerous than the pterosaurs that ruled the skies of the later Triassic and Jurassic.

DATING PTEROSAURS

How can we know that pterosaurs lived in the Mesozoic, first appearing in the fossil record in the Late Triassic about 220 million years ago and then disappearing at the end of the Cretaceous 66 million years ago?

As gravels, sands, and fine silts are deposited by water, and sometimes by landslides and wind, they build up in sequence atop the previous layer, so the higher in a column of deposits a pterosaur is, the younger it is relative to pterosaurs lower in the sediments. Over time, sediments form distinctive stratigraphic beds that are called formations. For example, *Pteranodon* and *Nyctosaurus* are found in the Niobrara and Pierre Shale Formations of the American Great Plains. Both were deposited in what was then the bottom of a shallow seaway that connected the Arctic Ocean with the Gulf of Mexico. The Niobrara was laid down first, from 87 to 82 million years ago, and the Pierre Shale was then deposited on top of the Niobrara. The oldest and therefore first known *Pteranodon* specimens, of the species *P. sternbergi*, are found in the middle of the Niobrara. From the upper Niobrara comes the next species, possibly descended from the first, the apparently larger *P. longiceps*. Also present in the upper Niobrara is *Nyctosaurus*. The last genus has not been found, at least so far, in the following Pierre Shale, which does include the youngest known *Pteranodon*, *P. mayesi*, which may be the descendant of *P. longiceps*.

Geological time is divided into a hierarchical set of names. The Mesozoic is an era—preceded by the Paleozoic and followed by the Cenozoic—that contained three progressively younger periods, Triassic, Jurassic, and Cretaceous. These are then further divided into Early, Middle, and Late, except the Cretaceous is split into only Early and Late despite being considerably longer than the other two periods (this was not known when the division was made in the 1800s). The periods are further subdivided into stages. The Niobrara began to be deposited during the last part of the Coniacian stage and continued to form through the Santonian and into the very beginning of the Campanian, when the Pierre Shale began to be laid down.

The absolute age of recent fossils can be determined directly by radiocarbon dating. This method, which depends on the ratios of carbon isotopes, works only on bones and other specimens going back around 50,000 years, far short of the pterosaur era. Because it is not possible to directly date Mesozoic pterosaur remains, we must instead date the formations in which specific species are found. This is viable because a given pterosaur species found as a fossil lasted only a few hundred thousand to a few million years.

The primary means of absolutely determining the age of pterosaur-bearing formations is radiometric dating. Developed by nuclear scientists, this method exploits the fact that radioactive elements slowly decay in a very precise, constant, cumulative manner over time. The main nuclear transformations used are uranium to lead, potassium to argon, and one argon isotope to another argon isotope. This system requires the presence of volcanic deposits that initially set the nuclear clock. These deposits are usually in the form of ashfalls, similar to the one deposited by Mount Saint Helens over neighboring states, that leave a distinct layer in the sediments. Assume that one ashfall was deposited 144 million years ago, and another one higher in the sediments 141 million years ago. If a pterosaur is found in the deposits in between, then we know that it lived between 144 and 141 million years ago. If the fossil is just above the 144-million-year-old layer, then it is probably closer to that age than to 141 million years, and so on. As technology has advanced and the geological record has become increasingly better known, radiometric dating has become increasingly precise. The further back in time one goes, the greater the margin of error, and the less exactly the sediments can be dated.

Volcanic deposits are often not available, and other methods of dating must be used. Doing so requires biostratigraphic correlation, which can in turn depend in part on the presence of "index fossils." Index fossils are organisms, usually marine invertebrates, that are known to have existed for only geologically brief periods, just a few million years at most. Assume a pterosaur species is from a formation that lacks datable volcanic deposits. Also assume that the formation includes marine deposits, or that the latter were laid down at the same time near the formation's edge. The marine sediments contain some small organisms that lasted only a few million years. Somewhere else in the world the same species of marine life was deposited in a marine formation that includes volcanic ashfalls that have been radiometrically dated to between 84 and 81 million years. It can then be concluded that the pterosaur in the first formation is also 84 to 81 million years in age.

Because it was not possible to accurately date geological deposits when the time divisions were mapped and named in the 1700s and 1800s, they later proved to be very irregular in the amount of time that each covers—the Norian stage is about ten times longer than the Hettangian stage, for instance.

Most known pterosaurs lived in coastal and marine habitats, so the correlation method is usually effective, and radiometric dating is often also accessible. But some pterosaur-bearing formations contain neither volcanic deposits nor marine index fossils. It is not possible to accurately date the pterosaurs in these deposits. We can only broadly correlate the level of development of the pterosaurs and other organisms in the formation with faunas and floras in better-dated formations, and this produces only approximate results. The reliability of dating therefore varies. It can be very close to the actual value in formations that have been well studied and contain volcanic deposits; these can be placed in specific parts of a stage. At the other extreme are those formations that, because they lack the needed age determinants, and/or because they have not been sufficiently well examined, can only be said to date from the early, middle, or late portion of one of the periods, an uncertainty that can span well over 10 million years.

EVOLUTION OF PTEROSAURS AND THEIR WORLD

Pterosaurs appeared in a world that was both ancient and surprisingly recent—it is a matter of perspective. The human view that the time of the pterosaurs, which largely corresponded with the Age of Dinosaurs, was remote is an illusion that results from our short life spans, as well as the recent appearance of our genus and species in the last few million and couple of hundred thousand years, respectively. A galactic year, the time it takes our solar system to orbit the center of the galaxy, is 200 million years, so the earth is a mere two dozen galactic years old. And just one galactic year ago, the pterosaurs had just appeared on planet Earth. When pterosaurs first evolved, our solar system was already well over four billion years old, and 95 percent of the history of our planet had already passed. A time traveler from our time arriving on the earth when pterosaurs first flew in the skies would have found it both comfortably familiar and marvelously different.

As the moon slowly spirals out from the earth because of tidal drag, the length of each day grows. When pterosaurs first evolved, a day was about 22 hours and 45 minutes long, and the year had 385 days; when they went largely extinct, a day was up to 23 hours and over 30 minutes, and the year was down to 371 days. The moon would have looked a little larger and would have more strongly masked the sun during eclipses—there would have been none of the annular eclipses in which the moon is far enough away in its elliptical orbit that the sun rings the moon at maximum. The "man on the moon" leered down at the pterosaur planet, but the prominent Tycho Crater was not blasted into existence until toward the end of the Early Cretaceous. As the sun converts an increasing portion of its core from hydrogen into denser helium, thereby raising its core temperature, it becomes hotter by nearly 10 percent every billion years, so the sun was about 2 percent cooler when pterosaurs showed up and around 0.5 percent cooler than it is now when they disappeared.

At the beginning of the great Paleozoic era over half a billion years ago, the Cambrian Revolution saw the advent of complex, often hard-shelled organisms, of which the trilobites are the best known. Also appearing were the first, very simple vertebrates. As the Paleozoic progressed, first plants and then animals began to invade the land, opening a whole new world to exploitation by multicellular organisms.

With aquatic and terrestrial environments well on the way to being conquered, still left untapped by animals was the most difficult habitat to use, the atmosphere. Although evolving to move through thin air is difficult, there being no buoyancy or terra firma for support, being a flier has many advantages. These include the ability to escape nonaerial predators (even some fish glide-fly to escape their pursuers); to hunt prey in the water, on the land, and in the air from the air; to quickly move short and long distances; to cover a given unit of distance at less energy cost than walking; to rapidly migrate enormous distances that

land animals cannot match; to readily breach geographical barriers from oceans to deserts to mountain ranges; to spot food items at long distances; and to breed in isolated locations free of nonaerial predators. The first, nonflying insects were not particularly successful. It was only after the first flying insects appeared, possibly by the Silurian or Devonian, that they became numerous in the Carboniferous, when gigantic dragonfly-like insects evolved with wings spanning two-thirds of a meter (2 ft). Those superbugs may have been able to prey on small-bodied vertebrates. Below these invertebrate fliers the first tetrapods were evolving, leading to a brief Age of Amphibians, followed by the classic Age of Reptiles in the late Carboniferous and much of the Permian. By the last period of the Paleozoic, the Permian, the continents had joined to form the supercontinent Pangaea, which straddled the equator and stretched nearly to the poles north and south. The megacontinent was roughly C shaped, with the grand, subtriangular Tethys Ocean jutting into the landmass from the east. The rest of the world consisted of the Panthalassic superocean, which was almost 5,000 km (3,000 mi) farther across than is today's Pacific, totaling nearly 25,000 km (15,000 mi) east to west. Northern Pangaea was Laurasia, the southern half Gondwana. India was on the southern shore of the Tethys, wedged in with Africa, Antarctica, and Australia. As seas increasingly swamped Europe, it was developing into an island archipelago off the northeastern coast of North America. With most of the land far from the ocean, the majority of terrestrial habitats were harshly semiarid, ranging from extrahot in the tropics to sometimes glacial at high latitudes. Some of the dragonfly-like aerialists at the beginning of the Permian were as big winged as those of the Carboniferous, but they died out, perhaps because of the increasingly harsh climate. Insect fliers would never again be so large, so big flying bugs would never compete with flying vertebrates, none of which had evolved by the end of the Paleozoic. The major land vertebrate groups had evolved by the Permian. Among synapsids, the mammal-like therapsids, some up to the size of rhinos, were the dominant large land animals in the Age of Therapsids of the Late Permian. These were apparently more energetic than is normal for reptiles, which may have been true of some other tetrapods of the latest Paleozoic. Toward the end of the period, the first archosaurs appeared. These modest-sized, low-slung, vaguely lizard-crocodilian creatures were a minor part of the global fauna. The conclusion of the Permian saw a massive extinction that was probably driven by the extreme volcanism that formed the vast Siberian Traps and that, in many regards, exceeded the extinction that killed off the terrestrial dinosaurs 185 million years later—among the losses were the last trilobites.

So, at the beginning of the first period of the Mesozoic, the Triassic, the global fauna was severely denuded. As it recovered, the few remaining therapsids enjoyed a second evolutionary radiation and again became an important part of the wildlife. Also doing well were diapsids. A number of examples of this group became part of the first wave of vertebrate aerialists. Some were relatives of lizards, with gliding membranes stretched over long, sideways-splayed ribs, similar to the modern *Draco* lizards. Also using the rib membrane was a relative of the archosaurs, also lizard-like, and another near archosaur used a membrane stretched out mainly by the hindlimbs. These were all gliders that could not power fly, and they did not survive for long as far as we know.

Down on the ground, the main competition of the stoutly built therapsids, none of which became aerial, came from the archosaurs, which also underwent an evolutionary explosion, first expressed as a wide variety of thecodonts. Some of those became fairly large. But at least one branch had downsized by the Middle Triassic—which is actually fairly early in the period—becoming the lightly built, long-limbed, hinge-ankled forms ancestral to perhaps both dinosaurs and pterosaurs. It is very possible that these lithe runners had elevated metabolic rates and protofeathers. Among them, the little lagerpetids show signs of being both protodinosaurs and protopterosaurs, and they may include the ancestors of pterosaurs or at least be very close relatives. Lagerpetids appear to have possessed some of the sensory adaptations found in pterosaurs (Ezcurra et al. 2020), in which case the latter inherited some of the features that primed them for the rigors of flight. Later protopterosaurs are predicted to have been quite small archosaurs with gracile heads, bodies, and limbs, including somewhat elongated arms that sported an unusually long outer finger supporting an incipient membrane airfoil, hindlegs that also helped support the membrane and had the ability to sprawl to the sides, a simple hinged ankle, plantigrade feet, and very possibly a covering of fuzz. That the hindlimbs of pterosaurs may have been able to sprawl and were connected to the wing membrane, and that the feet were plantigrade, imply that the protopterosaurs were more arboreal creatures rather than primarily ground runners like the protodinosaurs (Witton 2013). The grasping, clawed fingers are compatible with climbing. At the same time, pterosaurs lacked the ability to reverse the ankle that allows some small climbing mammals, such as squirrels, to quickly progress down tree trunks headfirst. If so, then pterosaur flight may have begun as a means to extend leaps between branches and perhaps rock faces and get down to the ground swiftly, saving considerable travel time and avoiding predators in the process. The same is probably true of protobats, and of protobirds according to some researchers. One possibility we can dismiss is that pterosaurs began to fly in pursuit of flying insects, because reliably catching aerial invertebrates while on the wing requires well-developed flight performance that can appear only among highly evolved powered fliers.

The absence of protopterosaur fossils suggests that the transition from the gracile archosaurs with no flight attributes to incipient flying protopterosaurs to fully aerial pterosaurs occurred quite rapidly in geoevolutionary terms. The evolutionary selective advantages of flight may have been so strong that, once initiated, the process operated at a high rate, and the advantages of well-developed flight may have allowed early pterosaurs to quickly outcompete their less aerially competent protopterosaur relatives. If so, the transition was more rapid than in birds—protobirds were in existence for 20 million or more years, helping them leave behind a substantial fossil record. The rapid pterosaur evolutionary experience appears similar to that of the later bats.

After half the Triassic was over, in the Norian, the first pterosaur fossils appear. These were a few species of small, long-tailed, basal rhamphorhynchoids. Because they were already fully pterosaurian, it is likely that the very first pterosaur species appeared a few million years earlier. With large sternal plates to help anchor the powerful flight muscles needed to operate fully developed wings, these were powered fliers with basic aerial abilities not dramatically inferior to those of the last pterosaurs over 150 million years later. Birds would not achieve a similar level of flight function for another 100 million years, and it would take bats 170 million years. All these Late Triassic pterosaurs are known from Europe, Greenland, and North America, where suitable coastal or lake deposits happen to have preserved them, but they must have been much more widespread considering their ability to travel via air, and the fact that the continents remained unified at that time. The short inner-armed and short-legged early eopterosaurs appear to have had the squirrel-like proportions that go along with good climbing ability, but they should also have been adept on the ground. The latter was especially true of caviramians, which had long legs. It can be speculated that the first wave of pterosaurs ate small creatures of the land. It is tempting to presume that they also fed while on shorelines and in the shallows of fresh and brackish waters, but a startling lack of their trackways implies otherwise, so eopterosaurs were not true shore pterosaurs, nor were they highly marine. Yet their habitats varied from coastal woodlands to deserts. It is very doubtful that they hunted while flying. Choice plant items like large seeds and small fruits may have gone down their gullets, and some of these early pterosaurs had fairly complex teeth that would have allowed them to cut up and even chew food items. Although the long-tailed Triassic pterosaurs were not particularly diverse, there was some significant variation—especially in the heads, which ranged from simple and low in preondactylians to large and deep in dimorphodonts; and in the teeth, which varied from simple spikes to the complicated tricusped teeth—and some had already developed impressive head crests, showing how important the display devices would be to the group. There is little doubt that these energetic beasts had the high metabolic rates that would remain with the group to the end.

Among terrestrial archosaurs the protodinosaurs were losing ground to their dinosaurian descendants at about the same time pterosaurs appeared. At the start, dinosaurs were small and few, but these high-energy animals rapidly diversified into both predaceous and herbivorous forms that gave thecodonts increasing competition. Just 15 or 20 million years after the evolution of the

The Late Triassic *Preondactylus* (top), *Eudimorphodon* (center), and *Longisquama*

first little protodinosaurs, prosauropods and sauropods weighing a couple of tons had developed. In only another 10 million years, sauropods, which were as big as elephants and were the first truly gigantic land animals, were extant. By the last stage of the Triassic, dinosaurs were becoming the ascendant land animals, although they still lived among thecodonts and some therapsids, a few of which reached elephantine bulk. From small theropods evolved the first small mammals at this time. Some of the latter may have gone down the gullets of pterosaurs, and they may also have been vulnerable to predatory dinosaurs when caught on the ground, a situation that would remain true throughout the Mesozoic. In the oceans, a remarkably diverse array of marine reptiles appeared, some of which looked like aquatic lizards, and some of which had heads like that of a duck-billed platypus. Others were armored forms, some with turtle-like carapaces. Protoplesiosaurs were also about, and the fish-mimicking ichthyosaurs were in force, some being colossi.

With the continents still collected together as the Triassic came to its end, the climatic conditions over most of the supercontinent remained harsh. It was the greenhouse world that would prevail through the Mesozoic. The carbon dioxide level was two to ten times higher than it is currently, boosting temperatures to such heights—despite the slightly cooler sun of those times—that even the polar regions were fairly warm. The low level of tectonic activity meant there were few tall mountain ranges to capture rain or interior seaways to provide moisture. Hence, there were great deserts, and most of the vegetated lands were seasonally semiarid, but there were forests in the few regions with substantial rainfall and groundwater created by climatic zones and rising uplands. It appears that the tropical latitudes were so hot and dry during the summer—although winter nights could be quite cold—that the larger dinosaurs, with their high energy budgets, could not dwell near the equator and were restricted to the cooler, wetter, higher latitudes. Like small dinosaurs, the first little pterosaurs may not have been so limited. The flora was in many respects fairly modern and included many plants we would be familiar with. Wet areas along watercourses were the domain of rushes and horsetails. Some ferns also favored wet areas and shaded forest floors. Other ferns grew in open areas that were dry most of the year, flourishing during the brief rainy season. Taller trees included water-loving ginkgoids, of which the maidenhair tree is the sole—and, until widely planted in urban areas, the nearly extinct—survivor. Dominant among large plants were conifers, most of which at that time had broad leaves rather than needles. Some of the conifers were giants rivaling the colossal trees of today; such formed the famed Petrified Forest of Arizona. Flowering plants were completely absent.

As the Triassic came to a close, pterosaurs had made an evolutionary mark by becoming the first flapping-flight vertebrates. But they were very limited in species diversity and lifestyle and were consistently not big—that no Triassic rhamphorhynchoid footprints are yet known when those of other creatures, including small dinosaurs, are abundant further

records the limits of their faunal impact. Pterosaurs had not made as big a splash in the air as had dinosaurs on the ground. The end of the Triassic about 200 million years ago saw another extinction whose cause has remained obscure. A giant impact occurred in southeastern Canada, but it was millions of years before the extinction; again, supervolcanoes related to the coming opening of the Atlantic may have been responsible. Many of the early sea reptiles bought the paleofarm, but plesiosaurs and the increasingly hydrodynamic ichthyosaurs survived and thrived. Up above sea level, the thecodonts and therapsids suffered the most: the former were wiped out, and only scarce remnants of the latter survived along with their mammal relatives. In contrast, crocodilians, pterosaurs, and especially dinosaurs sailed through the crisis into the Early Jurassic with little apparent disruption. The sauropods, whose high-held heads the pterosaurs flew past, just got bigger. For the rest of the Mesozoic, dinosaurs would enjoy almost total dominance on land except for a few semiterrestrial crocodilians; there were simply no major flightless competitors above a few kilograms in weight. In the air pterosaurs had no vertebrate competitors—at this time—and archosaurs would always rule the daylight skies up to today. From the Late Triassic on is thus the Age of Aerial Archosaurs.

As the Jurassic progressed, the prosauropods appear to have been unable to compete with their more sophisticated sauropod relatives and were gone by the end of the Early Jurassic. Pterosaurs pretty much had the skies to themselves. Although flying insects were very abundant, none were that much larger than those present today, and although some small lizard-like diapsids were gliding about, only the pterosaurs could power fly. The fossil record of the group is rather scanty in the Early Jurassic, pterosaurs being particularly scarce in sediments from toward the middle of that time because the deposits were ill suited for preserving the gracile aerial beasts, and pterosaur trackways have yet to be discovered. We do know that complex chewing teeth apparently disappeared, and why these sophisticated dental batteries were lost is perplexing. From now on, it would be only single-pointed teeth, which were sometimes small and sometimes gnashingly large. Enormous-headed dimorphodonts were largely terrestrial, but the first, spiky-toothed rhamphorhynchids indicate that they were shifting more toward fishing habits. The supercontinent was beginning to break up, creating African-style rift valleys along today's eastern seaboard of North America that presaged the opening of the North Atlantic. For the rest of the Mesozoic, the increased tectonic activity in the continent-bearing conveyor belt formed by the mantle caused the oceans' floors to lift up, slowly but persistently spilling the oceans onto the continents in the form of shallow seaways that began to isolate different regions from one another, encouraging the evolution of a more diverse global fauna. The expansion of so much water onto the continents also raised rainfall levels, although most habitats remained seasonally semiarid. The moving landmasses also plowed up more mountains, which were able to squeeze rain out of the atmosphere.

The fossil record of pterosaurs improves substantially in the Middle Jurassic, with finds from a variety of Eurasian locations. Their diversity was up, and flight performance improved to varying degrees (Witton 2013; Venditti et al. 2020). The last was especially true of the anurognathids, which begin to show up in the fossil record around the Middle/Late Jurassic boundary, although the less specialized members of the group must have evolved earlier. Sporting short, wide-mouthed frog heads well suited for collecting bugs on the wing, wings elongated by lengthening the radius and ulna of the lower arm, and very short tails (the first pterosaurs to have them), they were high-performance fliers that set the standard for later aerialists such as swifts and swallows, as well as fast bats. These supreme fliers are the first vertebrates thought to have specialized in hunting aerial insects, which until the Middle Jurassic had only the dragonflies to fear when flying during the day, and the dragonflies were themselves safe when in the air. It is probably not a coincidence that a few fossil insects from the later Jurassic show wing damage compatible with failed pterosaur attacks, including a dragonfly that typically flew during the day, and a lacewing that was usually nocturnal. The smallest known pterosaurs had adult wings with spans as short as half a meter, and no other pterosaurs would match the sheer flapping-flight ability of the anurognathids, which were the molossid bats of their day. Not entirely clear is whether the short tails of anurognathids as well as some other features mean that they were the rhamphorhynchoids closest to the pterodactyloids (as per Andres et al. 2010, 2014; Baron 2020; Wei et al. 2021), or whether they were more basal rhamphorhynchoids that independently lost their long tails as they evolved into high-performance fliers (as per Witton 2013). If, as seems likely, anurognathids had conjoined nasal and preorbital openings, then they were probably protopterodactyloids. Most other Middle Jurassic pterosaurs were more standard long-tailed rhamphorhynchoids not dramatically different from those of the Triassic and Early Jurassic. Pterosaurs in particular had not yet much exceeded 1.5 m (5 ft) in wingspan as far as we know.

In the Jurassic, evolutionary selective processes pushed some theropods, more specifically the three-toed avepod dinosaurs, to begin to challenge their fellow archosaur pterosaurs for dominance of the air without trying. The first fossils of protobirds begin to show up in the Middle Jurassic, in the same fine-grained Lagerstätte lake sediments of northeastern China that produce pterosaurs. Some were the bizarre scansoriopterygians that appear to have had membranous wings that were even more bat-like than those of pterosaurs. These quite small, arboreal creatures appear to have lacked the flight musculature needed for significant powered flight, and they are not known from the later fossil record, so they apparently did not offer much in the way of competition for the far more aerially sophisticated pterosaurs. More promising were anchiornians, found from the same place and time as the scansoriopterygians. Being more typical bipedal avepod dinosaurs, anchiornians were beginning to develop miniwings, with airfoils made of feathers. These protowings were too small for full flight, but they were a beginning. Interestingly, although the rhamphorhynchoids from these beds are somewhat more numerous in species and specimens than are the dinosaurian fliers, the pterosaur diversity advantage is surprisingly modest. Also notable in the Middle Jurassic is that small, highly developed gliding mammals appear in the fossil record, not long after the evolution of the first mammals.

The Late Jurassic provides paleozoologists with an even better pterosaur fossil record, especially via the famed German Solnhofen lagoonal island deposits, which indicates that the period saw major developments in pterosaur evolution. Finally, although most were not large, some rhamphorhynchoids became fairly big, with wings reaching up to 2.5 m (8.2 ft). The diversity of the long-tailed pterosaurs reached its height at this time, although how much of that was real versus the result of improved fossil samples is not obvious. As the Middle Jurassic ticked along into the Late Jurassic, some rhamphorhynchoids, the wukongopterids, had tails shorter than your average rhamphorhynchoid, as well as conjoined nasal and preorbital openings, and oversized heads. Yet more importantly, the truly and consistently short-tailed pterodactyloids show up in the fossil record later in the Late Jurassic; a somewhat earlier appearance cannot be ruled out. It seems that the very short-tailed pterodactyloids displaced their medium- tailed, close wukongopterid relatives even as longer-tailed rhamphorhynchoids continued to thrive. Also having the conjoined nasal-preorbital openings that first showed up in wukongopterids and perhaps anurognathids, pterodactyloids represented a major upgrade in general—as opposed to the specialized short-tailed anurognathids—as far as pterosaur aerial agility (Witton 2013; Venditti et al. 2020), which would not be matched by birds until the Cretaceous. The pterodactyloids also represented a major expansion of pterosaur lifestyles. Sporting longer walking limbs, with increasingly specialized snouts and teeth, and often found in shoreline habitats, a portion of the pterodactyloids were shoreline pterosaurs that inhabited freshwater and marine coasts. That perhaps explains why pterosaur trackways finally start to show up on what were ancient mud and sand flats as the pterodactyloids searched for small water-loving creatures to pick up—we know they were doing this because they left beak peck and scour marks as well as footprints. Prints attributable to long-tailed pterosaurs are also known from this time. There is, however, no evidence that Jurassic pterosaurs became truly marine, like petrels and albatrosses, over oceans filled with marine crocodilians, dolphin-like ichthyosaurs, and plesiosaurs, some of which sported very small heads at the ends of very long necks, while others had very big heads on short necks. Nor had any pterosaurs become gigantic by this time; the largest Jurassic pterodactyloids were no bigger than their rhamphorhynchoid relatives. Back on land, the dinosaurs were reaching their own new level of variety, with the sauropods becoming astonishingly enormous—it is possible that small pterosaurs could have ridden on their high, broad backs. Toward the other end of the dinosaur

The Late Jurassic *Rhamphorhynchus* (flying at top), *Anurognathus* (center right, front-on view), **and** *Pterodactylus* (bottom)

size range, little *Archaeopteryx* had wings as large as those of flying birds. But lacking a large bony sternal plate on which to anchor really powerful flight muscles, the classic protobird was far from offering direct competition to the far more common and diverse, coastal-adapted rhamphorhynchoids and pterodactyloids that shared its tropical, arid island European habitat, which was then not far off the northeastern coast of North America, and on the northwestern edge of the then-grand Tethys Ocean.

Both the rhamphorhynchoids and pterodactyloids of the Late Jurassic had teeth—although barely in the case of the cycnorhamphians, with their bizarrely twisted mollusk-cracking jaws—and they often, but perhaps not always, showcased head crests from small to large. Among the new pterodactyloids, some were evolving dense rows of sideways-flaring, comblike, superslender teeth for straining small organisms out of shallow waters. As for the air that pterosaurs flew in, during the Middle and Late Jurassic, carbon dioxide levels were incredibly high, with the gas making up between 5 percent and 10 percent of the atmosphere. As the Jurassic and the age of sauropods ended, the young North Atlantic was about as large as today's Mediterranean. Vegetation had not yet changed dramatically from the Triassic.

What happened to the global fauna at the end of the Jurassic is not well understood because of a lack of sufficient deposits. Some researchers think there was a major, fairly sudden extinction, including dinosaurs—the plated and spiked stegosaurs may not have made it into the Cretaceous—and pterosaurs, but the information needed to determine the situation is absent. Judging by what has not yet been found in Early Cretaceous sediments, the rhamphorhynchoids large and small, except the superflier anurognathids, were lost, leaving no long-tailed pterosaurs. Did a sharp extinction event wipe out the long-tails some 90 million years after they evolved, while the short-tailed anurognathids and pterodactyloids made it through in either small or large numbers? Or did the rhamphorhynchoids gradually lose out over some millions of years, perhaps in an evolutionary struggle with the more aerially adept short-tails? Unless suitable deposits straddling the Jurassic-Cretaceous boundary are found, we may never know the answers to these deep evolutionary questions.

The Cretaceous began 145 million years ago. This long period would see an evolutionary explosion of pterodactyloids, and even more spectacularly of dinosaurs and birds, surpassing what had gone before as the continents continued to split, the South Atlantic began to open, and seaways crisscrossed the continents. Greenhouse conditions became less extreme as carbon dioxide levels gradually edged downward, although never down to the modern preindustrial level (0.02 percent, now 0.04 percent). Early in the Cretaceous, the warm Arctic oceans kept conditions up there balmy even in the winter. At the other pole, continental conditions were more friendly toward winters frigid enough to sometimes form permafrost, and too cold for low-energy reptiles (Paul 2012, 2017a). General global conditions were a little wetter than earlier in the Mesozoic as air-humidifying seas developed between the sections of the once united supercontinent, but seasonal aridity remained the rule in most places, and true rain forests continued to be scarce at best. Although conifers and other nonflowering plants remained dominant, the flowering angiosperms began to appear late in the Early Cretaceous, albeit initially limited largely to shrubs lining watercourses, and fully aquatic plants such as water lilies. On the ground, sauropods remained abundant and often enormous, but they were less diverse than before, and to a fair extent the Cretaceous was the Age of Ornithischians, a diverse group of dinosaurian beaked herbivores. Avepod theropods small and large remained the dominant land predators, with the avian versions sporting increasingly effective wings. In the oceans, over which pterosaurs were increasingly venturing, sea reptiles, now including sea turtles, as well as increasingly modern bony fishes and sharks, prevailed. A proliferation of small near-surface fish in the open oceans, often driven up to the surface by their undersea pursuers, provided fine dining for flying fishers, as did the squid-like belemnites and true squids.

With the rhamphorhynchoids no longer in major force, pterodactyloids became diverse and started to become large, in some cases really large. The last was not true of all Cretaceous pterodactyloids, but none were as short spanned as the smaller of the Jurassic pterodactyloids, or the little anurognathids. The continued presence of the latter bug-eating pterosaurs may have hindered the evolution of similar insect-intercepting pterodactyloids, as well as birds of that type. Neither did pterodactyloids again evolve complex teeth for chewing as the early rhamphorhynchoids had. As always, pterosaur head crests were common—in some cases incredible in proportions as in the large thalassodromids and tapejarids—but not necessarily universal. Nonmarine pterosaurs of the time are arguably best recorded in northeast China lake deposits formed in highlands with chilly winters. Competitive with the Asian pterosaur fossils are those from South America. The most extraordinarily specialized filter-feeding pterosaurs evolved strongly curved, ultrafine, comb-toothed beaks functionally comparable to those of flamingos. At an opposite feeding extreme, the cutting-toothed istiodactylids look as if they were specialized land scavengers, while the deep-beaked tapejaromorphs are the only known pterosaurs that appear to have been highly herbivorous, chowing down on fruits like very oversized parrots. Exactly what moganopterans were doing with their delicate but widely spaced teeth at the front end of superelongated—two-thirds of a meter (1 ft)—and ultraskinny jaws is not at all clear. Most Early Cretaceous pterodactyloids were toothy, but the first toothless pterosaurs—the largely terrestrial azhdarchoid tapejaromorphs, thalassodromids, and chaoyangopterids—finally began to appear, finally doing what birds were already doing.

Also beginning was the development of the biggest pterosaurs up to that time, the enormous pterodactyloids (Witton 2013). Immense skulls show that some tapejarids sported a wingspan of up to 8 m (26 ft), larger than the latter pteranodonts; these were the first pterosaurs to achieve 200 kg (400 lb). Even larger pterodactyloids may have been recorded by Asian trackways that suggest individuals with wingspans up to 10 m (30+ ft) (Kim et al. 2012). Over the oceans, the first wave of truly marine pterosaurs, the ornithocheirid branch of the ornithocheiroids, also evolved toward the middle of the Early Cretaceous. Some were specialized for flapping and especially soaring long periods over the waves, where they hunted for fish and swimming invertebrates such as the squid-like belemnites and true squids. To a fair extent paralleling the pelagornithids and albatrosses of the Cenozoic, these saltwater superpterodactyloids achieved wingspans as great as 9 m (nearly 30 ft) and weights of over 100 kg (200 lb), exceeding the biggest seabirds of the Cenozoic, as well as the later Mesozoic pteranodonts. When floating on deep waters, ornithocheiroids had to be very concerned about being snatched under by the bevy of ichthyosaurs, plesiosaurs, sea crocodilians, and mosasaurs that cruised the warm Cretaceous seas, as well as big fish and sharks.

As major as the radiation of the early Cretaceous pterosaurs was, the dinoavian competition was ramping up as well. The lake deposits in northeastern China have produced a diverse array of dinosaurian protobirds and birds with flight abilities well beyond those of the Jurassic protobirds. Most of the Cretaceous dromaeosaur theropods bearing big sickle-clawed toes were fairly large ground predators. But they have a number of features of fliers, including birdlike shoulder girdles that included large sternal plates attached via ossified sternal ribs to ribs that bore ossified side processes called uncinates (which you can see on chicken breasts); folding arms; and, very interestingly, a very slender but long tail with a mobile base that was largely stiffened by long tendons that became ossified. The latter is remarkably like the tails of the now extinct rhamphorhynchoids. The arms of these dromaeosaurs were too small for them to fly; the many flight features suggest that the earliest dromaeosaurs were small-bodied, big-winged fliers. Indeed, such dromaeosaurs of the air have been found in the Chinese lake deposits; in fact, these microraptors were tandem biplanes that had leg wings about as large as the arm airfoils. Having the dromaeosaurs' large sternal plates and related flight features, as well as flattened hand bases to better support the outer primary wing feathers, this side branch of air-capable dinosaurs was somewhat better adapted for powered flight than the Jurassic archaeopterygians. Dromaeosaurs therefore stand as an example of the unpterosaurian propensity of flying dinosaurs to lose flight after having evolved it. What these rhamphorhynchoid-like microraptor dromaeosaurs did not constitute was a common form of serious competition for the much more aerially sophisticated short-tailed pterosaurs. Among other things, they could not achieve a combination of large size with aerial ability.

Largely revealed by the Chinese lake deposits, what was posing a swiftly growing challenge to pterodactyloids was mainstream early avian evolution. Some early birds lacked ossified sternal plates, but even these had much larger wings than the archaeopterygians. All other birds of this time had big sterna, some with strong bony keels anchoring very large flight muscles. Other

The Early Cretaceous
Anhanguera

sophisticated flight features were showing up, including highly flattened, fused hands with smaller and lost claws to better anchor the big, outer primary feather fans. A few Early Cretaceous birds had long bony tails, but most did not, so despite a much later start, birds were faster in losing long stabilizer tails in favor of more agile flight than pterosaurs. While many of the first great wave of birds had teeth, a number lacked them, so in another major regard birds were again evolving more rapidly than had pterosaurs. For that matter, the birds of the Early Cretaceous ranged from predators to herbivores, and from arboreal forms to ground dwellers, eaters of freshwater fish, and coastal and even marine forms. Among the water-loving birds, some were moving into the liquid in a way pterosaurs never did, as specialized swimmers with reduced or lost wings and flight performance.

With the largest known Early Cretaceous bird sporting wings that were a substantial 1.5 m (5 ft) across, none of the early birds were either gigantic or likely to soar over the oceans, so pterosaurs remained unchallenged in those aerial arenas. But even so, pterosaurs were far from being the uncontested vertebrate masters of the air they had long been. Unlike in the fine-grained

Jurassic deposits in which pterosaurs outnumbered protobirds by a modest to large degree, in Early Cretaceous lake-bottom sediments birds were many times more numerous than pterosaurs in types, species, and especially in individual specimens. Some of the Chinese birds are found in distinct layers in enormous numbers, apparently the victims of sudden kills by nearby volcanic eruptions. It seems that some of the first birds were flying about and roosting in mass superflocks, for which we have not found comparable evidence for pterosaurs. Birds were also winning the competition, such as it was, in leaving lots of footprints on wet soils. Because of competition from small and modest-sized birds, pterosaur diversity in the same size classes was suppressed in the Cretaceous. This is often cited as a reason that pterosaurs underwent a shift toward gigantic forms, but mindless evolution does not work like that—giant pterodactyloids would still have evolved for other intrinsic selective reasons even if Cretaceous birds did not exist. But the development of superpterosaurs meant they were occupying niches that the early birds could not, giving pterosaurs space to remain an important group of late Mesozoic aerialists. Also left to pterosaurs in the Early Cretaceous was hunting insects in the air, which small anurognathids were still doing much like those of 40 million years before.

During the Late Cretaceous, which began 100 million years ago, carbon dioxide levels continued to drop, so the dark Arctic winters sometimes became cold enough to match the seasonal conditions seen in today's high-latitude northern forests—whether the pterosaurs wintered over or migrated out is not known—and glaciers crept down high-latitude mountains (Paul 2017a). The continents were separating fast by geological standards, to the degree that by the end of the Mesozoic they were assuming a fairly modern configuration—in part because India raced (in terms of tectonic speed) north across the Tethys to eventually collide with Asia. The tectonic separation and the continuation of numerous interior seaways resulted in increasing division of continents, often into subcontinents, and Europe remained a complex of islands large and small. In the oceans, the ichthyosaurs declined and disappeared, perhaps because the similarly highly hydrodynamic sharks were assuming fully modern forms, and some lizards went entirely marine in the form of the mosasaurs, as had sea turtles. On land, mammals were increasingly modern yet remained small. Dinosaurs saw their ultimate radiation, including horned giants, the duckbills, and the great tyrannosaurs.

Hindering fuller knowledge of what was happening with flying archosaurs, both pterosaurian and dinoavian, in the Late Cretaceous is yet another scarcity of fine-grained Lagerstätte deposits. One resulting mystery is whether the anurognathids made it into the Late Cretaceous. In principle, the absence of fine-grained sediments means there could have been an array of small Late Cretaceous pterodactyloids that we do not know about, but this is very unlikely because we do have a fairly decent record of bird fossils, which, although fragmentary, leave no doubt that there were numerous small avians at that time. Back in the Early Cretaceous, birds experienced a major

split. A large branch, the enantiornithines, had diverged from the euornithines that would lead to today's birds. As the more archaic birds basal to the split declined from the Early to Late Cretaceous, enantiornithines became much more abundant than the euornithines. And enough has been found to understand that enantiornithines were in the main an array of small birds of diverse forms, ranging from arboreal to highly aquatic. Evidence that pterosaurs were doing as well in that size range is lacking, including trackways. It is likely that the well-developed, more fully erect hindlegs of birds, not connected to the wings, helped give the small birds an evolutionary leg up in locomotion in the air, trees, land, and water. Bird wings are also harder to damage— if a few feathers are torn or lost it makes little difference; indeed, major wing feather loss is normal when birds are molting. A simple, significant tear in a pterosaur membrane could be unrepairable and permanently ground and doom a pterosaur. The ways in which the feathery wings of birds, especially those of smaller birds, flap are very complex and aerodynamically felicitous in ways that the simpler airfoils of pterosaurs may not have been able to match, including skilled maneuvering in forested environs. On the ground, the avian ability to tuck wings up nice and tight makes it easier for them seek refuge in dense vegetation and underground when the weather goes seriously bad, including cold snaps and storms. And their feathery folded wings provide some protection to the body against physical injuries and low temperatures. For Cretaceous pterodactyloids, with their awkwardly folding wings that inhibited getting out of the way of meteorological conditions, more exposure to nasty weather is likely to have led to more mass mortality events. Further avian advantages included more flexible necks for better feeding and aerodynamics, kinetic heads that allowed bird beaks to manipulate food items more precisely (Larson et al. 2016), and bigger brains that one would think made birds better thinkers than their pterosaur rivals. The flying dinosaurs' hard-shelled eggs were another improvement. Being able to sit on and incubate robust eggs sped up their hatching and allowed Mesozoic birds to become more sophisticated parents that better cared for their eggs and perhaps their young (Erickson et al. 2017). One way or another, the flourishing evolution of small flying dinosaurs was not doing the now-always-large pterosaurs any good during the Late Cretaceous.

Meanwhile, some small euornithines, the toothed ichthyornids, were small seabirds. Another group of toothed euornithines went entirely marine by becoming completely flightless swimmers comparable to big penguins. On land, some birds went flightless. A South American example was about the size of a chicken. Dwelling on a European island was quite a big bird about the size of an ostrich, the first known example of avian loss of flight plus gigantism resulting from being isolated on a small plot of land. Neornithines, the modern birds, appear in the fossil record shortly before the end of the Cretaceous and were not abundant. Some looked like and very possibly sounded like ducks. Absent as far as we know were avian raptors comparable to eagles, hawks, falcons, and owls. The same seems to be true of swift aerial insect interceptors—were anurognathids still filling that aerial niche? Apparently still present up to the end of the Cretaceous were some of the small, long-tailed, winged dromaeosaurs.

The pterosaurs that were doing well were the gigantic pterodactyloids, all of which belonged to the eupterodactyloids—the late Early and Late Cretaceous can be labeled the Age of Giant Eupterodactyloids. These superpterosaurs came in two distinct flavors: marine as in the late Early Cretaceous, and continental. Oceanic toothed ornithocheirids made it into the early Late Cretaceous, but the Late Cretaceous marine giants of most note were their ornithocheiroid relations, the famed and toothless crested pteranodonts. Although best known from the interior seaway that split North America in two from the Arctic to the Gulf of Mexico, gigantic pteranodonts were much more widespread if not global. The smaller nyctosaurs were even more extremely adapted for oceanic soaring, to the degree that they had lost their three small fingers. It is a fairly common yet serious mistake for paleoartists to show *Pteranodon* soaring over the head of *Tyrannosaurus*. That is as ecologically nonsensical as illustrating an albatross flying over an elephant, all the more so because *Pteranodon* specifically lived millions of years before *Tyrannosaurus*. New fossils indicate that both pteranodontids and nyctosaurids did make it all the way to the end of the Mesozoic (Longrich et al. 2018).

The other group of toothless eupterodactyloids that made it to the Mesozoic/Cenozoic boundary experienced a very different, terrestrial path to extreme gigantism. While pteranodonts are limited to marine deposits, an array of the azhdarchoids that first appear in late Early Cretaceous deposits come from land or nearshore sediments, so the two groups were not much more likely to mix than albatrosses and storks. The hindlegs of the marine pterosaurs were short, at least relative to the wings, being just long enough to allow them to get around on the ground when breeding in isolated locations. Azhdarchoid legs were much longer for a given body size, up to the height of a human in the biggest species. Along with their long inner arms, and the onset of fully erect limbs that made very big ground walkers possible, these ultimate pterodactyloids were adept on the ground, giving them a new lease on evolutionary attainment that allowed them to fill niches Late Cretaceous birds were not yet able to enter and compete in. The tapejarids made it into the Late Cretaceous, but it was the azhdarchids that became the dominant nonmarine pterosaurs of the last portion of the Late Cretaceous (Witton 2013). Usually sporting exceptionally long necks even for pterosaurs, the azhdarchids began with modest dimensions but increased in size during the last millions of years of the Mesozoic, with wingspans of well over 10 m (33 ft). They also became as heavy as bears and horses while matching or exceeding the masses of the largest flightless birds. Most of them may have been ground, shoreline, and shallow-water walkers and waders, picking up small and medium-sized creatures for meals such as oversized

storks, herons, and egrets. Some scavenging may have occurred, but dart-beaked azhdarchids were not specialized for doing so like hook-beaked vultures and condors. A notable exception comes from a European island of the time, where a giant azhdarchid's robust head and neck suggest it was an archpredator (Naish and Witton 2017). It is possible that the most terrestrial azhdarchoids could experience permanent aerially disabling damage to a wing and survive for an extended period.

Some researchers see azhdarchids as condor/vulture-like soarers (Habib 2010; Witton 2013). But their wings were short relative to their heavy bodies, so they would have had poor gliding performance, and the inner wing bones were exceptionally stout for power exploitation rather than streamlined for drag suppression. It is more likely that these were short-range, burst-flapping fliers, like turkeys and bustards (Paul 1991, 2002; Goto et al. 2020). Notably, the smaller predaceous avepod theropod dinosaurs such as the dromaeosaurs were not doing as well during these times as they had been earlier. Perhaps competition from azhdarchids, which could be both big and aerial when the avepods could not, was a factor. If so, this would have been a rare case of nondinosaur vertebrates successfully outcompeting dinosaurs, and the only known case in which pterosaurs did so in a marked way.

While the big Late Cretaceous pterosaurs are well known, there is evidence that some azhdarchids were medium sized, with adult wingspans around 2 m (6.5 ft) (Prondvai et al. 2014; Martin-Silverstone et al. 2016), but more, better-quality specimens are needed to be sure.

Evolution accomplished quite a lot without trying with pterosaurs over their 150 million years. Ranging in size from cardinals to human-carrying sailplanes, some may have spent most of their time on the ground and only a little in the air, while others seem to have wave soared for months on end, landing only to breed. Diets ranged from fish to fruits to insects caught on the wing. Some had jaws bristling with big teeth, others none. Heads were sometimes short, broad, and frog-like, others long and superslender. Habitats ranged from the poles to deserts to the middle of oceans. Many preferred daylight, others moonlit and starlit nights. But pterosaur success and diversity had serious limits. Pterosaurs never became really small, as have countless thousands of bird species down to sparrows and hummingbirds, as well as bats. The smallest pterosaurs, anurognathids, had wings six times larger across than those of the smallest hummingbirds, and twice the size of those of sparrows and the littlest bats. As far as we know, there were no hovering, nectar-feeding pterosaurs. Or singing and talking pterosaurs. They did not become as arboreal as some birds have, and none appear to have pecked wood. Nor did any become sophisticated swimmers or fleet ground runners, much less flightless, it appears. None were raptor equivalents hunting other vertebrate fliers or ground creatures from the skies. Very probably no pterosaurs matched the intelligence of crows or parrots. None had the echolocation of bats. Or sucked blood, as some of the latter do.

The latest Late Cretaceous *Quetzalcoatlus* (top), *Tyrannosaurus* (left), and *Triceratops* (right)

Conversely, over their similarly long 150 million years of Darwinian evolution, birds have paralleled, sometimes broadly and in other cases closely, most or all of what pterosaurs did, ranging from swift hunters of flying insects to filter feeders in shallow water to gigantic wave soarers. On the other hand, pterosaurs did many things bats have not, including mudflat grub hunting, aquatic filter feeding, open ocean soaring, hunting on land, and becoming gigantic. Overall, cumulative pterosaur evolutionary achievements appear to have been intermediate between those of bats and birds. Specific to the Late Cretaceous, it appears that pterosaurs were the dominant big fliers that birds could not yet compete with, but little birds had probably entirely excluded pterosaurs from being small fliers, so only the big pteranodonts and azhdarchids remained, as far as we know. And that was a troubling sign of high evolutionary vulnerability.

Near the end of the Cretaceous, a burst of uplift activity and mountain building helped drain many of the seaways. Flowering plants were fast becoming an ever more important part of the flora, and the first hardwood trees evolved near the end of the period. Conifers remained dominant, however, and ginkgos were fairly widespread. Grasses had evolved, but they tended to be water-loving types and did not yet form extensive dry grassland prairies.

Then things went catastrophically wrong.

EXTINCTION

The mass extinction at the end of the Mesozoic is generally seen as the second most extensive in the earth's history, after the one that ended the Paleozoic. However, the earlier extinction did not entirely exterminate the major groups of large land animals. At the end of the Cretaceous all nonavian dinosaurs, the only major land animals, were lost, leaving only flying birds as the sole survivors of the Dinosauria. Even among the birds, all the toothed forms, plus the most abundant birds at the time, the enantiornithines, as well as the flightless birds, were also destroyed. Only neornithines survived, which means that if those birds had not evolved by that time, no dinosaurs would have made it into the Age of Mammals. And of course all pterosaurs were lost, leaving the skies to those avians, all sophisticated neornithines as far as we know, that were left over. The pterodactyloids had lasted 100 million years, not that much longer than the long-gone short-tailed rhamphorhynchoids.

The extent of the extinction at the Cretaceous/Paleogene (K/Pg)—formerly the Cretaceous/Tertiary (K/T)—boundary, which also saw the destruction of most of the sea reptiles (some marine crocodilians and turtles excepted) as well as many oceanic invertebrates, was such that the loss of the pterosaurs is not especially surprising. Pterosaurs had never been nearly as diverse as land dinosaurs, whose extinction is correspondingly harder to understand. Even birds at the end of the Age of Dinosaurs and Eupterodactyloids were much more diverse than the pterosaurs. Even so, many questions surround the extinction of a group that had existed for over 150 million years before vanishing at the same time as many other organisms.

A changing climate has often been offered as the cause of the demise of the pterosaurs and others. But the climatic shifts near the end of the Cretaceous were neither strong nor greater than those already seen in the Mesozoic—the world remained largely tropical and subtropical. And pterosaurs inhabited climates ranging from tropical deserts to icy polar regions, so it is questionable whether yet another general change in the weather would have posed such a lethal problem. On the other hand, pterosaurs were more linked to atmospheric conditions than were nonflying creatures. Perhaps changes in winds, which may have become more vigorous as mountains rose and big internal seas receded, were a problem for soaring azhdarchids and even more so for the lightly loaded pteranodonts and nyctosaurs. But there is no clear evidence this was true, in part because we do not fully understand giant pterosaur aerodynamics, and if azhdarchids were short-range power fliers, then stronger winds should not have been critical.

The withdrawal of the seaways may have been a factor for the marine pterodactyloids. Possibly pteranodontids and nyctosaurids were specialized for flying over these shallow seas between major landmasses. In that case, the reduction of these seascapes would have been a serious problem. But it is quite possible that species of nyctosaurs and/or pteranodonts were adept at flying and feeding over all the world's oceans, including the great Pacific, which was somewhat bigger than it is now. In that case, the loss of the shallow seas should not have been more of a problem for the wave-soaring pterosaurs than it is for today's albatrosses, and pteranodontids and nyctosaurids are known from the latest Cretaceous. It is possible, however, that some of the marine pterosaurs such as *Pteranodon* and *Nyctosaurus* were restricted to interior seas while other, unknown members of their groups were deep-ocean specialists. In that case, the loss of the shallow marine waters would have reduced pterosaur diversity somewhat, albeit not critically.

Mammals consuming pterosaur eggs are another proposed agent. But pterosaurs had been losing eggs to mammals for nearly 150 million years, as had reptiles and birds, without long-term ill effects, and pteranodonts probably bred on mammal-free islands in any case. Diseases spread as retreating seaways allowed once-isolated pterosaur faunas to intermix, but this would not have applied to marine species. Epidemics had failed to crash populations over a vast span of time because they were too diverse to be destroyed by one or a few diseases, and they would have developed resistance and recovered.

The solar system is a shooting gallery full of large rogue asteroids and comets that can create immense destruction. There is widespread agreement that the K/Pg extinction was caused largely or entirely by the impact of at least one meteorite, a mountain-sized object that formed a crater 180 km (over 100 mi) across on the Yucatán Peninsula of Mexico. The evidence supports the object being an asteroid rather than a comet, so speculations that a perturbation of the Oort cloud as the solar system traveled through the galaxy and its dark matter are problematic. The explosion of 100 teratons surpassed the power of the largest H-bomb detonation by a factor of 20 million and dwarfed the total firepower of the combined nuclear arsenals at the height of the Cold War. The blast and heat generated by the explosion wiped out the fauna in the vicinity, and enormous tsunamis cleared off many coastlines. On a wider scale, the cloud of high-velocity debris ejected into space glowed hot as it reentered the atmosphere in the hours after the impact, creating a global pyrosphere that may have been searing enough to bake animals to death as it ignited planetary wildfires that destroyed most of the world's forests. The initial disaster would have been followed by a persistent dust pall that plunged the entire world into a dark, cold winter lasting for years, combined with severe air pollution and acid rain. As the aerial particulates settled, the climate then flipped as enormous amounts of carbon dioxide—released when the impact happened to hit a tropical marine carbonate platform—created an extreme greenhouse effect that baked the planet for many thousands of years. The crash of marine life could have done in any marine pterosaurs that managed to get through the above. Similar shortages of food on land and extreme weather conditions would have crippled the continental azhdarchids. A critical factor may have been body size. While there were so many species of little birds that some made it through the crisis, the failure of pterosaurs to maintain a similar array of small species left the group with a low number of big-bodied pterosaurs that were statistically vulnerable to being wiped out. Those big wing membranes that could not be tightly tucked away from the weather during the postimpact cold pulse probably did not help. Such a combination of factors appears to solve the mystery of the annihilation of the pterosaurs (Witton 2013; Brusatte et al. 2014–15).

Even so, some problems remain. It is not certain whether the pyrosphere was as universally lethal as some estimate. Even if it was, heavy storms covering a modest percentage of the land surface should have shielded a few million square kilometers, equal in total to the size of India, creating scattered refugia. Birds and amphibians, which are highly sensitive to environmental toxins, survived the acid rain and pollution.

The way pterosaurs reproduced may have played a role in their loss. The only vertebrate powered fliers that survived, the neornithine birds, use mostly body warmth to incubate their strong-shelled eggs, so they were able to hatch their eggs during the postimpact cold snap (Erickson et al. 2017). Unincubated because they were soft shelled, pterodactyloid eggs would not have hatched in the chilly soil or too-cool vegetation mounds. And better parental care may have given bird babies a survival boost. The surviving birds were also non–forest dwellers that could get along in the largely tree-free postimpact world.

If the impact was the only exceptional big event that occurred in association with the extinction, then the latter could be readily and fully assigned to the former. But unfortunately for earth history simplicity, there was also another, longer-running matter as the Mesozoic transitioned into the Cenozoic that may have complicated the situation. Massive volcanism occurred at the end of the Cretaceous as enormous lava flows covered 1.5 million square km (over 579,000 square mi), a third of the Indian subcontinent. It has been proposed that the massive air pollution produced from the repeated supereruptions as they damaged the global ecosystem so severely in so many ways caused pterosaur populations to collapse in a series of stages, perhaps spanning tens or hundreds of thousands of years (Schoene et al. 2019). The result may have been a depopulated K/Pg pterosaur population rendered terminally vulnerable to an impact event. Others disagree. This hypothesis is intriguing because extreme volcanic activity also occurred during the great Permian-Triassic extinction, and similar volcanism may have been behind the extinction at the end of the Triassic. As the solar system orbits within the galaxy, it is possible that periodic encounters with a postulated thin plane of dark matter heats the earth's interior enough to initiate such bouts of supervolcanism (Rampino 2015). Although the K/Pg Deccan Traps were being extruded before the Yucatán impact, evidence indicates that the latter—which generated earthquakes of magnitude 9 over most of the globe (11 at the impact site)—may have greatly accelerated the frequency and scale of the eruptions. If this is correct, then the impact was responsible for the extinction not just via its immediate, short-term effects but also by sparking a level of extended supervolcanism that prevented the recovery of the pterosaurs. It is also possible that the Yucatán impactor was part of an asteroid set that hit the planet repeatedly, further damaging the biosphere.

AFTER THE AGE OF PTEROSAURS

The complete absence of nonavian dinosaurs, as well as most sea reptiles, freed up enormous evolutionary space for mammals to evolve into similarly large animals that have dominated the Cenozoic continents and the oceans, although it took about two dozen million years for therians to fully begin to do so. The lack of any pterosaurs left the surviving dinosaurs, the neornithine birds, unhindered by competition from other archosaurs in their domination of the skies, especially when the sun was above the horizon, and entirely when it came to the air above the big lakes and oceans. Most of the birds from the earliest

Cenozoic appear to have been waterbirds, and specialized oceanic examples rapidly evolved to replace the now absent giant marine pterosaurs, with an array of terrestrial birds following soon after, some of which were entirely flightless and quite big. The land bird explosion would culminate in the Late Cenozoic radiation of a massive diversity of small passerine songbirds, many with marvelous musical talents. Also supporting an array of avian species were islands, where some flightless birds would grow as heavy as horses.

The same lack of fine-grained Lagerstättes that hinders our understanding of pterosaur and avian evolution before and after the end of the Mesozoic also hides the beginnings of bats. All that can be said is that protobats may have been developing shortly before the end of the Cretaceous, and well-developed chiropterans show up in lake-bed deposits just a dozen million years after the last pterosaurs. They have become the primary night fliers in search of fruits and insects, the latter pursued with sophisticated acoustics available only to mammals with their exceptional hearing apparatus.

What if neornithine birds had not evolved by 66 million years ago, or had not made it through the crisis? Then we would live in a bird-free world, likely with bats dominating the daytime as well as the nighttime skies. Presumably the day-loving bats would be a remarkably diverse lot, with some examples filling niches currently occupied by birds. But it is unlikely that there would be mammalian equivalents of peregrines, woodpeckers, and songbirds. Probably few people would be putting out feeders for bats, and sales of field guides for bats would be mediocre.

BIOLOGY

General Anatomy

Most pterosaurs are known from their bones alone, but we know a surprising amount about their soft tissues from a small but quickly growing collection of fossils.

Pterosaurs were a fascinating mix of features found in their archosaurian relatives, the flying dinosaurian birds, as well as features of the flying mammalian bats, so the anatomy and function of these three groups are often compared in order to better understand pterosaurs and their often unique attributes. In this section, "bird" is usually used to refer to modern forms.

While it was not true of all pterosaurs, and some birds are big headed—toucans come to mind as well as hornbills—pterosaurs, the pterodactyloids especially, were prone to have large heads disproportionate to their relatively small bodies, and even to the span of their wings, much more frequently than birds, much less bats, none of which have particularly large heads. In most cases a large snout or beak was mainly responsible for the great size of pterosaur heads. No pterosaur had a proportionately very small head like some dinosaurs, including birds. Pterosaur heads were usually low and subtriangular in side view. Some were incredibly shallow. A few were strongly bowed downward at the midsection in side view. A few were deep, and some of these were more rectangular. Most were narrow in top view, others were broader, and those of anurognathids were astonishingly broad.

All birds have beaks covered in keratin sheaths. No bat has a beak, and neither did the blunt-snouted anurognathids. The jaws of many pterosaurs, even some with teeth, were beaked. In a few cases the keratin sheath that extended the length of the beak was preserved; sometimes the extension was modest, and in other examples it was considerable. In most restorations, the length of the beak extension is therefore a guess, and others are free to restore it differently.

Bird skulls tend to be lightly built and are notable for including kinetic heads in which the beak can be flexed up and down

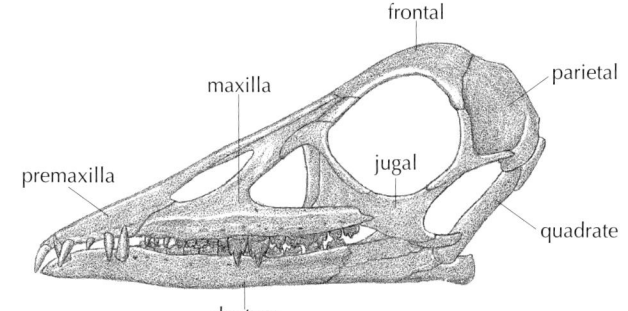

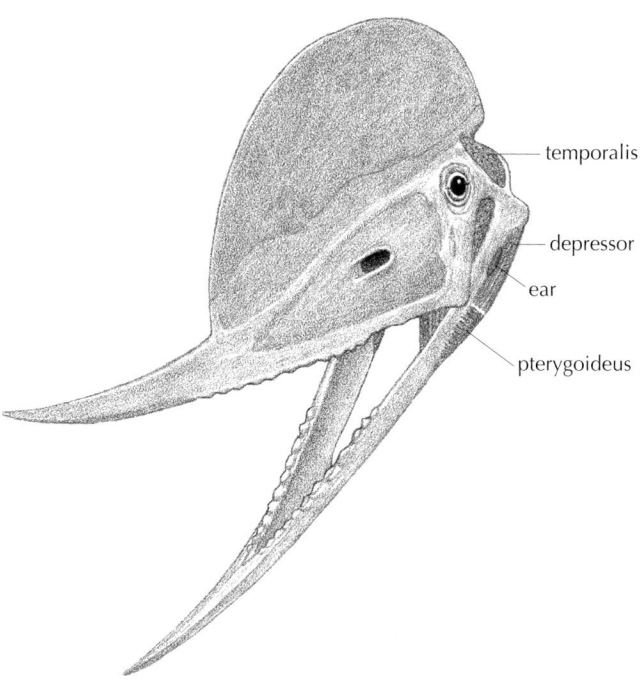

Pterosaur skull and muscles

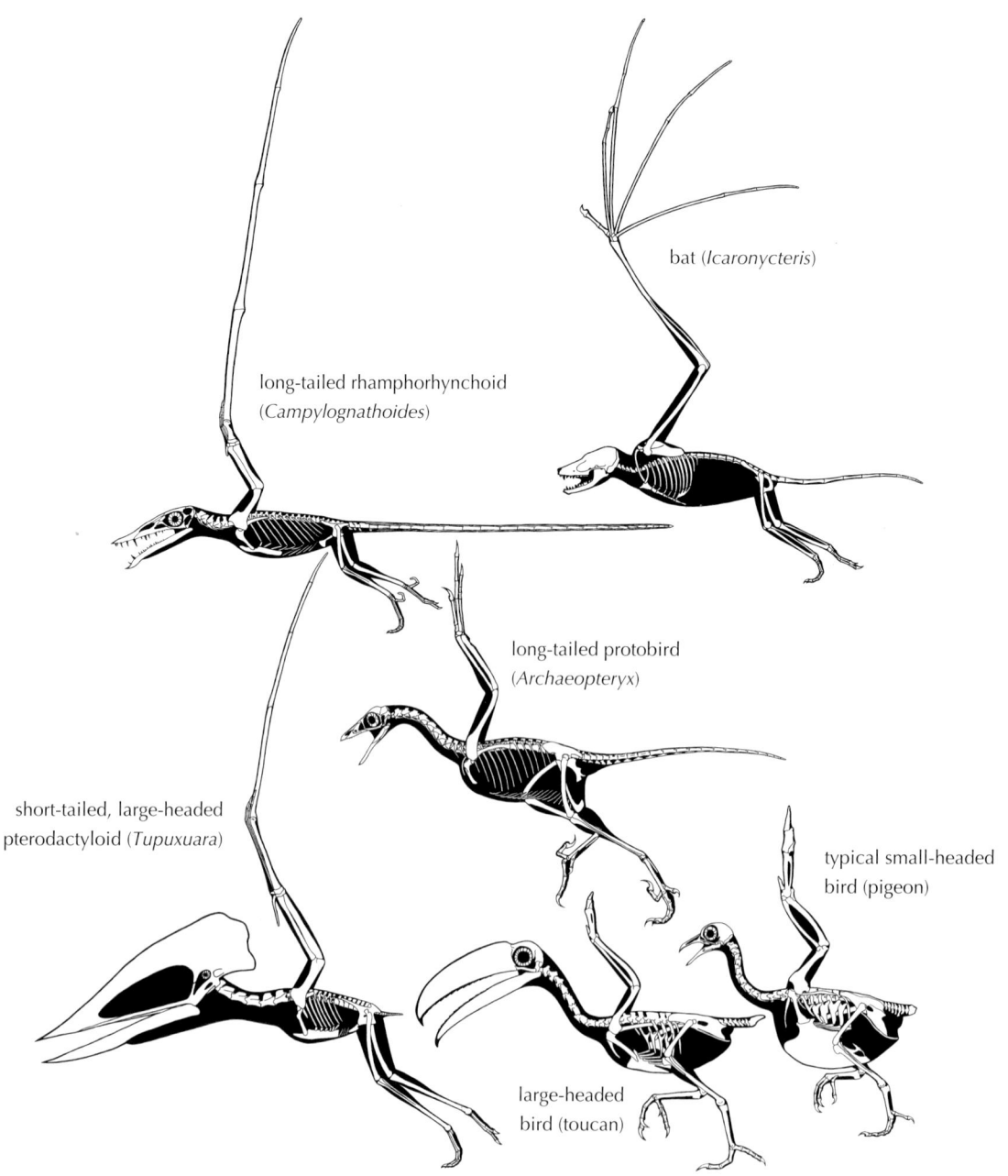

Pterosaur types compared with birds and a flying mammal

relative to the back of the skull—this is most obvious in parrots when they manipulate an object with their fingerlike beaks. Some pterosaur heads were remarkably delicately constructed, and with a few notable exceptions they were not heavily constructed. Even so, pterosaurs, like bats, had stiff, akinetic heads, although none of the former had the nearly solid-surfaced skulls of the latter.

In nearly all pterosaurs, the nasal passages or sinuses or both were very well developed, a feature common to most archosaurs in general. In particular, pterosaurs retained at least a fairly large opening immediately in front of the orbits that was part of this nasal complex. Except in the anurognathids, in which the beakless snout was severely abbreviated like that of a bulldog, pterosaur nostrils were never close to the tip of the upper snout, as they always are in bats. The backward shift of pterosaurian external nares was similar to that of similarly beaked birds, and the migration toward the orbits, leaving a long, solid snout or beak, became more extreme over time. Eventually the nasal opening began to merge with the preorbital opening until they formed a single large fenestra in front of the orbits. Such a merger does not occur in birds. Where the soft nostrils were set in these openings is not certain. In most animals the nostrils are very far forward in the bony external nares, but in some birds with large external nares the nostrils are in the center. In birds it is sometimes possible to see straight through the nostrils from one side to the other, and that could have been true of at least some pterosaurs.

When head crests were present, they ranged from small to incredibly large. In some cases the crests were entirely bony, while in others they were composed entirely of soft tissues such as keratin, and in others they consisted of both (Witton 2013). That some pterosaurs are known to have had soft crests even though there is no bony trace leaves open the possibility, but does not establish, that all species had a crest of some sort. Many pterosaur hard dorsal crests extended along most of the length of the skull, which was possible because the skull was akinetic—the bony top crests of some birds are set either in front of or behind the point of head flexion to avoid interfering with the mobility of the beak. Apparently, pterosaur crests were always fairly or very thin midline structures. In the relatively few birds that have them, crests, or casques or combs, are sometimes inflated from side to side even in fliers. No bat has anything close to a cranial crest.

Unlike the heads of mammals with their extensive facial musculature, and like those of reptiles and birds, the heads of pterosaurs lacked facial muscles, so the skin was directly appressed to the skull. Along with the close configuration of the keratin beak sheaths, this feature makes the life appearance of pterosaur heads easier to restore than that of mammals. The skin covering the large openings in front of the orbits probably bulged gently outward—these were *not* shallow depressions, as some artists show. Jaw muscles likewise bulged gently out of the skull openings aft of the eye sockets—in nearly all pterosaurs the muscles on the head were not powerfully developed. It is possible but by no means certain that the fairly short teeth of early pterosaurs were covered by lips, as in lizards, but in many toothed pterosaurs the teeth were much too long for that, so both the upper and lower teeth were fully exposed even when the mouth was closed, as in crocodilians. In pterosaurs with especially long teeth the result was a visually spectacular intermeshing of upper and lower teeth when the mouth was closed.

Pterosaur teeth varied from simple, short, and conical, to complex in a few early pterosaurs, to very long, slender spikes that gave many pterosaurs a freaky snaggletoothed look, to enormous numbers of fine bristles that formed a filtering comb. In a number of pterosaurs the teeth projected out to the sides to varying degrees; in some young rhamphorhynchids a tooth began on the side of the upper jaw and gradually migrated down until it was in line with the rest of the teeth in the adults. As is typical of reptiles including archosaurs, teeth were repeatedly replaced through life, a given tooth lasting some months or around a year or so. In most cases the exposed tooth was replaced simply by being pushed out from below by the replacement, which meant that for a substantial period the new tooth was too small to be fully functional. Pterosaurs developed a unique system in which the replacement tooth developed immediately behind the one it was replacing and grew to nearly two-thirds its final length before the old tooth fell out—which is why it sometimes looks as if two teeth are set right next to one another even when the teeth are overall widely spaced—so there was less irregularity in the rows of teeth.

Bony (sclerotic) eye rings were probably present in all species. Also found in birds but not bats, eye rings record the actual size of the eye, both in the diameter of the entire eyeball, and indirectly in that the diameter of the inner ring tends to match the area of the visible eye when the eyelids are open. Pterosaurs were prone to have the large eyes expected in visually oriented fliers. But some of the comb-toothed feeders were not. Relative eye size decreases as animals get bigger, and many pterosaurs had oversized heads; one or the other of these factors can give the false impression that the eyes were not large. So can how the eyelids cover most of the eyeball. One group that did have very large externally visible eyes was the anurognathids, in which the eyeballs were almost as tall as the head, and the very large sclerotic hole indicates that the visible eye was half the height of the head or more. Whether the pupils of pterosaur eyes were circular or slits is not known. The latter are most common in nocturnal animals, which the anurognathids may have been, and either may have been present in different species. The eyes of birds and reptiles are protected by both lids and a nictitating membrane, and the same was presumably true of pterosaurs.

The outer ear was a small depression between the quadrate and jaw-closing muscles at the back of the head. The eardrum was set in the depression and was connected to the inner ear by a simple stapes rod. The orientation of the semicircular canals of the inner ears has been used to try to determine the posture of pterosaur heads. The results seem to indicate that some pterosaur heads were normally tilted downward more than would be expected. However, the orientation of the canals is not consistent in living animals and can differ even among individuals. The latter issue is especially pertinent since ear canal orientation can usually be measured in only one specimen. It seems that the posture of the semicircular canals is determined at least in part by the orientation of the braincase with the rest of the skull and does not reflect the orientation of the head as well as has been thought.

Made of multiple elements—the jaw of mammals is unique in having just one bone—pterosaur lower jaws were never very deep, even in those species that had deep skulls. In some examples they were very slender, sometimes to an extraordinary degree. In a number of examples the front end of the jaws was adorned by a modest-sized, downward-projecting crested keel. In most pterosaurs the lower edges of the front portion of the lower jaws were joined together along the midline for a long distance, forming a narrow Y shape with an arced internal surface to the split portion in bottom view, and often a concavity ranging from subtle to obvious in side view. This anchored the rim of the throat pouch, which continued under the front portion of the neck—the farther forward the joining of the jaws, the longer the pouch, which is preserved in a few cases. Modest-sized hyoid bones indicate that pterosaur tongues were modestly developed, less so than is usual in birds, and fairly mobile (Li et al. 2018). Despite a fair amount of preserved soft tissue, no pterosaur is yet known to have had lower jaw or neck wattles similar to those of birds, although these cannot be completely ruled out.

The necks of many dinosaurs, including birds, tend to articulate in an S curve. Those of pterosaurs appear to have followed a shallow U curve, yet there are problems in accurately restoring neck posture when the cartilage and other soft tissues that help create the curvature are not present, and standing and walking animals often tend to carry the head higher than the vertebrae might indicate (Paul 2017b). But when flying, pterosaurs presumably kept the neck fairly horizontal in order to decrease frontal profile drag. Bird necks are extremely flexible because the articulations between the centra have a saddle shape that maximizes rotation in all directions, as do the extralarge auxiliary articulations, and there are many neck vertebrae, from 13 to 25. This allows some large, long-necked birds such as pelicans and herons to pull the head back and rest it on top of the trunk, including when flying. Pterosaurs lacked the extraflexible articulations and had only 7 to 9 neck vertebrae. So their necks were surprisingly stiff (Witton 2103), and some of the taller, longer-armed forms may have had some trouble reaching their head down to ground level to drink. Pterosaurs probably lacked the ability to suck—a trait limited among tetrapods to mammals and pigeons—in which case they had to scoop up water like most birds.

Pterosaur neck length varied from very short to very long. As in mammals, the variation resulted largely from extreme variation in the length of the vertebrae, from short to hyperelongated—unlike birds, in which individual vertebrae are never particularly long, so a high number of elements combine to make a neck long. Neck robustness was also widely divergent. In some short-necked examples, the neck was quite robust; when the neck was highly elongated it was always slender boned, even in giant pterosaurs. Shallow sets of nuchal ligaments have been preserved in a few specimens. These were necessary to help support the longer necks but may have been absent in shorter-necked species. The neck muscles of pterosaurs with very long, slender necks would not have been especially powerful, in part to keep body mass to a minimum. It is very possible that the individual neck vertebrae would have been visible as subtle bulges in long-necked pterosaurs, as in giraffes. Conversely, pterosaurs with robust heads and short, stout necks should have had fairly strongly developed neck muscles.

The bodies of pterosaurs are small relative to the head-neck complex and the wings, in some cases to an almost perplexing extreme—neither birds nor bats are prone to having such small bodies. The marine soarers were likely to have had quite small trunks relative to their wings, and giant azhdarchids, for all their great weight, were small bodied relative to their heads, necks, and legs.

Pterosaur trunk and hip vertebrae articulated in a dorsally convex arch that varied from fairly strong to perhaps nearly straight, as is true of birds. The nature of the vertebral articulations indicates that pterosaurs, again like birds, had stiff backs, to the degree that fusion of the shoulder vertebrae occurred in some pterodactyloids. The neural spines of the trunk vertebrae were not particularly unusual in rhamphorhynchoids and basal pterodactyloids, but in derived versions of the latter they cofused to form a continuous solid notarium that anchored the top end of the scapula—interestingly, the notaria appear to have evolved a number of times in pterodactyloids (Witton 2013; Aires et al. 2020). As in lizards, crocodilians, and dinosaurs including birds, the front ribs are strongly swept back in articulated pterosaur skeletons of all types and are not vertical as they are in many mammals. All the rib heads were mobile in all rhamphorhynchoids and many pterodactyloids, but the chest ribs commonly became fixed to their vertebrae in big pterodactyloids. Relative to the length and depth of the body, the rib cage was fairly broad—sometimes considerably so—and was widest at or just behind the shoulders and tapered backward. The belly was never very large because no pterosaurs were leaf-eating herbivores that needed large digestive tracts to process tough plant materials—this is like bats, but unlike some herbivorous birds such as grazing geese. The small volume of pterosaur bellies helps explain their small trunks. The ribs lacked the backward-projecting bony uncinate processes found in most birds. On the bottom of the belly between the sternum and the pelvis was a short series of about half a dozen gastralia, a series of flexible bony rods in the skin of the belly. The gastralia were flexible, each segment being made of multiple pieces of thin bone. Gastralia are fairly common in tetrapods, including crocodilians, and in some dinosaurs, including protobirds and early birds, but are absent in derived avians. The arrangement of pterosaur gastralia varied—in some they formed a typically simple series of subparallel rows, while in others the inner ends near the midline were clustered together and the outer ends were much more separated, forming a radiating pattern—both schemes are found among both rhamphorhynchoids and pterodactyloids.

Pterosaurs lacked clavicles, which are common in tetrapods, including dinosaurs. In some of the latter, including flying birds, the clavicles fused to form a furcula. In rhamphorhynchoids and basal pterodactyloids, the shoulder girdle was quite similar but not identical to that of avian fliers. The scapula blade was a fairly long, slender strap that sat nearly horizontally on top of the rib cage and met the coracoid at a sharp angle. The coracoid was also a strap-like element, somewhat shorter and more robust than the scapula, that ran vertically and a little back and down and inward to where it articulated near the body midline with a socket in the sternum. Interestingly, the articulation of the right and left coracoid with the sternum did not always exactly match up in side view, one being a little forward of the other.

In some pterodactyloids the back tips of the scapulae were close to but did not meet at the spines of the vertebrae. In derived pterodactyloids the scapula blade became shorter and stouter and was directed subvertically upward in side view, as well as strongly inward, so that the tip articulated with the notarium atop the spinal column, fixing the shoulder girdle firmly in place by bony contacts at both the top and bottom ends. The right and left scapula-coracoids now formed a near-vertical circle around the chest.

The sternum had two sections. In the front it was a modest-sized vertical keel. Behind the articulation with the coracoids it flared out sideways and became a broad, subhorizontal plate, curving somewhat upward toward the sides. There may or may not have been cartilaginous ventral extensions of the center line keel in a few cases, such as pteranodonts. Along the side edges were articulations for a series of short, ossified sternal ribs that connected the sternum to the main ribs. This is similar to birds, but while the sternal ribs of the latter are smooth edged, those of pterosaurs usually had sets of small, bony, lobed projections on both sides—these sternocostapophyses appear to be absent on the earliest pterosaurs. Farther to the back, in at least some cases, the sternal ribs became somewhat longer, in which case the sternum was pitched a bit upward at the front, rather than being completely flat in side view.

The bat shoulder girdle could hardly be more different. As is typical of mammals, the coracoid barely exists. Also typical of mammals, including humans, is that the scapula is a broad, subtriangular plate attached to the sternum by a long, slender clavicle via loose joints. Unlike the rigid avian and pterosaurian scapula, that of bats is highly mobile, and its movement makes a major contribution to the wing stroke. The dramatic movement of the wing at its base may contribute to the agility of bat flight. The sternum is merely an elongated rod attached to the ribs by simple ossified sternal ribs.

Although not as large as that of most birds, the pterosaur pelvis made up from a quarter to nearly half the length of the body, mainly because of the long, slender forward projection of the ilium. The bat hip is not particularly large. The sacrals are a fairly long series of fused vertebrae attached to the pelvis, which

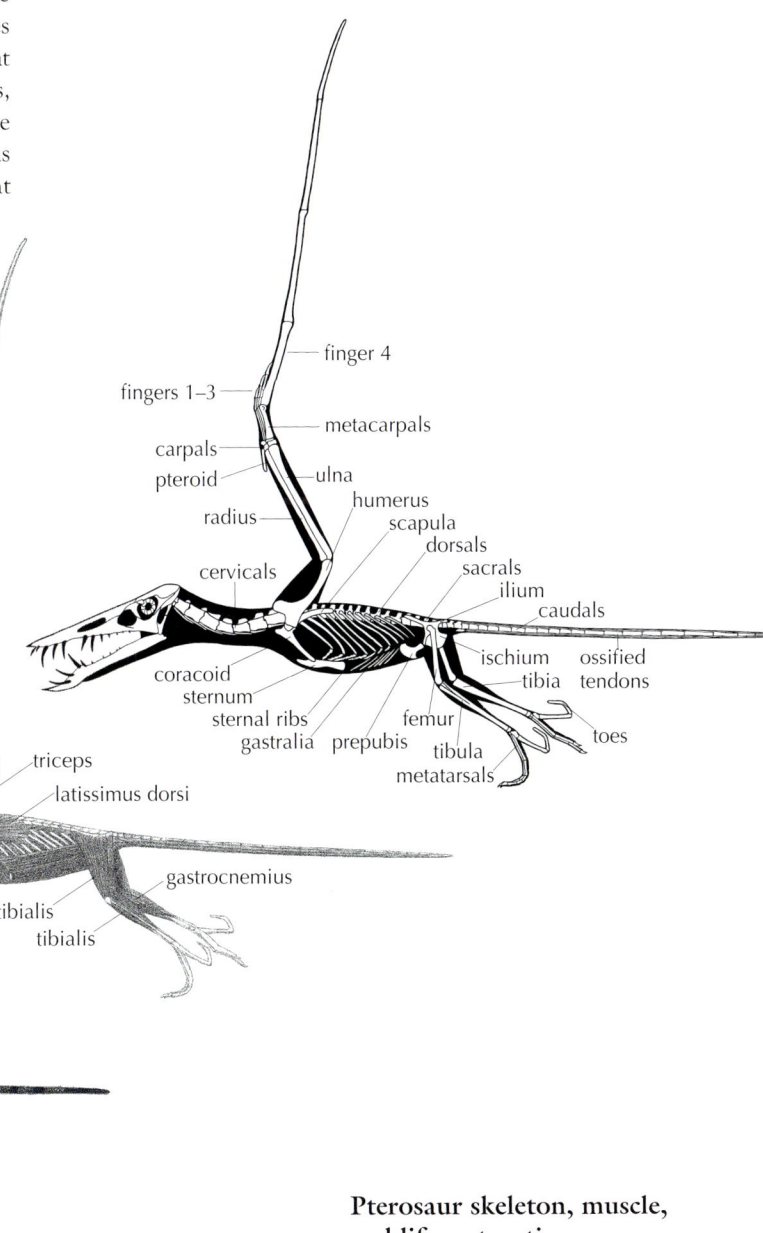

Pterosaur skeleton, muscle, and life restoration

in rhamphorhynchoids and pterodactyloids remained broadly similar in configuration. In top view the pelvis was fairly broad up front, narrowed toward the back, and shallow in side view, continuing the streamlined teardrop shape of the body. The pubes were vertical, short, curved plates. Articulating via a mobile joint with the front lower edge of the pubis was the prepubis, an unusual flattened bone of somewhat variable shape just behind the gastralia not found in other archosaurs.

Because dinosaur trunk vertebrae and ribs formed a short, fairly rigid body in which the shoulder and hip girdles were close together, the trunk musculature was rather light, like that of birds.

The tails of most rhamphorhynchoids were highly flexible at the base, and the rest of the structure was stiffened by the ossification of very long, very thin, largely forward-projecting tendon extensions of the vertebrae. Preserved soft tissues show that at least some of these long tails had adornments along part of the length or at the tip. In some cases these may have been vertical aerodynamic stabilizer rudders, while in others they were additionally or alternatively display adornments. The anurognathids and pterodactyloids had short tails that were fairly flexible along their length, except in pteranodontids, which sported a distal rod somewhat similar to the fused but short pygostyle at the end of bird tails.

The pterosaur shoulder joint faced primarily sideways so the arms could be held out to the side and raised high for flapping. This is similar to birds and bats, but important differences are present because while bird arms, lacking a suitable hand, are never used for moving on the ground, the pterosaur arm, which in almost all cases had a hand with clawed fingers, was used for such.

In the pterosaur arm the humerus was short, often very short, making up only a twelfth to less than a sixth of the bone span of the wing. This is markedly less than in most birds and all bats, in which the upper arm bone makes up more of the bone span—often a fourth, and well over a third in very long-spanned birds such as albatrosses in which the humerus is extremely slender as well as elongated. At the other extreme, some small birds with high flapping rates such as swifts and hummingbirds have short, thick inner arm bones that make up just a fifth of the bone span. The pterosaur humerus tended to be a stout element, often very much so, with a quite large, rather rectangular pectoral process projecting forward and somewhat downward near the shoulder joint. In birds the pectoral crest is longer along the shaft and less developed, except in basal birds, in which it is more prominent; the crest is very shallow in bats. The elbow hinge joint allows nearly 180 degrees of rotation of the lower arm in bats, birds, and pterosaurs, from folded nearly flat against the upper arm to nearly straightened out. The pterosaur radius and ulna were always at least moderately elongated, as in birds. And as in birds and other dinosaurs—but unlike in many mammals, including humans—the two outer arm bones were not able to rotate much relative to one another along their long axis. The same is true of

bats because all but the inner portion of the ulna is absent, as it is in a number of running ungulates. In pterosaurs the radius was in front of the ulna, and a little higher when the wing was held out horizontal.

In birds and bats the wrist is where the outer wing folds against the inner wing. In the flying dinosaurs this was via a prominent pulley joint, with the pulley roller being part of the hand so the roller surface faced inward. The primary wrist carpals of adult pterosaurs were simplified into two large, interlocking blocks, which severely limited wrist rotation along its long axis. How much fore-and-aft hinge motion was possible at the wrist is not entirely clear. Flexing the wrist a modest degree was allowed by sliding articulations, but going farther would have required disarticulating the carpal blocks. This occurs in animals such as horses when the forelimb is not in contact with the ground. There may have been a few degrees of up-and-down movement outside that simple fore-and-aft plane.

The three petite inner fingers, which were probably one (the thumb) through three, varied some in length, becoming progressively and markedly longer proceeding outward. The number of joints increased progressing outward as the fingers became longer. The variation in the position of the fingers recorded in articulated fossils and handprints shows that the digits were highly flexible. They could move inward for gripping items, and outward so they could be used to walk with a posture that was like that of dinosaurs, dogs, and cats in being digitigrade—the wrist does not contact the ground—but more similar to that of bats in that the fingers were directed sideways—and in the case of the third finger backward when on the ground—rather than forward. And the fingers could move fore and aft to a limited extent so they could be tightly appressed to the main finger during flight to minimize drag. It has been claimed that a derived rhamphorhynchoid had an opposable thumb adapted for climbing (Zhou et al. 2021), but this is not certain.

In aerial vertebrates something is often superelongated to support a large airfoil. In some reptiles it is the ribs. In birds it is the outer wing feathers; the hand is not especially long. Among archosaurs the outer, fifth finger was typically quite short, so it was probably lost in pterosaurs, which probably had the fourth finger hyperenlarged into a superfinger. The resulting outer wing was folded against the inner at the finger base via a prominent pulley joint that was oriented in the direction opposite that of birds. This allowed nearly 180 degrees of motion of the large wing finger, from flat against the inner hand as shown by some articulated specimens, to entirely or nearly in line with it. Movement of the joint other than fore and aft was probably limited to a little passive motion. The great length of the pterosaur finger is similar to the bat's four superlong wing fingers, which also fold up at their bases, but it is quite unlike the short main, central finger of birds, which is very rigid relative to the inner hand. The elongation of the pterosaur wing finger is more extreme than that of bats because the shorter inner arm of pterosaurs means the big finger had to make up a greater portion of the total span. In nearly all pterosaurs the wing finger had four elements, but in a couple of known examples there are only three finger bones. The bones were usually a flattened triangle in cross section, with the point at the leading edge to keep the frontal profile at a minimum to reduce drag. The exception were azhdarchoids, in which the cross section was a very shallow, inverted T shape. The wing finger was nearly straight but generally had a gentle convex forward arc that seemed to vary in its degree. The outermost finger bone was sometimes straight, but more often it arced backward to some degree. In one known rhamphorhynchoid, the wing-tip bone instead arced strongly forward. None of the joints in the pterosaur superfinger were mobile, as they are in bats, but neither were the joints ossified, and the very long, slender bones were probably somewhat bendable, so the great digit remained passively flexible to a modest degree.

That all the major arm joints including the elbow outward—the elbow, wrist, wing finger base, and other joints—were either simple fore-and-aft hinges or immobile means that the pterosaur arm was a fairly rigid structure as far as rotation along its long axis. So its pitch as a wing, or its rotation when walking, could be strongly adjusted only at the shoulder joint, as is true of birds. With their many wing fingers, bats can alter wing pitch to varying degrees along their airfoils, rather similar to the wing warping of some early light-winged aircraft such as the Wright Flyers.

Birds have a complex means of tightly folding their wings at the elbow and wrist, which is not found in pterosaurs. In most animals, at the elbow the condyles of the humerus that articulate with the radius and ulna are fairly symmetrical, and both are set about an equal distance from the shoulder joint. So when the elbow is closed, the radius and ulna do not slide alongside one another. Birds are distinctive because the outer condyle for the radius is set substantially closer to the shoulder joint than is the inner condyle for the ulna. This means that when the elbow folds up, the radius slides forward along the ulna, which rotates the wrist in a manner that improves the ability of the wrist pulley to in turn fold the hand against the outer arm. Because the pterosaur outer wing folded up primarily at the finger base while the wrist remained primarily straight, pterosaurs did not need to slide the radius along the ulna to help create extreme wrist flexion. So the elbows of most pterosaurs small and large had a more typical tetrapod structure like that of bats, with symmetrical humerus condyles for the radius and ulna.

In the hindlimbs, the pterosaur hip joint is most like that of many mammals, including humans. The socket is a cup without an internal opening. A correspondingly spherical femur head caps the end of a somewhat narrower shaft that connects it at a modest angle to the rest of the femur. In dinosaurs, birds included, the hip socket is an internally open horizontal cylinder into which fits a cylindrical head of the femur that projects inward at a strong angle, up to 90 degrees, from the main shaft of the thighbone. The dinosaurian system is a hinge joint that strongly limits femoral action outside the prescribed near-vertical fore-and-aft arc. However, in no theropod, including birds,

is the knee as close to the body midline as is the hip joint. The knee is bowed out to some degree in order to keep it clear of the deep body just forward of and under the hips. In birds with especially big bellies between the knees, such as geese and the big ratites, the femur angles strongly out from the hip joint as much as 40 degrees, so the knees are strongly bowed out. The shank then slopes strongly inward, so the gauge of the hind feet is always narrow in birds. The upper leg posture of bats is quite different. The hip socket faces strongly sideways and even a little upward, so their femora sprawl strongly to the sides, which aids their legs in supporting the aft membranes.

The sphere-in-a-cup hip socket of pterosaurs may have given their thigh much more freedom of movement, more like many mammals, including humans, than birds. That could have allowed the hindlegs to splay out horizontally to support and control wing and leg membranes during flight, and to be directed downward when walking and running (Witton 2013). In most pterosaurs this wide swing from the legs being sprawled during flight to the feet being under the body was achieved by the hip socket facing strongly sideways, even upward to a small degree. Because the femoral head was not directed strongly outward, the femur could not point directly downward as it can in humans. If so, the standard pterosaur femur and knee had to be bowed out strongly, despite the absence of a broad belly between the knees. In some strongly terrestrial pterosaurs such as the azhdarchids, if not the azhdarchoids they were nested within, the hip socket faced somewhat ventrally. Combined with the angle of the head shaft to the rest of the femur, this should have allowed the femur to be held more vertically downward than was usual for pterosaurs. Other researchers disagree with the above scenario. They argue that the ligaments that held the femur head in the hip socket did not allow the femur to splay out horizontally (Manafzadeh and Padian 2018). It is possible that these factors varied among pterosaurs.

The pterosaur knee and ankle were hinge joints, and the tightly bound tibia-fibula complex in which the latter bone did not connect with the ankle could not be rotated along its long axis, so movements of the lower leg were not especially flexible—as in birds and some mammals, bats included, but unlike many mammals, including humans. The knee was probably permanently flexed and could not be fully straightened out, as in birds but not as in we humans. The knee could be tightly folded. The ankle could be strongly folded, and nearly straightened in order to streamline the foot directly backward during flight; it could also aid in pushing off the ground when walking, running, or taking off.

Pterosaur feet sported both avian and bat-like features; they were plantigrade, with the ankle contacting the ground as in bats, bears, and humans, rather than digitigrade as in cats, dogs, and dinosaurs, including birds. Bird feet are prone to being long and narrow. The three inner metatarsals that make up the bulk of the upper foot are at least moderately long and slender and fused into a solid unit. The outer toe is lost, the inner toe when present is a short, reversed hallux, and the three main toes are fairly long, with the central toe the longest. In many birds the toes are flexible and in many cases provide a firm grip. Although the feet of some highly aerial birds like swifts, swallows, and nightjars are somewhat shorter, they retain the basic avian configuration. Bird toe claws are extremely variable in form. Bat feet are uniform in general design—all are short and broad; the short metatarsals are not tightly appressed to one another, much less cofused; all five digits are similar in length and have large, hooked claws; and the shortish toes are subequally proportioned. Fairly uniform in configuration, pterosaur feet were moderately long and markedly narrower than those of bats yet broader than those of birds. The four main slender metatarsals were not fused but did not splay out as much as in bats. The toes were medium in length, with the inner two a little longer than the outer two. They were not especially flexible, and the claws were not particularly large. It has been claimed that an anurognathid had a reversible inner toe well adapted for perching (Lü et al. 2017), but this is not certain. The rhamphorhynchoid outer toe was a large divergent digit specialized for supporting aeromembranes. In form and function, it was rather similar to the slender calcar spur that emerges from the outer side of the ankle of many bats. The pterodactyloid outer toe was reduced to a nubbin.

Walking, Running, Climbing, Swimming

The way pterosaurs moved on the ground, or whether they even did so much, has long been controversial. Aside from flightless marine examples, and some small aerial insect hunters and hummingbirds, most birds are competent to excellent ground walkers and runners, some to the point of being flightless. Ostriches are so specialized for speed that they have only two toes, only one of which is large, and the back part of the toes is held clear of the ground, helping make them among the fastest land animals. Bats are largely the opposite. Aside from vampires, most spend their nonaerial time hanging from tree branches or cave ceilings. Some researchers have seen pterosaurs as more analogous to the latter in avoiding much contact with flat ground. If some or most were competent ground creatures, then questions arise as to whether they were predominantly or entirely bipedal, like their avian archosaur relatives, or largely or always quadrupedal, perhaps with a wide-gauge sprawling gait like that of membrane-winged bats. There is less dispute these days, but there are still some serious issues concerning restorations of pterosaurs on the ground.

One thing that is clear is that, like the trunk of dinosaurs, birds included, the pterosaur trunk was too rigid as well as too short to contribute to ground locomotion by flexing laterally as in lizards and crocodilians, or vertically as in galloping young crocodilians and many mammals.

Fossil trackways are crucial to solving the pterosaur land loco-motion problems. For one thing, their existence demonstrates that some pterosaurs did walk about and appear to have been reasonably well adapted for doing so. The trackways further put constraints on how they did so by establishing where the hands and feet were placed, and their orientation. Once very scarce and of uncertain identity, pterosaur prints are increasing in number, but they remain much rarer than those of dinosaurs, of which enormous numbers are known. The relative scarcity of pterosaur prints is a potential problem, because it leaves open the possibility that varying ways in which they moved around are not yet known from our limited sample.

The difficulty has been particularly acute when it comes to rhamphorhynchoids, because there were no unambiguous prints attributable to the long-tails until shortly before completion of this book (Mazin and Pouech 2020). For that matter, pterosaur trackways are not yet known from the Triassic or the Jurassic until near its end despite the footprints of other small creatures being common. Therefore, we do not know whether rhamphorhynchoid ground locomotion changed over time, or whether most or all rhamphorhynchoids always moved about on all fours, as the few Late Jurassic rhamphorhynchoid tracks indicate, or bipedally, at least sometimes in some species.

Otherwise, the pterosaur prints found so far appear to be those of later Jurassic and Cretaceous four-toed pterodactyloids. It is said that there are thousands of pterodactyloid prints, but only a few have been published. Nearly all these trackways show the pterodactyloids that made them clearly walking on all fours, leaving no doubt they could be quadrupedal. But a few examples of large Cretaceous Asian pterodactyloids seem to show only the hindprints and may have been laid down by bipeds.

Regarding the forelimb, by being bipeds, birds avoid the many anatomical difficulties and resulting oddities that stem from using arms primarily adapted for cutting through the air for a very different type of movement on terra firma. Bats are much more informative because they show how one group of arm-winged fliers employ their winged forelimbs for ground locomotion, particularly the vampires. When they walk, their humerus is held horizontally and moderately swept back, and the outer arm is nearly vertical in front view, with the hands falling on a fairly wide gauge, and the palm of the hand faces inward while the one free finger is directed straight out to the side. No other modern animal walks like this.

The known pterosaur trackways do not immediately settle the question of their arm posture because the gauge of the handprints is extremely inconsistent. Lateral hand separation can be fairly narrow, only a couple of hand widths in the case of some of the largest known pterosaur tracks, those of a gigantic azhdarchid with a wingspan of about 6 m (20 ft) (Hwang et al. 2002). Gauge is moderate in most pterosaurs, with a separation of four to eight hand widths—these are small animals, and there is no consistent difference between rhamphorhynchoids and pterodactyloids in this regard. The separation is an extreme

two dozen hand widths in one of the smallest examples, which happens to be a pterodactyloid.

Handprint orientation goes a long way toward solving the pterosaur arm posture problem. The free fingers were widely divergent, and in pterodactyloids, as in bats, they were splayed out predominantly sideways, albeit in some cases also somewhat backward, with the inner digit pointed forward or sideways, and the outer of the three almost straight behind. The free fingers projected sideways and a little backward, meaning that the palm of the inner hand faced inward and a little forward. That in turn means that the wing finger, folded via the big pulley in the same plane as the rest of the inner hand, was directed a little inward rather than directly behind the rest of the hand, a fact verified by how traces of the wing finger preserved in the most sprawling of the pterodactyloids point backward and somewhat inward. As for the rhamphorhynchoid handprints, the fingers

look as though they were—rather surprisingly—directed more strongly but not entirely forward, an orientation more like that of most animals.

As noted above, the pterosaur radius was fixed in about the same plane as the ulna, so the two bones could not be rotated to cross one another, as we supple-armed primates can easily do to swivel the hand along its long axis relative to the inner arm. The pterosaur radius and ulna were in turn in the same plane as the flat of the hand, with the radius in line with the metacarpals of the free fingers and in front of the ulna, which anchored the wing metacarpal, and the humerus condyles for the radius and ulna were about the same distance from the shoulder joint in most examples. Exceptions are the small dimorphodonts and the big azhdarchids, in which the radius articulations were placed somewhat closer to the elbow, forming a more angled elbow articulation as in birds. The orientation

Pterosaur and other trackways drawn to same hind stride length

of the hand in trackways indicates that the radius was indeed ahead of the ulna when pterosaurs were moving on the ground on all fours, and apparently somewhat inside the radius in the rhamphorhynchoids.

A number of researchers and artists who have illustrated pterosaur forelimbs being used for walking have neither paid sufficient attention to the above and other anatomical details, nor illustrated them in all views with the hands and feet fitting properly into the trackways, the joints all correctly articulated, and the bones in proper orientation. It has become fairly common, but not universal, to show the humerus positioned as it is in walking ungulate mammals and quadrupedal dinosaurs—directed back and down from the shoulder joint, with the elbow tucked in near the body (Bennett 1997; Unwin 1997; Witton 2013). Meticulous detailed restorations based on complete, uncrushed, properly articulated pterosaur arms showing that this works have not been presented, and this very narrow, parasagittal inner arm posture is contrary to the wider gauge

of most pterosaur hand trackways and entirely incompatible with the most splayed-out examples. This arm posture also puts the outer condyle of the humerus and the radius with which it articulates directly to the outside of the ulna, resulting in a number of anatomical difficulties.

To start with, with the pterosaur humerus in the above ungulate-like position, the palm would face directly forward. You can see how this happens with your own arm and hand. Begin by placing one of your upper arms so it slopes down and a little forward from your trunk, with the elbow flexed 90 degrees, as you do when sitting and using a keyboard—this is the equivalent of the erect posture some researchers use for the pterosaur arm, with the proviso that the pterosaur's trunk is of course tilted more horizontally when standing and walking. In that case, with the palm facing the keyboard, your radius and ulna are crossed nearly as far as they can be, something pterosaurs could not come close to doing. Now, rotate your hand so the palm faces upward. That uncrosses the radius and ulna, which was how

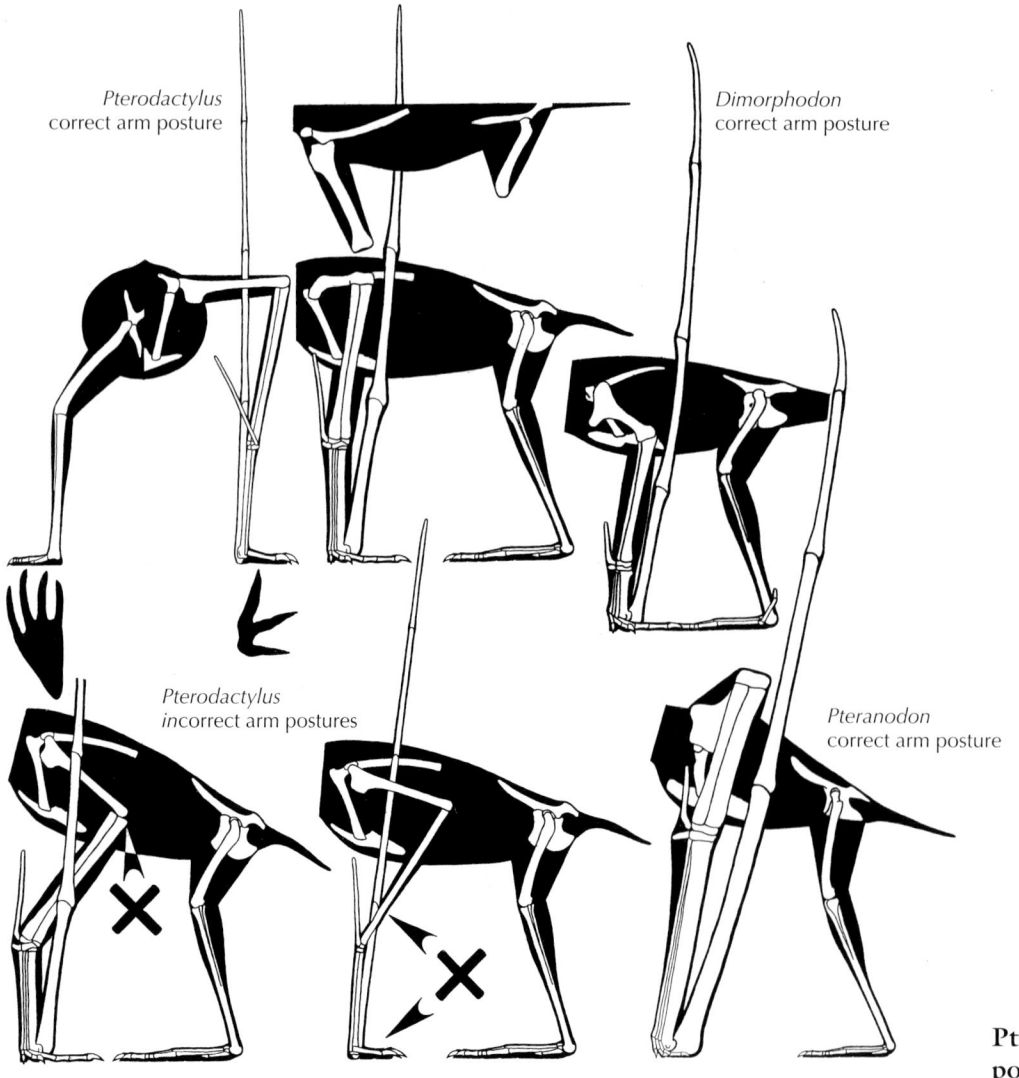

Pterodactylus correct arm posture

Dimorphodon correct arm posture

Pterodactylus incorrect arm postures

Pteranodon correct arm posture

Pterosaur standing postures

the pterosaur outer arm bones were permanently fixed. Your upward-directed palm is now equal to the forward-facing palm of the pterosaur. The obvious problem is that no animal walks with its palm facing forward and fingers pointing directly backward. Also in error is that the free fingers would be directed backward with the outermost digit pointed directly inward, contrary to the trackways. And the folded wing finger would be pointing too directly inward. Not only is that not shown in the illustrations with tucked-in elbows, but even when tightly folded, the outer wing is at high risk of scraping alongside the body, and perhaps even wedging and jamming up between the inner arm and body wall. If the pterosaur radius was placed outside the ulna, then when the wrist was flexed fore and aft it would have bent in the wrong plane of action for these fliers with their hinged wing joints.

Some have tried to get around the above defects of tucked-in elbows by vaguely illustrating the elbow end of the radius of walking pterosaurs as articulating with the humerus much closer to the shoulder than the ulna, and therefore the radius head is placed too far inward relative to the elbow end of the ulna. This is actually what occurs in quadrupedal dinosaurs, but pterosaurs do not possess the special, laterally expanded proximal ulna head that allows this in the dinosaurs. Or the illustrators show the radius and ulna crossing one another, which occurs in many mammals but not in pterosaurs. Or they fudge the illustration by including what is in effect a long-axis twisting of the main metacarpal of the inner hand so the palm faces inward rather than forward, as the rest of the arm articulation indicates it should. The pterosaur skeletals shown in takeoff position in Witton (2013) feature the radius head too close to the shoulder joint, and the small fingers directed too strongly backward—although the latter may be an illustration artifice to allow viewers to compare the dimensions of those digits.

Yet another problem with directing the humerus strongly downward in walking pterosaurs is that it places the elbow well below the shoulder joint. This is not a problem in most animals because the length of the arm from the shoulder joint to the walking fingers is fairly similar to the length of the hindleg from the hip joint to the flat of the foot. But in many pterodactyloids with elongated inner wings, the distance of the arm from the shoulder joint to the hand is much longer, up to over two times longer, than the distance from the hip joint to the ankle. Placing the elbow so low only exaggerates the length disparity, by a sixth to a quarter, forcing many pterosaurs to be restored with the arms awkwardly and impractically splayed forward, with the hands far forward of the elbow and the wrist bent strongly downward. This produces a bizarre, praying mantis–like pose not seen in any vertebrate, including bats, which avoid that extreme pose by keeping the humerus horizontal—a pterosaurian version of a shopping cart; and/or pitching the body strongly upward.

The way to eliminate or at least reduce all these problems is by positioning the humerus sprawled well out to the side, with the elbow at about the same level as the shoulder joint, or somewhat lower in the few examples with short outer arms and inner hands, or higher when the arm from elbow to wrist is considerably longer than the leg from hip to ankle. Among pterodactyloids the humerus is swept back to some degree, not very different from its posture during flight at midstroke. The more angled elbow articulation of dimorphodonts and azhdarchids suggests that their humerus would have been most strongly swept back. This bat-like, horizontal, swept-back posture of the inner arm places the radius ahead and a little to the outside of the ulna when the outer arm is directed downward about 90 degrees from the elbow so that the radius-ulna unit is subvertical. In front view, the hands are then almost directly beneath the elbows. Because the pterosaur humerus was short, this in turn causes the hands to follow the modest-gauge handfall pattern preserved in most pterodactyloid trackways.

To mimic this arrangement, start by putting your arm in the position described above, with your upper arm parallel to your body, the elbow flexed about 90 degrees, and your hand facing upward so the radius and ulna are uncrossed. Then, while holding everything from your elbow on down the same, swing the upper arm so it is nearly horizontal, being careful not to start crossing the radius and ulna in the lower arm. The palm of your hand will now face inward and just a little forward relative to the long axis of your trunk, just as the pterosaur trackways record. And the hand is well out to the side, as most of the trackways show.

The rhamphorhynchoid trackways with their forward-facing fingers suggest that their upper arms were sprawled straight out to their sides, as is common in reptiles and in some primitive mammals. With the humerus subhorizontal, the gauge of the handprints was again not narrow.

Because the azhdarchid outer arm and inner hand from the elbow to the finger base was exceptionally long relative to the length of the humerus, a small inward slope of the arm below the sprawled elbow on down would have placed the hands near the midline, resulting in their narrow-gauge handprints. If instead pterosaur outer arms and inner hands were sloped more outward as vampire bats can do, then the result would be a wider-gauge handfall pattern, all the way to the extreme seen in the most sprawling pterosaur trackway.

In side view, the hands of pterosaurs with arms of moderate length are now normally and elegantly under the shoulders and elbows as in normal animals, rather than gawkily directed far forward as in many restorations. The wrist was held nearly straight in accord with the natural articulation of its blocky carpal bones, and as is normal when the hands of animals are in contact with the ground. Meanwhile the free fingers were splayed forward and a little outward in rhamphorhynchoids, and outward and a little backward in pterodactyloids—as in the trackways—while the wing finger was mainly behind the inner hand and lower arm and not directed strongly inward, so it did not bang into the body.

As the pterosaur walked, the big pectoralis muscles that produced most of the thrust-generating downward flap during flight held up the front end of the animal by holding the humerus constantly close to horizontal. That would use a fraction of the muscular work produced by the pectoralis during powered flight. Most of the fore-and-aft arm rotation occurred at the shoulder joint using the same wing-pitch-altering muscles used during flight to rotate the humerus along its long axis. The large, downward-projecting pectoral crest of the humerus was used as a back-and-forth muscle lever near the shoulder. Swinging the elbow end of the humerus a little back and forth could have added a corresponding amount to the length and power of the propulsive stroke. Adjusting the horizontal angle of the humerus would have placed the elbow at the changing heights needed as the lower arm and inner hand swung from forward at the beginning of the propulsive stroke, to vertical at the middle, and backward at the end. At the end of the propulsive stroke the downward motion of the humerus would have added to the power of the final push-off. It is possible that the wrist helped compensate for the elbow height changes that occurred during the propulsive limb swing by flexing in the middle of the stroke. On the other hand, in tetrapods the wrist is typically held straight during the propulsive stroke and is flexed only to help keep the hand clear of the ground during the recovery swing. The latter would also be facilitated by lifting the elbow higher during recovery than during the propulsive stroke.

The shoulder joints of most pterosaurs were more open at the back than are those of birds. This is a major reason that some argue that pterosaurs had an erect arm posture, but it may instead reflect other differences between the two groups. Being quadrupedal, pterosaurs could probably take off with a push from their arms (Habib 2008; Witton 2013), which probably required the humerus to swing farther back than the humerus of bipedal birds, which never push off the ground with their wings. Also, when folding their wings, birds have to point the stiff inner wing feathers downward rather than inward to keep them from being compressed against the body wall. In order to do that, a very unusual and complicated system evolved in which, as the arm folds against the body, the humerus rotates along its long axis so its back edge and the feathers anchored on it are directed downward. What is the lower side of the humerus when the wing is held out for flight now faces to the side, and the radius and ulna fold up alongside that. During this operation, the long-axis rotation of the humerus allows its head to remain in articulation with the sideways-facing shoulder joint. Because the pterosaur inner wing membrane was highly flexible, there was no need for the humerus to rotate along its length, so the wing folding was much more straightforward, consisting of just the arm being folded, with the lower surface of the humerus continuing to point in that direction. That required the pterosaur shoulder socket orientation to include a strong backward component.

As far as the prints left by pterosaur hindlegs, the variation in the lateral gauge of the foot trackways is not as extreme as that of the hands. That is because there is no example of extreme leg sprawling. The foot gauge width was usually moderate in rhamphorhynchoids and pterodactyloids alike, the feet being laterally separated by five to eight foot widths even when the hands were splayed much wider. Exceptions were the giant azhdarchid prints, as well as the even larger Early Cretaceous pterodactyloids in Asia, with wingspans of up to perhaps 10 m (30+ ft) (Kim et al. 2012), whose feet were separated by one-half to two foot widths, a fairly narrow gauge similar to that of the hands in the case of the quadrupedal azhdarchid tracks. In some other cases, the hands and feet also follow much the same gauge, albeit broader, while in others the handprints are well outside those of the feet.

The typical pterosaur combination of semierect femora with a vertical shank posture kept the feet from sprawling far out to the sides, but the result was the considerable distance between the left and right feet preserved in most pterodactyloid trackways. In azhdarchids, if not in the azhdarchoids they were within, the more erect leg posture allowed by their more downward-oriented hip socket is in line with the narrow gauge of their trackways.

Pterosaur hindprints show a very long heel pad pressed into the ground behind the toes, key evidence that their feet were plantigrade rather than digitigrade, as were the feet of their avian archosaurian relations, which also have a simple hinge-joint ankle. This is somewhat perplexing because, aside from how carrying the ankle high off the ground would reduce the common disparity between the length of the arms and legs, pterosaurs lacked the backward-projecting Achilles heel that provides leverage for the flat feet of bats, humans, bears, and crocodilians, but flat-footed lizards also lack such a heel lever. The trackways also affirm that the foot was directed predominantly forward, often with a sideways splay to varying degrees, a normal tetrapod orientation quite unlike that of bats, whose feet are oddly directed straight to the side like their hands.

Being front heavy, the hindlegs bore less of a pterosaur's weight when walking than did the more robustly boned and powerfully muscled arms. That was most true in the often big-headed, sometimes long-necked, and always short-tailed pterodactyloids. Because pterosaurs were not bipedal the way birds are, pterosaur leg muscles were correspondingly probably not as bulging and powerful as those of avepods and were more like those of humans. With the rather short pelvis, the thigh was fairly narrow, although the platelike ischium below and behind the hip socket supported strong leg-pulling muscles. Pterosaurs lacked the very prominent forward projection of the tibia at the knee found in avepods, including birds, so the shank muscles did not form as prominent a drumstick shape. The feet were operated largely by tendons coming down from the shank.

Trackways and anatomy reveal that pterosaurs were not outright fore-and-aft sprawlers like lizards, nor were the tracks generally as broad gauged as those of the bats best adapted for getting around on the ground, the vampires. Neither were most pterosaurs as erect legged and narrow gauged as dinosaurs, including birds, that sport very narrow-gauge trackways—this being true even of waddling ducks, geese, and pigeons—or many mammals, humans among them. Any who disagree need to produce detailed restorations of undistorted pterosaur limbs with fully articulated joints that actually fit into the trackways. Most pterosaurs were somewhat like crocodilians, which often use a semierect high walk that produces a trackway of similarly moderate breadth. The pterosaurs most like erect-gaited quadrupedal dinosaurs and mammals were the exceptionally terrestrial azhdarchids, although even they do not appear to have had tightly tucked-in elbows and knees. Note that differing limb postures and designs—sprawling versus erect, flexed versus straight jointed, short versus long limbed, heavy versus light footed, bipedal versus quadrupedal—have remarkably little effect on the energy cost of moving a given distance for any given body mass (Paul 2002, 2012). So pterosaurs were likely to have been typical in this regard. Nor does walking or running a given distance make much difference as far as energy loss. The cost is similar regardless of speed, although the effort per unit time rises with speed. What does make a difference is moving on land versus in air—the former is rather costly on a distance basis, burning about three times as much energy as does power flying the same distance.

In pterodactyloid trackways, the hindprint is set a little ahead of the foreprint, while for dinosaur trackways, the opposite is true. This is probably because pterodactyloid limbs were so long relative to their short bodies that as the hand lifted off the ground just before the forward-swinging hindfoot on the same side was set down—the normal footfall pattern of walking animals—the hindfoot overstepped where the hand had just been. The footfall pattern of rhamphorhynchoids was less consistent, and sometimes the foreprint was placed ahead of the hindprint. That may have been because the short-limbed long-tails had fewer problems with leg interference.

The length of the stride compared to the height of the hips—two to four times the length of the foot in pterosaurs—can be used to approximate the speed at which the track maker was moving via a formula that indicates that the known pterosaur fossil trackways were laid down at speeds of 4–14 km/h (2.5–9 mph).

That pterosaurs were at least strongly quadrupedal makes sense because, as in bats, their arms were more powerfully muscled and stronger boned than their legs and usually bore hands, which exclusively bipedal birds lack. There is no known pterosaur that could not go on all fours. But with a long tail to better balance a usually smaller head, an always short neck, and the body, it is possible that rhamphorhynchoids were more prone to progressing bipedally, including when running without the intention of taking off. It is possible that the more front-heavy pterodactyloids were able to go on two legs by standing erect, as well as or better than apes and bears. The extremely marine nyctosaurs, lacking even the free fingers found in all other pterosaurs—though these were very small in their fellow oceanic ornithocheiroids—are particularly likely to have gone about on their hindlegs alone. They may have spent at least a

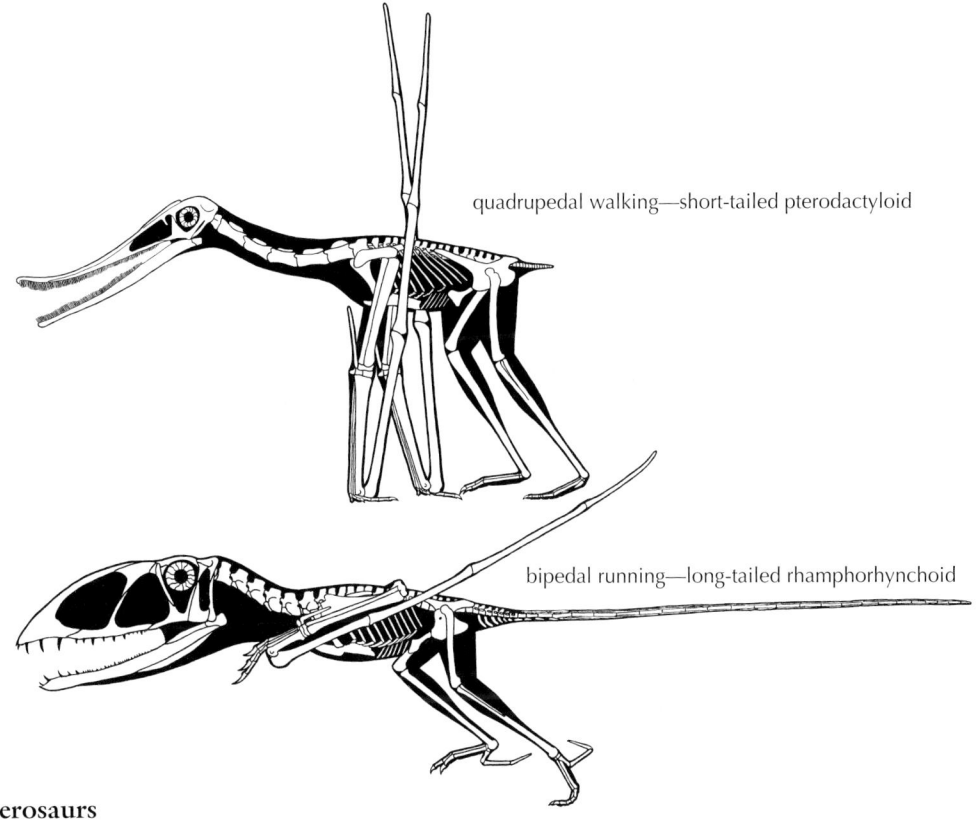

quadrupedal walking—short-tailed pterodactyloid

bipedal running—long-tailed rhamphorhynchoid

Quadrupedal and bipedal pterosaurs

portion of the limited time they moved on the ground during breeding season walking on the big knuckle of their wing fingers, rather like the knuckle walking of big apes. Potential direct evidence for bipedal pterosaurs comes from the hindprint-only trackways—that the hindfeet stepped on the foreprints cannot be ruled out, but no handprints are known in any of the sets of tracks of the giant Asian pterodactyloids (Kim et al. 2012).

Most land animals can run, as can many birds on their hindlegs, including many fliers from strong to weak, but a few such as tortoises can only walk. The ground agility of bats that are specialized for hanging under branches and rocks varies from often minimal to well developed in the prey-stalking vampires, which can even achieve a bounding run to flee discontented victims. Lacking terrestrial speed specialization, pterosaurs could not match fast-running birds, but some should have been able to achieve a fairly good clip. The pterodactyloid trackway with the sprawling arms has the longest recorded stride length, so the animal was either walking fast or barely running on all fours and was doing so by striding rather than bounding, although it is possible that faster-running pterosaurs bounded. The wide gauge of the arms may have been a means to prevent the un-splayed legs from getting tangled up with the folded wings, which is the practice of running bats. High-speed trackways are always scarce because animals make the intense effort needed to run usually only when necessary.

Pterosaur ground locomotion abilities varied widely. The scarcity of rhamphorhynchoid foot traces leaves open the possibility that they spent very little time on the ground, which is perplexing because many of them look as though they should have been competent on terra firma. Certainly they were not out there feeding on the sand and mud flats that most readily preserve prints, as were the later pterodactyloids and especially shorebirds. It is notable that the earliest pterosaurs, the preondactylans, had long legs that suggest good terrestrial abilities (Witton 2013). Conversely, the classic rhamphorhynchoid *Rhamphorhynchus* had exceptionally short legs compared to its relatives, indicating notably limited ground performance. Also interesting is that the most aerially adapted continental pterosaurs, the anurognathids, look more terrestrially capable than swifts, nighthawks, and the like, which either avoid ground contact altogether or do not walk about.

As for the pterodactyloids, it makes sense that the gigantic oceanic ornithocheiroids had limited ground locomotion abilities because they spent their time either soaring over the waves or breeding on isolated islands, neither of which required exceptional ground performance. The ornithocheirids and pteranodonts were not awkward on the ground as much because their legs were exceptionally short relative to the body as because the legs were so much shorter relative to the arms that even with the humerus held horizontal, the body would have been pitched very strongly upward with the outer arms and inner hand directed downward. The difference could have been reduced, however, by holding the elbow a little above the level of the shoulder. The fore-aft disparity was taken to an extreme in the nyctosaurs, which also lacked free fingers. Perhaps the sheer awkwardness of being all on fours caused the long-armed ornithocheiroids to sometimes walk on two legs, although this too would have been rather awkward, as seen in rearing bears or apes. Alternatively, they may have splayed their arms well out to the sides with the hands following a wide gauge, as the trackway of a running pterosaur shows it did.

Most pterodactyloids had less disparate fore-aft limb ratios and should have been able to progress readily across the ground, in some cases across wet flats in search of prey items, in other cases across dry land in search of food items, as in the assorted azhdarchoids up to the biggest giants. Perhaps the pterosaurs best adapted for walking about were the dsungaripterids. Their legs were not only the longest relative to their bodies among the great group but were also as long as the inner wings, eliminating any awkward slope of the trunk when the humerus was held horizontal. That makes biosense since dsungaripterids appear to have been dashing about on shoreline flats and streambeds as they searched for mud- and sand-boring invertebrates to dig up with their heavy spiked beaks.

Because many pterosaurs were small and lightly built and had grasping, claw-tipped fingers and toes, they had the potential to be arboreal to some degree. The long limbs, webbed feet, and rather short toes typical of pterodactyloids indicate they were not spending much time scrambling about within bushes and trees. The combination of shorter upper arms and inner hands, as well as legs with fairly long fingers and toes bearing large hooked claws, suggests that at least some of the low-slung rhamphorhynchoids with squirrel-like proportions were adept climbers—which might help explain why they did not leave lots of footprints on flat ground. If, as is uncertain, some wukongopterids had the opposable thumb ascribed to them (Zhou et al. 2021), then that is evidence for arboreality in a few derived rhamphorhynchoids. The anurognathids are particularly interesting because the tucked-up posture in which some specimens have been found suggests that they lay flattened out along the tops of broad branches, or on the ground like some nightjars, to remain inconspicuous when resting (Witton 2013). If some anurognathids had a reversed inner toe similar to those of birds, they could have used the digit to help grasp branches when climbing (Lü et al. 2017). The toes of a hanging bat all share the same subequal proportions, allowing them to readily wrap around a perch, but the toes of pterosaurs were too asymmetrical to do that (Witton 2013). There is currently no evidence that any pterosaurs lived or nested in burrows or caves like some dinosaurs, including a number of birds, and many bats. This may not have been feasible because of their gangly folded wings, but the possibility cannot be ruled out.

Many pterosaur hindprints are webbed, which is in line with the soft tissue webbing preserved in some fossils. This may have been true of all pterosaurs or at least all pterodactyloids, but it is possible that some pterodactyloids and/or rhamphorhynchoids were not web footed; the few known rhamphorhynchoid footprints seem to lack webbing (Mazin and Pouech 2020). The highly terrestrial anurognathids were apparently web toed. The webbing of the hindfeet helps explain why some pterosaur trackways show only handprints. Because the fingers were short and webless, they bore more weight than the much bigger-surfaced webbed feet, which therefore enjoyed a much higher surface-area-to-weight load. As a result, the hand fingers tended to sink deeper into soft sediments than the flat feet, and the arms bearing more weight than the legs enhanced this effect. If the sediments were barely soft enough yet sufficiently firm, then only the fingers would leave an impression. Or it is possible that the trackways that have been found are an underimpression rather than a surface impression, and only the fingers sank far enough to disturb the muds or sands a short distance under the surface. Another possibility is that the pterosaur was floating in water to at least some degree and poling along the bottom with its inner arms, which were longer than the legs and so could reach the bottom while the hindfeet could not. A few trackways appear to record floating pterosaurs scraping with feet and/or hands along the shallow bottom (Lockley and Wright 2003).

Whether pterosaurs, especially the giants, could readily dive and swim underwater is problematic. Partly because they lacked a dense, smooth covering of feathers, their heads, necks, and bodies were not streamlined in a hydrodynamic manner. Nor do the gangly wings appear suitable for underwater propulsion any more than those of bats, which never dive, or for being smoothly tucked out of the way. Water is nearly 800 times denser than air, and big, flat beaks and especially head crests would have caused steering issues. The slightest deviation from the intended course would have resulted in a strong hydrodynamic deflection off course. And while the modestly muscled legs and webbed feet were sufficient for surface paddling, they lacked the power and other specializations seen in birds that swim mainly or entirely with their hindlimbs.

Pterosaur Pneumatics

Further contradicting deep-swimming pterosaurs, among the pterodactyloids especially, is their light, airy, and correspondingly buoyant construction, including sometimes very thin-walled pneumatic bones and air sacs. These should have especially precluded the high-velocity splash diving performed by some birds such as gannets and boobies, but some small pterosaurs might have been divers like the big-headed kingfishers. Having descended from nonpneumatic mammals, bats show that internal air spaces other than lungs are not necessary for high-performance flight. Birds, having descended from already pneumatic avepod dinosaurs, integrated internal air voids into their flight systems. Deep-diving birds are notably less pneumatic than the avian norm. Because the preflight ancestors of pterosaurs are poorly known, it is not certain

how much of their pneumaticity they inherited—it may have been none, little, or substantial—versus how much evolved in the group independently. In basal rhamphorhynchoids, the pneumatic structures were limited to the skull and vertebrae. These expanded somewhat to the shoulder girdle and inner wing in more derived rhamphorhynchoids, tended to be further expanded in pterodactyloids, albeit with a few exceptions, and were taken to an extreme in some pterodactyloids such as azhdarchids, and especially pteranodonts and nyctosaurs, in which most of the arms and the upper hindlimbs were air filled (Witton 2013; Larramendi et al. 2021). That marine soarers are so exceptionally buoyant is specific and good evidence that pterosaurs were not underwater swimmers as opposed to floaters.

The pneumatic bones of pterodactyloids could be amazingly thin walled, just a few millimeters even in the gigantic forms. Strength was maintained in part by internal struts in an evolutionarily selective parallel to stress analysis, in which a supporting structure is placed only where stress loads required its presence. Some birds have the same. But whether pterosaur skeletons were exceptionally lightweight in order to reduce overall mass is open to question. The pneumatic skeletons of birds are not lighter than those of other tetrapods relative to overall heft, according to some accounts (as noted by Witton 2013), and the quality of the data used in those studies is questionable (Larramendi et al. 2021), so whether simple weight reduction is or is not a primary selective factor behind the evolution of nontrachea/lung internal air spaces is not certain; respiratory functions may have been the driving evolutionary factor. Filling bones with air does balloon the surface area available for adding muscles without adding weight, so that may be a selective factor. And filling big beaks with air allows them to be very large without making the animal front heavy.

In addition to bones filled with air, pterosaurs probably had air sacs filling parts of the head, neck, and trunk similar to those present in avepod dinosaurs, as well as the highly pneumatic sauropod dinosaurs, and those in the main body could have been particularly large. Most land animals float, albeit barely enough to allow them to swim. Their density, or specific gravity (SG), is just a little less than that of water because the air inside the lungs counters the density of their soft tissues and bones,

Longitudinal section of large pterodactyloid humerus

which are denser than water. Bats have SGs barely below 1.0. In birds, all the combined air spaces can make up to around a third of internal body volume when at maximal expansion during the peak of inhalation, but this can be misleading because the respiratory tract is only half-filled with air on average during the inhalation-exhalation cycle. In many flying birds normal SG is around 0.9, ranging from about 0.85 to over 0.95, higher than has usually been thought. Large-beaked birds such as toucans, sporting big bills with SGs of around 0.1, have overall densities of around 0.75. This is not as surprising as it might seem—mammalian fliers do not have reduced density, and fliers can actually degrade flight by being overfilled with air because the resulting ballooning of the body threatens to unduly increase frontal area, which increases drag. Back in the last century when it was assumed that pterosaurs had to have been soaring ultralight air beings, SGs were estimated to have been absurdly low, 0.2 to 0.5. This is abjectly impossible because it is not biologically practical for animals to consist largely of air-like inflated balloons, nor is it aerodynamically efficacious. Current restorations of pterosaur SGs indicate they were even denser than has recently been thought, ranging from about 0.95 in rhamphorhynchoids, to 0.9 in the more pneumatic small pterodactyloids, down to 0.75–0.85 in giant pterodactyloids that sported the most pneumatic and biggest heads and arms (Larramendi et al. 2021).

Skin, Feathers, and Color

So far, the only scales on pterosaurs have been found on the underside of the hindfeet, where they formed a pavement of small, polygonal scales (Witton 2013). It is possible but not certain that the tops of the feet also bore scales. Where other bare skin is preserved, it is fairly smooth.

A modest number of pterosaur specimens record the presence of filamentary body coverings (Witton 2013; Yang et al. 2020). These are not fur, the fibers not being the same as the hair that adorns the unrelated mammals. What they appear to be are feathers. The filament shafts are hollow, which is true of feathers but not of normally solid-shafted mammalian fur. And pterosaur filaments are in at least some cases branched (Yang et al. 2018, 2020), a characteristic of feathers but not fur. The branching is fairly simple, like the feathers adorning some nonavian dinosaurs as well as birds, although the ultrasophisticated contour feathers common to many birds are not seen in pterosaurs. These pterosaurian pycnofibers, or pycnofeathers, were usually short, at 5–10 mm, but were sometimes longer atop the necks of some pterosaurs. Aside from bare horny beaks, the fibers covered much or all of the head—except in the beakless anurognathids, which were pycnofibered from the tip of the snout to the neck, body, wings, and hindlegs down to the ankles. So far, no example of pycnofeathers forming display structures such as cranial crests or anything elsewhere on the body has been discovered.

Examples of pterosaur fibers

The body coverings of flying birds are almost always smoothly streamlined to minimize drag, but in slow-flying bats the body fur is sometimes more erect and fluffy, and that may have been true of some of the aerially less capable pterosaurs such as *Pterodaustro*. Drag minimization is always critical for soarers to maximize their glide ratios, so all soaring pterosaurs should have had smooth pycnofiber coverings. Being as aerodynamically sleek as possible is also important for any aerial giant whose ability to get its massive body into the air might be marginal, as in the azhdarchids. Flying birds with short necks can pull the highly flexible neck back into a strong U curve that brings the head close to the body. The broad contour feathers emerging from the back of the head and neck then form a smooth aeroshell that blends the head into the body. Pterosaur necks were not flexible enough to do that, nor could the short, simple pycnofibers form a continuous aeroshell. So pterosaurs were more comparable to long-necked birds, in which the slender form of the neck remains obvious during flight.

Because hollow, branching fibers also covered a variety of dinosaurs, it is a reasonable scientific bet that ornithodiran fibers evolved once (Paul 2002, 2012, 2017a, Yang et al. 2018, 2020). If so, then pterosaurian pycnofibers and dinoavian feathers are truly the same thing. The absence to date of fibers from Triassic and Early Jurassic protodinosaurs and dinosaurs is the kind of negative evidence that is no more meaningful than the lack of fossil scales, and it is likely to be corrected by the eventual discovery of fibers in basal examples if sufficient fine-grained sediments that could preserve them are found. However, it cannot be ruled out that fuzzy body coverings evolved more than once in ornithodirans.

One question is why pterofeathers, and for that matter any external insulating fibrous body covering, appeared in the first place. The first few bristles must have been too sparse to provide insulation, so their initial appearance should have been for non-thermoregulatory reasons. One highly plausible selective factor is display purposes. As the pycnofibers increased in number and density to improve their exposition effect, they became thick enough to also help retain the heat generated by the increasingly energetic archosaurs. Also plausible is an initial sensory function, as in the whiskers of mammals. Pycnobristles adorning the jaws of anurognathids may have been for tactile purposes. As insulation became the primary function of pterofuzz—because hollow-feathered pycnofibers enjoyed a significant heat retention advantage over solid-fibered fur—the air contained within the feathers made them as much as twice as efficient as fur at trapping warmth for a given mass of insulation. And insulation can work in both directions; under certain conditions feathers and fur can help keep an animal cooler than it would otherwise be in a very hot place. The thick pycnofeather coats that have been preserved adorning the bodies of some specimens are fully characteristic of thermal insulation.

Apparently because pterosaurs had bat-like main wing membranes, it became an arbitrary convention for paleoartists to color them rather like bats, largely in drab, fairly solid blacks, dark grays, browns, or gray browns. There was never a good reason to do so in such a near-universal manner, because the largely daylight-flying archosaurian pterosaurs were not close lifestyle analogues to or phylogenetic relatives of the more nocturnal, mammalian bats. Being more closely related to their fellow ornithodiran birds, and largely sharing daytime skies with them, pterosaurs more likely were often or always colored more like birds. Because bird eyes can see ultraviolet light, their color patterns include ultraviolet patterns that humans cannot perceive, and presumably the same was true of pterosaurs. As for what we can see of avian coloration, it is highly variable, ranging from drab to brilliantly colorful—sometimes this extreme occurs within a species, the female being the former and the male the latter. That may have occurred among some pterosaurs if the males needed to stand out for reproductive competition, and the females needed to be able to hide as well as possible.

If the anurognathids were cryptic branch and ground huggers like nightjars, they may have been similarly camouflaged. This appears to be supported by the brownish coloration with a red component indicated by melanosome capsules within pycnofeather specimens that can be assigned to *Dendrorhynchoides*. Whether a speckled pattern was present is not apparent. One caveat is that some researchers question the ability to restore the colors of fossils via their pigment capsules.

Another example of preserved pterosaur color patterning records dramatic subvertical banding on the soft tissue head crest of the rhamphorhynchoid *Pterorhynchus* (Czerkas and Ji 2002). The observed banding records the pattern of the coloration, not specific colors. It is quite likely that such bold coloration was widespread among pterosaurs, especially in the crests, whose primary function was probably display. The deep snouts and big beaks of various pterosaurs also may have been bold patterned, as they sometimes are in birds, such as auks.

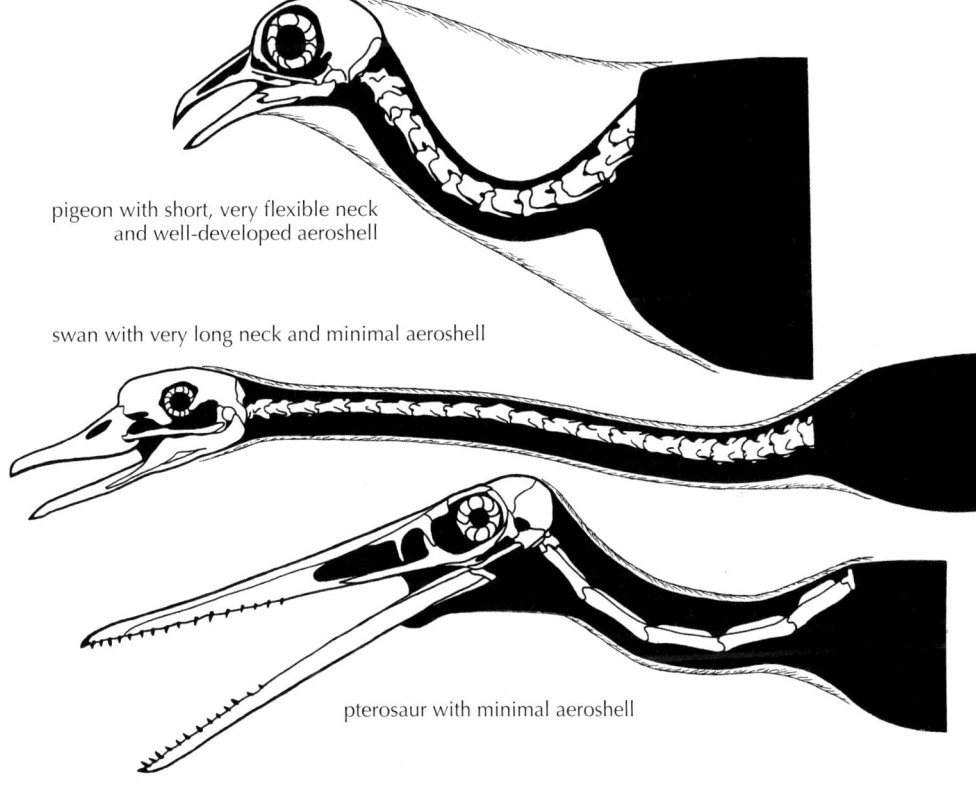

pigeon with short, very flexible neck
and well-developed aeroshell

swan with very long neck and minimal aeroshell

pterosaur with minimal aeroshell

Aeroshell necks

It is very possible that shore and marine pterosaurs were colored like birds with similar habits, in attractive patterns of pleasing whites, grays, and browns. Topsides could have been darker to protect against chronic exposure to ultraviolet radiation, undersides lighter to decrease their visual profile when silhouetted against the bright sky. It has recently been suggested that dark wing tops also increase lift by heating the air flowing over the wings, decreasing its density and further reducing air pressure, which helps create more lift without drag (Rogalla et al. 2019). The effect is enhanced by a light-topped inner and dark-topped outer wing; the differential seems to enhance lateral airflow in a manner that improves lift efficiency. Or, the wing tops may have been light toned in order to minimize heat overload from the sun in the living membrane tissues— overheating is not a problem for inert bird feathers. Other pterosaurs may have been a single solid color top and bottom, perhaps solid black in some examples, or solid white in others. It is common for the tips of bird wing feathers to be darker on the trailing edge than farther inward on the airfoil. This is at least in part because the dark pigment capsules that provide the color also serve to strengthen the feather ends against wear and abrasion. Whether the trailing edges of pterosaur membranes exhibited the same adaptation is not known. Because pterosaur eyes were like those of birds or reptiles, not mammals, they lacked white surrounding the iris. Pterosaur eyes may have been solid black or brightly colored, as in many reptiles and birds.

Flight

As important as walking and the like was for pterosaurs, they were adapted above all else for progressing through thin air, which is why they all had large wings. In order to understand pterosaur flight, we must first address a basic question.

How Wings Really Work

The popular explanation for how wings work goes as follows. Wing tops are more curved than the bottoms. As a result, the air traveling over the top has to travel farther in order to meet up with the air moving along the flatter and therefore shorter underside. Because it has to travel farther, the topside air has to travel faster. The faster a fluid moves along a surface, the lower the pressure, which is called the Bernoulli effect. This is why, if two boats are moving close alongside one another, they are in danger of being sucked into one another and colliding; the water being squeezed between the vessels moves faster than the water on the outer sides of the hulls, so the water pressure is lower between them than on their outer hulls. The boats will therefore be pulled together and collide along their inner sides if the helmsmen are not careful; this is a problem when ships are engaged in refueling and supply operations. Because the pressure is higher

on the bottom of the wing than on the top, lift is generated, and up it supposedly goes. Note that the same applies to horizontal helicopter rotors, which produce both lift and thrust, as well as vertical propeller blades, which produce forward thrust.

The above explanation cannot be and is not correct. At air shows when a plane flies upside down, it does not come crashing to earth. Nor do all wings have the standard shape. The stunt planes common at air shows have symmetrical wing cross sections that provide the same aerodynamics whether the plane is upright or rolled onto its back. The wings of some recent airliners, such as the gigantic Boeing 777 and the superjumbo Airbus A380, are actually flatter on the top than on the bottom because such supercritical wings reduce drag as they approach the speed of sound. Some wings—those of early flying machines, hang gliders, and many ultralights, bats, and pterosaurs—are arched sheets in which the bottom follows the same dorsally convex arc as the top. Paper airplanes tend to have flat wings, and it is easy to produce lift with a flat piece of cardboard.

In a standard wing, the extra distance the air needs to travel from the leading to the trailing edge on the curved top compared to the air flowing along the straighter bottom is just a small percentage more, not close to substantial enough to produce the dramatic speed differential needed to generate

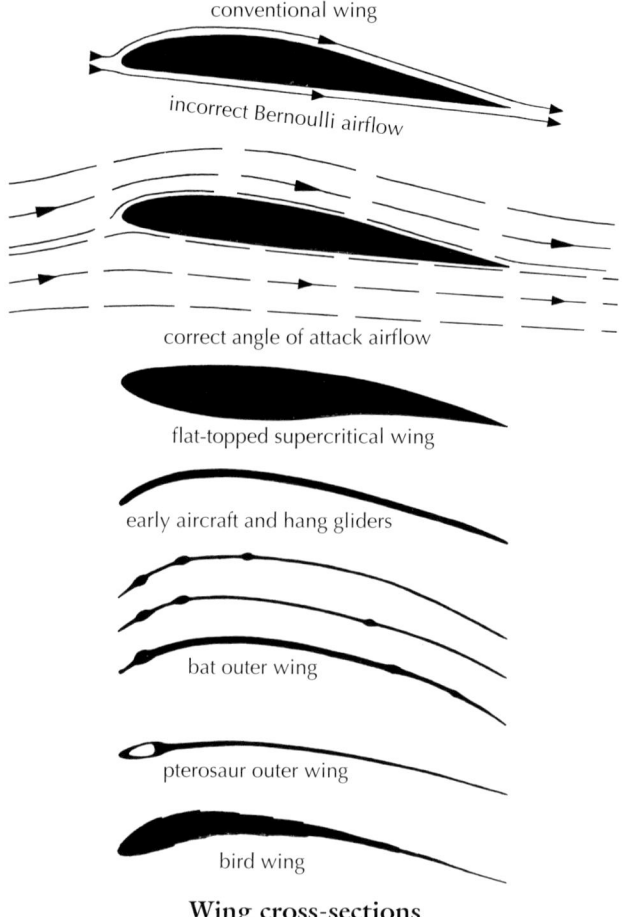

Wing cross-sections

the big pressure difference required to produce enough lift to allow flight. To produce lots of lift, the air going over the top has to move a whole lot faster than that going underneath the wing. Also, pressure differentials alone cannot produce lift. In the end, it all comes down to Newtonian physics. In accord with an action requiring an equal reaction, in order to sustain a body of mass denser than air in the air, enough air must be projected downward with enough velocity to equal the mass of the object. You can observe the latter effect by watching a bee or wasp flying just above grass or loose soil. In order to keep the insect off the ground, the buzzing wings must produce a notably strong downdraft that bends grass blades and scatters dirt and dust. Helicopters from small drones to large machines produce downdrafts that in the latter case can easily knock down a person. This happened to Julie Andrews near the beginning of *The Sound of Music* every time the camera-bearing helicopter roared directly over her head to shoot the alpine pasture scene.

A key requirement for a wing—including a chopper blade—to work is that it needs to have an angle of attack. If the aircraft, whether biological or artificial, is to maintain a constant altitude, the leading edge of the wing needs to be higher than the trailing edge. This helps produce the downward rush of air that keeps a B-52H or *Pteranodon longiceps* in the air, but it is much more complicated—and fascinating—than that.

As air is approached by the leading edge of a wing with the proper angle of attack, the air does not remain undisturbed until the wing's leading edge cuts into it, as one might expect. Instead, the air begins to lift upward when it is still about half a wing chord ahead of the leading edge. This happens because the bottom of the wing, which projects downward toward its trailing edge, acts like a bit of a dam, obstructing the airflow below the wing, slowing it down, and causing it to pile up in front of the leading edge. This also leaves a deficit of air behind the trailing edge of the wing. In order to compensate for the latter, the air flowing over the top of the wing has to speed up greatly, flowing along the wing's surface half again or more as fast as that along the bottom. The dramatic speed difference is most easily seen in online videos in which smoke streams showing the flow of air are pulsed to show how fast the two bodies of air are moving. Because the top air is moving so much faster than the underside air, the big speed differential produces the big pressure difference needed to deliver abundant lift via the Bernoulli effect. In addition, the topside air arrives at the trailing edge long before the bottom air. This creates a standing vortex above and behind the wing, rather like the standing wave downstream of a boulder in the rapids of a fast stream. Because the air at the trailing edge of the vortex is moving downward, and because the downward and backward bottom surface of the wing also pushes the air down, the resulting downwash of mass provides the Newtonian equal action and reaction that keeps the mass of the wing from dropping earthward.

This not-simple set of effects works regardless of the cross-sectional profile of the wing as long as it is sufficiently shallow relative to its chord length. The reason the bottom of standard wings is less curved than the top is partly because this causes the leading edge to be close to symmetrical relative to the airflow, reducing drag, and because the differential curves do produce a minor extra Bernoulli effect that increases lift efficiency a little at very high speeds when the angle of attack is at its bare minimum. The standard wing profile is a good generalized shape that, with appropriate modifications for specific needs, works well in a variety of types of aircraft. But the standard profile runs into problems near the speed of sound. When a wing is moving just below that speed, the faster-streaming air flowing over the top is racing at or above the speed of sound. That is a negative result because the fast-flowing top air produces minor shock waves that create extra drag, as well as potentially interfering with the control surfaces. By having a flatter top, supercritical wings slow down the top air enough so that it does not move at the speed of sound, eliminating the shock waves.

Producing lift always produces drag—the wing or blade that generates lift is a drag even when it is the thin blade of a supersonic machine such as the razor-winged F-104 Starfighter. This is exasperated by the angle of attack, which prevents the wing from presenting its minimal frontal profile to the air. In general, producing more lift produces more drag. High-camber wings consisting of strongly arced thin sheets produce a large amount of lift because the strongly downward-deflected aft portion of the curve directs air strongly ventrally, but there is also a lot of air resistance because the bottom air has to first flow hard up into the concavity before it continues aft. Plus, their thinness means that sheet wings cannot be as strong as thicker wings. They are therefore most suitable for slow fliers, such as many early aircraft, hang gliders, and bats. Because bird wings are sort of a cross between standard and sheet wings—fairly thick, curve topped, and flat bottomed forward, with sheetlike feathered trailing edges—and because they have less camber, they are somewhat better suited for higher speeds than are those of bats. That is why no bat can match the highway-like cruising speed of pigeons, much less the speedway velocity of diving peregrine falcons.

As complicated as the above aerodynamics are, there is yet more complexity involved with wing aerodynamics. When wings are viewed from top or bottom, the air flows straight back from the leading to trailing edge only if the wing is held straight out rather than swept back or forward, and only if there are no wing tips—in other words, the wing goes on forever laterally. But all real wings come to an end. In that case, the higher pressure of the air on the underside of the wing causes it to slide outward to the side toward the wing tip, while the lower-pressure air on the top slides inward. As the bottom air slides out from under the wing tip and the top air shifts in the opposite direction, a whirling wing-tip vortex is generated. This produces a large amount of drag without any compensating extra lift. The effect can never be entirely eliminated and is worst if the wing tip is simply squared off. One way to reduce the problem is with the winglets that adorn the wing tips of many airplanes. Another way, which is used by many birds, is to have wing tips consisting of multiple feather tiplets, as this helps break up the vortex to some extent. A third option is to have a sharp tip. Used in a number of airplanes, including the semielliptical-winged Supermarine Spitfire and the new Boeing 787 airliner, it is also found in some birds, the fastest-flying bats, and all pterosaurs. Having the top of the outer wing darker and correspondingly hotter from sunlight makes the air warmer and therefore less dense than it is over the lighter-colored inner wing. In flight, that causes the air to flow more laterally over the wing top than it otherwise would if the wing top was all the same tone. That in turn reduces the wing-tip vortex and the resulting drag, which may be a reason some birds are colored in that manner. A small pycnofiber tuft lining the trailing edge of the outermost portion of the wing membrane of at least some anurognathids probably helped suppress the wing-tip vortex.

As bad as they are, wing-tip vortices can be exploited by formations of fliers. This is done by flying in the classic V formation used by the likes of geese and pelicans, as well as military aircraft—experiments flying airliners this way are being conducted. In this arrangement, all but one of the fliers in the flock fly just behind one another, with one wing tip flapping in coordination with the flier ahead in order to keep the tip in the outer, upward-rotating portion of the vortex coming off the end of the wing of the bird just ahead. This gives the follower free lift and reduces the work it has to do to maintain speed and altitude. Direct measurements show energy savings of a substantial 10 to 15 percent. The flier in front does not experience any loss in performance from the freeloading trailing flier, but that individual is not getting anything out of the formation. So it tires more rapidly, and when it has had enough it drops back and begins to trail another flock member, as another in the group temporarily assumes the lead until it too tires. This is done only by fairly large birds that power fly substantial distances in flocks. If any of the larger pterosaurs regularly did the same, they may have adopted V formations. This would presumably exclude the big wave soarers such as pteranodonts because, like albatrosses, they did not constantly flap and probably never flew in flocks.

Pterosaur Wings

The configuration of the pterosaur wing was anatomically based on the posture of the arm bones that supported the airfoil membranes. Although the precise poses will never be known and presumably varied to some extent, the always short humerus of the inner arm should have been swept backward to some degree, with the elbow correspondingly flexed forward, and the long radius-ulna unit of the upper arm also swept forward. In front view, when the wing was held out horizontally in neutral flight posture, the humerus was probably tilted upward a little, with the outer arm a little less so, producing a slight dihedral, which is common in aircraft. All this is similar to bats and most birds, the partial exception being big birds in which the

humerus is very elongated along with the rest of the wing, as in the superwinged pelagornithids and albatrosses. In those, the inner and outer arms are nearly straight, with only modest flexion at the elbow, and it is unlikely any pterosaur was like this.

In birds and bats, the wrist, which is always the key rotation point for wing folding when not flying, is always flexed fairly sharply so that the hand is straight or swept backward somewhat during flight relative to the outer arm. In front view, this is where the wing may slope outward and downward some, producing an anhedral, which is found in some aircraft. The best-developed familiar avian example of an inner dihedral and outer anhedral is the gull wing. The situation with pterosaurs, whose flight wing configuration is not likely to be reliably preserved in the fossil record, and whose main wing folding occurred farther out at the finger base, is more ambiguous. It is possible that the wrist was flexed significantly backward and downward, causing the main wing flexion at this joint. The alternative is that the wrist was nearly straight, and, with the most flexible joint being the wing base, the main wing flexion was there, which seems more likely. The fossil big finger bones of at least some ornithocheirids appear to have been dorsoventrally curved to give the outer wing a shallow arch, and this may have been true of other pterosaurs. That any pterosaurs had a wing dihedral/anhedral is highly plausible but somewhat speculative. The rest of the wing finger followed the general direction of the innermost element in a gentle backward arc.

Operating the wing skeleton, held together with bone-to-bone ligaments and joint capsules, was a set of muscles, sometimes attaching to bones via tendons. Easily the biggest arm muscle was the pectoralis, which did more than any other to power flight, achieving that by doing the great majority of the work of the downstroke that produced most of the forward thrust. The pectoralis spread over and was anchored on the entire sternum, from the front vertical keel across the big chest plate. Its outer end inserted on the large pectoral crest of the humerus just lateral to the shoulder joint. By increasing the leverage of the pectoralis on the wing, the pectoral crest improved the muscle's ability to pull the arm down during the downstroke. It also improved the ability of the pectoralis and other, lesser muscles to adjust a given wing's angle of attack relative to the body. That system worked automatically when flapping. On the downstroke, because the crest was at the leading edge of the wing base, contracting the pectoralis rotated the leading edge downward and increased its pitch in addition to depressing the overall wing. On the upstroke, the crest being at the leading edge meant that contracting the wing elevators also pitched the leading edge of the wing up a little, resulting in the lift-generating wing also becoming a variable-pitch thrust-producing propeller. Pterosaurs could also voluntarily use this system to adjust wing pitch, symmetrically on both sides or differentially for maneuvering, to control flight. Because the shoulder joint was the only arm joint that could rotate in pitch many degrees, the otherwise stiff-jointed pterosaur wing could rotate extensively along its long axis. This wing pitch control point was vital to aerial power and maneuvering, as it is in birds and to a lesser degree in the more flexible-winged bats.

In modern flying birds the sternum is even larger than that of pterosaurs, with a very deep bony keel, but this does not mean that the avian pectoralis is correspondingly dramatically larger than that of pterosaurs. For one thing, the often very large sternal keel of birds helps support the supracoracoideus, a large muscle that loops up and over the shoulder joint to help elevate the wing during the upstroke. Pterosaurs, like bats, entirely lacked this unusual complex; the supracoracoideus was merely a small muscle that helped pull the wing down. Also, in today's birds the pectoral crest of the humerus is modest in size, and the bulk of the pectoralis volume is supported by the big sternal plate. But in basal birds, the sternum and pectoral crest were more like those of pterosaurs, with the first being a flat plate and the latter a very large hatchet-shaped blade. In pterosaurs and early birds, more of the mass of the pectoralis was supported by the pectoral crest than it is in derived birds, rather than by the sternum. This meant that the pterosaur shoulder region was flattened out and somewhat broader from side to side than those of deeper-bodied birds, somewhat like late-generation fighter jets.

Opposing the pterosaurs' downward-flapping pectoralis was the upward-flapping latissimus dorsi. Anchored on top of the rib cage and the neural spines of the chest vertebrae, it inserted on top of the humerus. Because it took much less work to lift wings that already wanted to rise up under the combined upward push of wing lift and the mass of the body suspended

between the wings, the latissimus dorsi was not nearly as large as the pectoralis, although it may have been the second most powerful wing muscle.

Aside from the big pectoralis and latissimus dorsi muscles, the rest of the important flight muscles were positioned largely either in front of or behind the wing bones. This had the advantage of keeping the frontal area of the arm minimal, reducing aerodynamic drag. This meant that, as in birds and bats, the top and bottom of the arm bones from the middle of the inner arm out were largely bare of muscles. On the inner arm the stout biceps did the most to flex the elbow downward, thereby making an important contribution to the total downstroke, and to also flexing the elbow forward. The longer but less powerful triceps, positioned behind the humerus and wrapped around the elbow, performed the opposite actions. Farther out, the upper arm and beyond were worked by a complex of long, slender muscles, anchored in part at the elbow end of the humerus, and also on the radius and ulna, which operated the wrist and the rest of the wing via tendons, often very long—beyond the wrist there may have been no muscles.

Although pterosaurs used their arms for nonaerial locomotion, the primary evolutionary purpose of the arm skeleton and musculature was to support and power the wings, the airfoils of which were formed by three membranes, two supported largely by the arms. One is also present in birds and bats, the propatagium, which spreads across the space between the shoulder and wrist, in front of the backward-flexed elbow. This leading-edge airfoil broadens the chord of the inner wing and helps streamline the inner and lower arms into the airfoil. In

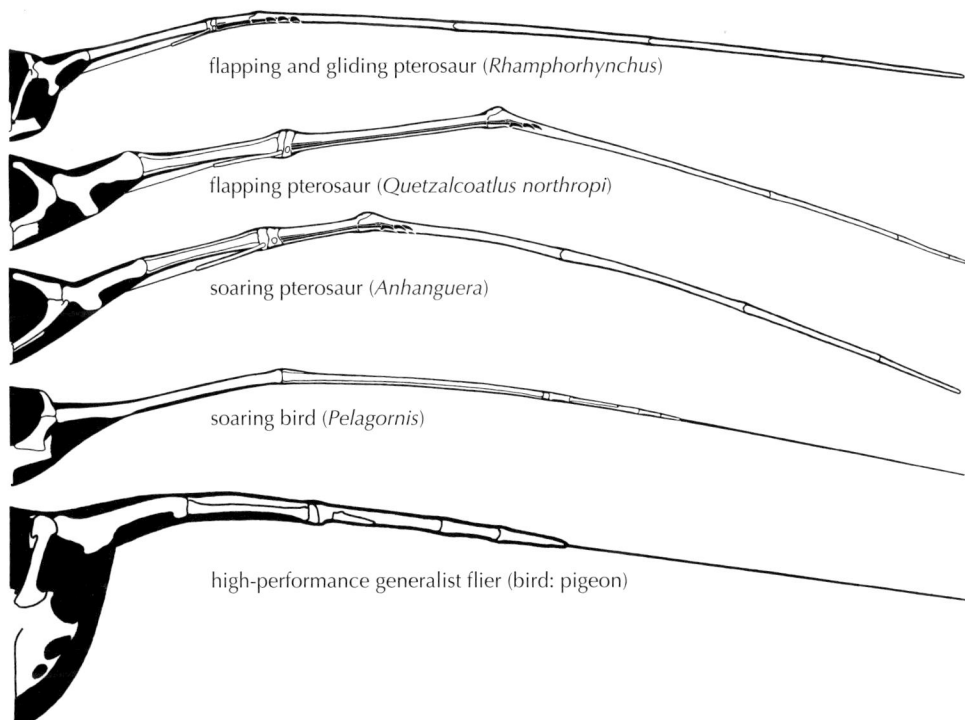

flapping and gliding pterosaur (*Rhamphorhynchus*)

flapping pterosaur (*Quetzalcoatlus northropi*)

soaring pterosaur (*Anhanguera*)

soaring bird (*Pelagornis*)

high-performance generalist flier (bird: pigeon)

Pterosaur chest and wing frontal profiles compared with birds

49

pterosaurs, the leading edge of the propatagium approaching the wrist was supported by the pteroid bone. This was an elongated inner carpal (Peters 2009); there is nothing similar in bird or bat propatagia. Perhaps the pteroid was used to modify the camber of the propatagium to alter the lift capacity versus streamlining the inner wing, especially in those pterosaurs in which the bone was long. But in a number of rhamphorhynchoids—including, interestingly enough, the agile anurognathids—the element appears too short to have strongly altered the leading-edge membrane. Another wrist element, the preaxial carpal, projected a little forward of the joint and may have supported the inner end of the shallow outer propatagium, which could have helped streamline the leading edge of the fairly thin inner metacarpal.

By far the largest wing surface of pterosaurs was the brachiopatagium, which made up three-quarters to nine-tenths of the total airfoil area. From the tip of the big finger, it ran along behind the arm to anchor on the body and apparently on the legs, in some if not most or all pterosaurs. In rhamphorhynchoids, the front edge of the outer membrane could be anchored in the shallow groove on the back of the wing finger, but this is not present in pterodactyloids. Probably because the bat outer brachiopatagium is supported by four splayed-out fingers, of which the leading-edge digit is not the longest, the membrane of the flying mammals is consistently made up of thin, supple, stretchable tissue similar to a latex sheet. Apparently, because it was not supported by a series of digits, most of the pterosaur brachiopatagium was a thicker, stiffer, multilayered structure (Witton 2013). The underside had a dense vascular network that supplied the tissues with blood. Next and in the center was a thin sheet of connective tissue and light muscle. This sandwich was topped by a continuous sheet pavement of slender actinofibers that were oriented directly backward behind the inner arms, rather like the shafts of inner wing bird feathers, and beyond the wrist they were directed increasingly outward, again like the feather shafts of bird primary feathers. In small pterosaurs the actinofibers were a fraction of a millimeter wide; how thick they were in the giants is not known. On the inner wing the fibers were rather short, apparently fairly flexible, and covered only the membrane immediately behind the arm, leaving the rest of the inner membrane consisting of highly stretchable tissue as in bat membranes. Outside the wrist the actinofibers were very long, sometimes 2,000 times longer than wide, running all the way to the trailing edge of the brachiopatagium, and were markedly stiffer. Sometimes the filaments split into two as they progressed to the trailing edge. The fibers did not make the outer membrane as permanently flat and fixed as a sheet of thick paper; the membrane could fold, but only in fanlike pleats paralleling the fibers.

Wing membranes add mass to animals that need minimal weight, so evolutionary selective pressure works to keep them as thin as possible. Bat membrane thickness ranges from just 0.02 to 0.15 mm (0.00075 to 0.01 in) from the smallest to the largest species, leaving even the thickest membranes translucent.

The thickness of bat membranes stretched between their fingers tends to be fairly uniform. Pterosaur membranes, not being supported by multiple fingers, having multiple layers, and stiffened by actinofibers, should have become progressively thinner aft of the supporting arm and finger, and should have been thicker on average at a given body mass than those of bats. But because wing membranes are so large, their mass as a percentage of the total adds up surprisingly fast, depending on their thickness and chord (Larramendi et al. 2021). If average (thicker at the front, thinner toward the trailing edge) brachiopatagium thickness ranged from approximately 0.2 mm (0.0075 in) in the smallest to 4 mm (0.15 in) in the most gigantic, then the membranes would have made up about a tenth of the total mass of pterosaurs. That is more than the wing feathers of birds, or the wing membranes of bats, which are half as heavy. On the other hand, the multiple wing fingers of bats weigh more than the one pneumatic pterosaur finger, so it balances out. And because the main pterosaur wing membrane was thicker than the thinner membranes of bats and was reinforced by the actinofibers, it should have been less vulnerable to being torn than those of bats. The main membranes of pterosaurs may have been translucent like those of bats, especially those of smaller examples, but dark pigmentation could have rendered them opaque.

Although very large portions of the brachiopatagium have been preserved in a few fossils of small pterosaurs, in no case is its exact profile known. That is because the postmortem events that happened to the deceased pterosaur's body and the fossilization process always prevented the wing from being stretched fully out as it would be in flight; all fossil membranes are folded to some extent. The preservation of membranes is often patchy and ambiguous, and it does not help that different researchers' interpretations of the extent of a given specimen's fossil wing tissues can consequently differ dramatically. It is possible to readily restore the wing profiles of a number of extinct birds because if enough of the wing feathers are preserved in place and complete, even if the wing is partly folded, the dimensions and profiles of the stiff feathers can be used to map the overall wing profile during normal flight. No giant pterosaur wing membrane has been found anywhere close to intact.

Among bats, the high stretchability of the thin membranes between the wing fingers means that the trailing edges form prominently concave, stretched-out curves between each fingertip, creating the classic bat wing profile frequently reproduced in often sinister logos. It has been fairly common for artists to portray pterosaurs with similarly concave trailing edges on their wings, starting at the wing tip, and some researchers contend this is correct because the trailing edge of the pterosaur membrane should also have been under elastic tension (Palmer and Dyke 2012; Hone et al. 2015). However, the un-bat-like stiffening of the pterosaur's outer membrane out to the edges by the actinofibers should have allowed the trailing edge of the outer membrane to assume a convex curve, as is present in their preserved membranes (Bennett 2000).

Pterosaur preserved wing membranes

Rhamphorhynchus

Rhamphorhynchus

Sordes

Dendrorhynchoides

Pterodactylus

The forward curvature of the tips of the wing finger of some pterosaurs should also require that the trailing edge of the membrane be convex. Some researchers propose that pterosaur finger bones were so flexible that the forward-curved wing bone was bent back and straight by the pull of the stretched-out membrane when the wing was under tension in flight (Hone et al. 2015). Whether ossified bones would be as bendable as fishing rods is dubious—if they were, the wing finger would be too floppy to produce an effective airfoil; it would be subject to severe flutter in the airstream. Also working against this idea is that bending the slender fingertip backward and only backward would work only if the cross section of the bone were shaped to limit flexion to that direction—the forward-swept tips of recurved archery bows bend only backward when the string is drawn because the bow's broad limbs are strongly compressed fore and aft. The front-to-back flattening relative to limb breadth prevents the tips of the arms from flopping to one side or the other as the bow is drawn, as the round cross-sectioned tip of a forward-curved fishing rod would do. Pterosaur wing finger bones, instead of being strongly flattened fore and aft—which would not be aerodynamic—were streamlined and transversely flattened, meaning that they would be twisted out of the flat plane of the wing if the forward-swept tips were pulled under the extreme tension of the membrane that would be needed to straighten the bone. It is concluded that the wing finger

was moderately flexible, that the actinofibers meant that the internally stiffened outer wing membrane did not need to be under high tension during flight, and that pterosaur wing tips were often and probably always convex on their trailing edge.

The biggest issue yet to be fully resolved is how the inner brachiopatagium was anchored. In all bats the membrane is connected all the way along the leading edge of the hindlegs down to at least the ankle, in a few cases even farther. The membrane appears to have been attached to the ankle in at least some rhamphorhynchoid specimens, and possibly in some pterodactyloids, but that evidence is more ambiguous. In a specimen of the pterodactyloid *Pterodactylus*, the attachment appears to be on the thighs on both sides, above the knees, but this has not been confirmed by other specimens. That the brachiopatagium anchored on the legs is in accord with the pterosaur's bat-like ability to splay out the hindlegs on the same plane as the airfoil, if that is correct. But the folding of the membrane, and the possibility of displacement during the process of decay and preservation, render the apparent connections problematic. Perhaps the membrane actually attached to the ankle in *Pterodactylus*, and the seeming connection above the knee is an illusion because of the folding of the membrane across the splayed-out legs. Or perhaps the seeming ankle attachments are the result of postmortem displacements. Also complicating the situation is the argument that the hip socket ligaments

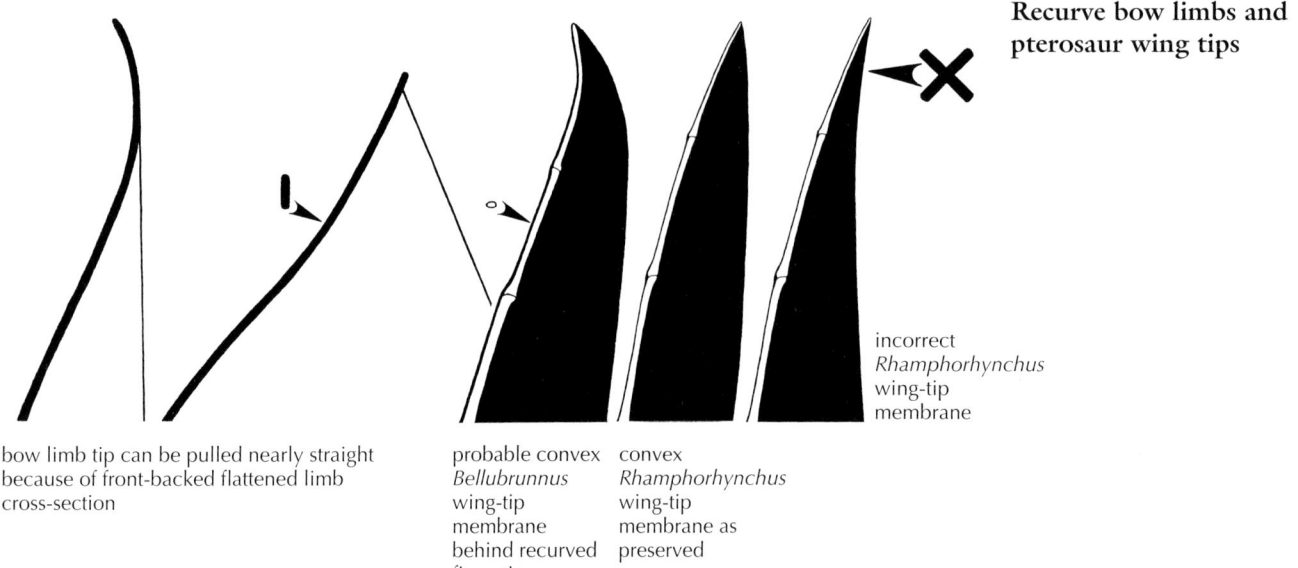

Recurve bow limbs and pterosaur wing tips

bow limb tip can be pulled nearly straight because of front-backed flattened limb cross-section

probable convex *Bellubrunnus* wing-tip membrane behind recurved fingertip

convex *Rhamphorhynchus* wing-tip membrane as preserved

incorrect *Rhamphorhynchus* wing-tip membrane

prevented pterosaur legs from being sprawled out horizontally. It is quite possible that the exact nature of the leg attachment varied among pterosaurs, perhaps between rhamphorhynchoids and pterodactyloids, or within one or both groups.

Pteranodontid tails may be significant because they ended with elongated rods, which suggests that some form of airfoil was attached, perhaps the inner trailing edge of a brachiopatagium that did not attach to the legs. Of related interest are azhdarchids, because the significant possibility that they could not splay out their hindlegs sideways implies that their situation was markedly different from that of other pterosaurs. They may have had a distinctive leg posture during flight to hold the membrane, perhaps with the legs stretched straight back as in fruit bats. Possibly the brachiopatagium-leg connection was lost and the membrane was attached to the tail, as it may have been in pteranodontids. If the latter was true, then what was the leg pose during flight? Were the legs directed straight backward from the hips, or did they fold up fully or partially, as is common in flying birds?

Bats, but not birds, feature a third, rear-end uropatagium membrane. Anchored on the hindlegs, it has quite a variable configuration in bats, ranging from a very large, broad sheet if the legs are held widely splayed out and the twin membranes run along the entirety of a fairly long tail; to narrow, subtriangular bands behind the legs if the tail is very short or much of the tail is free of the uropatagium, and if the legs are held nearly straight back, leaving little space between them. Pterosaur uropatagia appear to have been similarly variable, with the interesting proviso that they apparently did not attach at all to the long tails of at least some rhamphorhynchoids even at the base—this makes sense in that it allowed the tail to be entirely unlinked to the legs, so both could operate independently either on the ground or in flight (Witton 2013).

The splint-like fifth toes of rhamphorhynchoids helped support and manipulate the trailing edge of the interleg membrane, as does the calcar spur in bats that have it. Because the fifth toe was the outer digit, and it was on the topside of the foot when the legs were spread to support membranes during flight, the two outer legs formed inverted, airflow-channeling walls below the level of the interleg membrane; in bats, the legs are simply lateral to the uropatagium. In *Sordes* the trailing edge of the inner portion of the uropatagium was a concave aft shallow V; this may have been true of some or all other rhamphorhynchoids. Lacking the splint-like fifth toe, the pterodactyloid uropatagium was less extensive; that of *Pterodactylus* was a subtriangular sheet on each leg running from behind the hip to the ankle. How true this was of other pterodactyloids is uncertain, as it is possible that the specialized pteranodontid tail was somehow involved in the interleg membrane.

When rhamphorhynchoids stood on all fours, the membranes would have formed a sort of tentlike configuration, with the brachiopatagium making up the side walls much of the way to the ground, and the extensive uropatagium making up the back wall, while the front would be open. The much more limited and posteriorly open uropatagium of pterodactyloids should have produced a less pronounced tent effect.

With the three flight membranes described, it is time to look at the overall wing and tail section profiles in plan view. Supported along much of their length by a series of long wing fingers, bat wings have a broad chord that tends to promote turning more than speed even among the narrowest-winged examples, the molossids. As a result, the trailing edge of the inner brachiopatagium usually runs in a fairly straight line or a gentle curve out from the ankle to the similarly broad outer wing. Exceptions are some fruit bats, which hold the hindlegs straight back during flight, and the inner brachiopatagium sweeps strongly back toward the ankle.

More brachiopatagia are preserved for *Rhamphorhynchus* than for any other pterosaur, and some appear to be spread out close

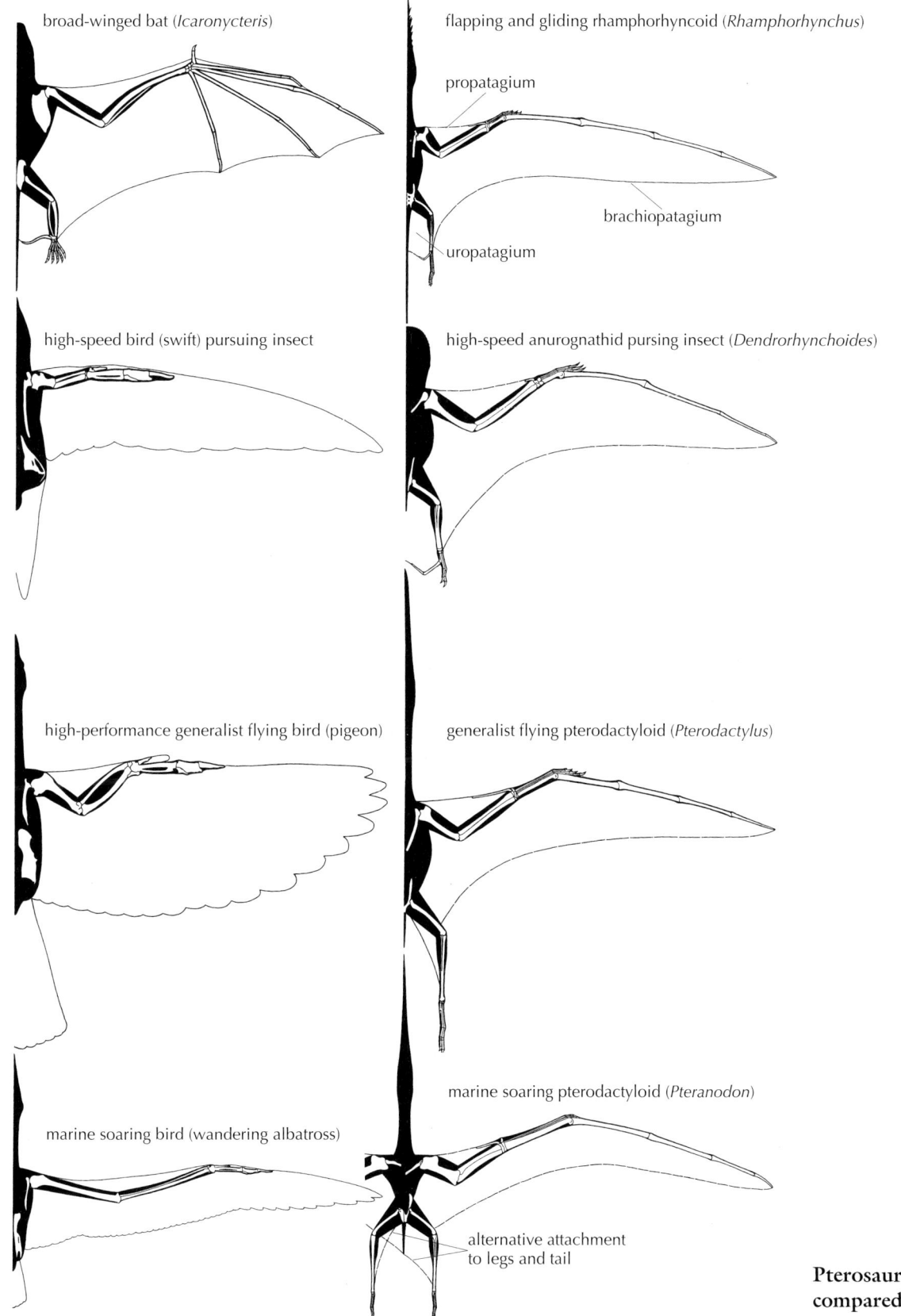

broad-winged bat (*Icaronycteris*)

flapping and gliding rhamphorhyncoid (*Rhamphorhynchus*)

propatagium

brachiopatagium

uropatagium

high-speed bird (swift) pursuing insect

high-speed anurognathid pursing insect (*Dendrorhynchoides*)

high-performance generalist flying bird (pigeon)

generalist flying pterodactyloid (*Pterodactylus*)

marine soaring pterodactyloid (*Pteranodon*)

marine soaring bird (wandering albatross)

alternative attachment to legs and tail

Pterosaur wing profile compared with birds and bat

to full chord breadth. If the wing membrane is drawn from the wrist out so that it is a little broader than that of the specimen, the chord is narrow along the outer wing from the tip to behind the wrist. That means that the main membrane was far from being so broad chorded that the inner trailing edge emerged straight out from the ankle. That in turn indicates that the trailing edge of the inner brachiopatagium swept strongly back to the ankle, which some specimens suggest it anchored to. In an anurognathid rhamphorhynchoid, the fore-aft chord of the membrane behind the elbow appears to be preserved, according to some researchers' interpretation. Again, the inner trailing border of the main membrane appears to have swooped dramatically forward like that of fruit bats from the ankle of the long leg out to the rest of the brachiopatagium (Larramendi et al. 2021).

In pterodactyloids the wing chord was fairly narrow at the base because the main membrane probably emerged well forward, off the lower thigh. If the brachiopatagium was not attached to the legs in some pterodactyloids, then the wing chord must have been narrow. It therefore appears that, from a gentle convex trailing edge at the wing tip, the main wing membrane of pterosaurs was fairly to quite narrow chorded, promoting speed over agility. Although broader wings cannot be entirely ruled out based on the limited data on hand, such wide chords would have significantly added to the total mass of the animal. The propatagium provided a gentle concave leading edge to the inner wing and broadened its chord, contributing to the overall wing area. Almost certainly the overall chord of the wings would have been the least in the very long-winged, short-bodied, and short-legged oceanic ornithocheiroids, nyctosaurs most of all. The sometimes even bigger azhdarchids had markedly shorter and correspondingly probably somewhat broader wings relative to their bodies. The wing chords restored herein produce wing area to mass ratios close to the avian norm, supporting their validity. Between the legs were the uropatagium sheets, apparently much more extensive in rhamphorhynchoids than in the pterodactyloids lacking the outer splint-like toe.

Viewed overall, the restored pterosaur wing was an evolutionary work of simple, clean elegance in top or bottom profile view, much more so than the creepily irregular rear edges of bat wings, and exceeding the attractiveness of even most birds' airfoils. The extraslender wings of the big marine pterosaurs may have been the most aesthetically graceful, being similar to those of albatrosses.

The camber of the bat brachiopatagium is strong because it is maintained by the two strongly arced midwing fingers, a configuration that favors agility via lots of lift over speed via low drag. The camber is the least in the faster bats such as molossids. The shallower feathery camber of bird wings favors speed over maneuverability. Presumably the cross section of the pterosaur brachiopatagium followed a dorsal arch maintained over most of the span by the stiffening of curved actinofibers. Lacking the long fingers to maintain a strong camber and ease the turbulence of airflow during slow-speed flight, pterosaur wings should have

had more modest camber than those of bats and should have favored speed over agility, although this balance probably varied considerably among pterosaurs.

Along the outer pterosaur wing, especially the wing finger, the aerodynamically flattened leading edge bone, tendon, and ligament struts were so slender that they would not have seriously interrupted the frontal streamlining of the airfoil. The three joints between the four wings formed minor bulges, which were reduced to two in the extra-high-performance nyctosaurs and an anurognathid to bring the drag of the outer wing down to a bare practical minimum.

Farther inward on the wing is a different matter. In order to provide sufficient bending strength and muscle power for flapping flight, the bones and muscles of the inner arm from the wrist inward have to be fairly robustly built in bats, pterosaurs, and birds, producing considerable frontal area that generates drag. In birds this is not as much of a problem because the progressive reduction of layers of feathers toward the trailing edge allows the inner wing arm to be smoothly integrated into a gently cambered teardrop shape well suited for minimal drag. The resulting wing cross section is similar to that of moderately fast planes like biplane fighters during and after the World War I. The leading edge of the propatagium further reduces the problem of smoothing airflow over the humerus, radius, and ulna and their muscles but does not entirely solve it.

For bats and pterosaurs, smoothing out the connection between the very thin fore and aft wing membranes and the thick inner arm was not so readily solved. Mostly rather slow fliers, bats usually accept the imperfection—in fact, the resulting irregular top surface of the outer wing created by the multiple fingers may improve airflow during slow flight by helping to keep airflow from breaking away from the top of the wing and creating an unwanted stall during slow flight. The exception are the fast-flying molossids, in which carpet-like strips of short fur are used to blend the top of the propatagium with the inner arm and both sides of the brachiopatagium. Although preserved pterosaur membranes usually lack evidence of pycnofiber coverings, there is some fossil evidence of such carpeting; hence it is possible that some pterosaur wings minimized drag via pterofuzz. Also possible is that air sacs filled out the wing membranes close to the wing bones to help smooth out the surface profile. Blending via pycnofeathers and/or air sacs may have been particularly necessary among giants in order to smooth out hefty inner wing bones, and such should have been an aerodynamic priority. Almost certainly pterofuzz helped smooth the base of the inner brachiopatagium, as is true in bats. Because pterosaur legs were slender, they would not have produced serious drag as they anchored the rear of the inner brachiopatagium and the uropatagium.

Pterosaur wings lacked a feature of bird wings, the leading-edge alula or bastard wing. This is a set of small feathers supported by the freely moving thumb splint. When a bird is flying slowly to turn hard or land, it can lift the alula a little above the leading edge of the main wing, allowing air to flow between the two airfoils. This creates a leading-edge slot, similar to those used on some airplanes, that alters airflow over the top of the wing in a manner that allows the wing to adopt a higher angle of attack and generate yet more lift, rather than accidentally stall out and suddenly lose all lift. The improvement in lift improves turning ability at low speeds and slows down landing speeds. Because pterosaurs lacked a similar structure, as well as the higher camber that can be adopted by bat wings, pterosaurs were probably not as maneuverable when flying slowly, nor were they able to fly as slowly, as birds and especially bats.

Flight Control

Now that we have detailed the anatomy and basic actions of the pterosaur wing, the next task is to look at how it generated and controlled flight.

During the late 1800s, when increasingly serious attempts were being made to produce powered, winged planes that could carry humans, it was widely assumed that controlling a flying machine would be a lot like steering an automobile, a boat, or even, more appropriately, a lighter-than-air ship or submarine. Turning right or left would be a simple matter of moving the rudder in the correct direction. Going up and down would require using a movable elevator. It was the Wright brothers who first realized that this simplistic view of aviation was errant, and that flying through thin air that is 800 times less dense than water is inherently dynamic. That the Wrights were mere bicycle makers is actually key to their realization. On a bike, one does not merely turn the front wheel as in an auto; one *banks* into the turn—cycling is dynamic. To stay in the air is more than a matter of sufficient lift and power; it requires control of orientation in upward-downward pitch, right-left yaw, and right-left roll. The latter realization in particular is what allowed the Wrights to first get into the air in 1903, albeit in a misdesigned Flyer 1 that they accidently made so extremely unstable that it was barely controllable. That is why their first flights were so short—there was fuel enough to go a few kilometers, but the pitch controls were oversensitive so the machine kept semicrashing, until on the fourth flight one of the brothers managed to squeeze a few hundred meters out of it. It was not until 1905 that they got the bugs worked out and could stay up in the air under full maneuvering control as long as the fuel did not run out. In 1906 Europeans started flying, but without dynamic controls they could barely steer in the air and were amazed when in 1908 one of the Wrights showed how their machine could fly like a giant bird, under complete command by the pilot.

What the Wrights realized is that, like bicyclists, birds dynamically bank into a turn. If a winged flier tries to turn by simply kicking over the rudder while keeping the wings level, it will skid and end up going partly sideways while barely beginning to make the turn. And in any case, few if any biofliers have a prominent vertical rudder. When banking, the wings are not just generating vertical lift; the lift is now directed partly away from the direction the flier wishes to go, producing a dramatic sideways thrust that pushes the flier into a tight turn—the steeper the bank, the harder the turn. Hang glider pilots bank by shifting their weight left or right. Most aircraft use ailerons or other hinged flap-like mobile panels on the outer wings. But the early Wright Flyers used a bat-like wing warping in which the lightly built, thin-sheeted, flexible wings were flexed along their long axis, so that the angle of attack was higher on one side than on the other. The greater the angle of attack, the more lift an airfoil produces—as long as the attack angle is not too steep—so the angled-up wing lifts up, the other tilts down, and banking is produced, which automatically generates the turn. Interestingly, many airplanes, including airliners, rarely use the rudder when turning; like a flying animal, they rely entirely on wing banking to produce smooth passenger- and cargo-pleasing turns. The rudder is used only in particular situations, including emergencies such as an engine going off-line. In the Wright Flyers and some other early planes, the wing warping caused the attack differential to gradually increase toward the outer wings. The more complex wing warping enjoyed by multifingered bats helps them make sudden hard turns. Because the main rotation of wings along their long axis is limited to the shoulder in flying birds and pterosaurs, the entire wing on a given side is pitched up or down as needed for banking. This aerodynamically simpler, stiffer scheme, though not used in any piloted aircraft, limits the turning ability of birds and presumably pterosaurs relative to chiropterans. On the other hand, the bat-like minimuscles that were probably within and helped finely manipulate pterosaur membranes should have improved their agility vis-à-vis birds to some degree. Another item that may have helped pterosaurs turn was their small free fingers. Presumably these were normally folded tight during flight to maximize streamlining, and extending or flexing the short fingers up or down to project into the airstream on just one wing would have created some drag on that side. The turbulence created by the splayed fingers would have disrupted the air flowing over the downstream area of the airfoil and enhanced the amount of drag on that wing, further turning the creature in that direction. Somewhat similar to the drag-inducing split flaps used for turning on some aircraft such as the B-2, this limited form of pterosaur yaw control would have worked better when the free fingers and their claws were relatively bigger.

Flying craft need to control orientation on all axes—roll along the long axis of the main body, left and right yaw, and up and down pitch. Although flying is dynamic, the degree of dynamism is quite variable. A priority for small private planes is to make them as easy and safe to fly by amateur pilots as is practical. To do that, they are made as stable as possible—so stable that if the pilot lets go of the controls the plane will naturally assume a steady, horizontal path. This is why when the pilot of a private plane becomes gravely disabled the plane often cruises along until the fuel runs out. One way to achieve high stability is to simply place flat wings atop the fuselage, so the low-placed mass of the latter provides stabilization. If, on the other hand, the wings are attached to the lower fuselage, then the wings are given a strong dihedral, with the tips markedly higher than the bases to form a shallow V. If the plane starts to tilt to the right, then the right wing now being more horizontal relative to the ground produces more vertical lift, while the left wing being more steeply angled away from horizontal, and with the high-pressure bottom air more easily spilling off the wing tip, produces less vertical lift, so the craft automatically rights itself. During World War II the P-51 fighter was given a little extra dihedral because it was designed to fly farther than other single-seat, single-engine fighters, and the improved stability reduced the fatigue imposed on the pilot by long missions.

A high degree of stability is not ideal for some aircraft because it makes it harder to maneuver, so stunt planes and most fighters are designed to be at most moderately stable. To do that, the wing dihedral is modest, absent, or reversed via an anhedral if the wing is mounted high on the fuselage, as seen in the Harrier jump jet. When stability is low, the pilot has to constantly fly the plane—unless autopilot is available—to keep it straight and level. Some modern jet fighters, starting with the F-16, are so aerodynamically unstable that a human pilot has no hope of keeping them in control for even a few moments, so computers are in constant use to keep the machines from tumbling through the air.

The neural networks of flying insects and vertebrates are de facto expert computer controls that allow them to constantly and without specifically thinking about it keep stable at all times regardless of the individual's aerodynamic stability or lack thereof at a given moment, within reason. None of these networks possess the high automatic stability of a private plane. Even so, a pterosaur would not have cruised far with its wings assuming so much anhedral that if it began to unintentionally tilt right, the right wing would become increasingly vertical and generate less lift, while the increasing horizontal left wing would produce ever more lift and threaten to flip the ancient flier over. To keep from rolling, the pterosaur would have had to struggle, with physical and mental difficulty, to constantly adjust its wing controls to keep on an even keel, wasting energy, tiring itself out, and risking loss of control. Instead, cruising animal fliers tend to put their wings in a posture that provides easy-to-adjust-for stability. A notable example is the way that many vultures, which soar with little or no flapping for hours at a stretch, habitually pose their wings in a distinctive shallow V dihedral.

Among animals that fly, only one group apparently had potential flat vertical tail rudders: the long-tailed rhamphorhynchoids. Because preservation of soft tissue vanes is very spotty, it is quite possible that some portion of rhamphorhynchoids lacked the vanes. Some, most notably *Rhamphorhynchus*, had a short deep vane adorning the tail tip. Others had a series of shallow vanes along a long portion of the tail. That the vanes are asymmetrical indicates that they were vertical in the living animals. The shallow vanes do not appear highly aerodynamic, so they are candidates for display rather than aerodynamic purposes. The same may have been true of the deep tail-tip vanes, but it is also possible that the use of a vertical rudder was variable in rhamphorhynchoids in accord with the different aerial needs of different species; we will likely never know. What is certain is that the long tails impacted the flight of the pterosaurs that had them one way or another, as did the lack of them in other pterosaurs. The long tails of rhamphorhynchoids made up a nonnegligible twentieth of their total mass. With such a significant mass placed well away from the body center, the tail could be used to help quickly change direction along all axes, especially in yaw if the vanes acted like air rudders. On the other hand, long tails, vaned or not, would also tend to provide more stability than in pterosaurs that lacked the distally placed inertial mass and vertical aerosurface. That most rhamphorhynchoids retained the long tail for some 90 million years indicates it was a successful adaptation for aerodynamic purposes, for reproductive display, or both. On the other hand, many of the exceptionally high-performance anurognathids lost the long tail, and the short-tailed pterodactyloids replaced all the long-tailed rhamphorhynchoids and often became oceanic and/or enormous. This indicates that lacking a substantial mass well away from the body center was overall superior to having it, probably by increasing dynamic agility by concentrating mass more toward the body center. The simple reduction in total mass was also an advantage. The absence of long tails in dinoavian fliers after the Mesozoic, and the similarly short tails of bats, support this dynamic flight hypothesis.

That the atypical rod at the end of the pteranodontid tail sported a vertical rudder surface is highly speculative but cannot be ruled out.

Many and perhaps all pterosaurs had another potential vertical rudder, the head crest. For that matter, the big beaks or deep snouts of some archosaurian fliers had aerodynamic steering potential. Very few aircraft have a rudder in front of the wings. A very large surface area well ahead of the central axis of the body would produce tremendous bending forces on necks, which would have been problematic, especially among pterosaurs with longer and more slender necks. Birds—including those with head crests, as far as we know—and bats turn without the use of vertical rudders, which favors pterosaurs not doing so either. A related matter is the little-considered issue of how pterosaurs oriented their head during banking turns. Birds tend keep the horizontal plane of their head and eyes level with the ground as the following body and wings roll into a turn, even a hard turn. Presumably this differential improves spatial orientation as the bird's flight direction rapidly changes—doing so is possible in birds because their neck vertebrae are so numerous and rotationally flexible relative to one another. Bats, with their shorter, stiffer necks, seem less prone to this action. Because pterosaur neck vertebrae were not highly rotationally flexible and were few in number even in long-necked pterodactyloids, they must have kept their head tilt much the same as their body tilt, hard banking both head and body into tight turns. This point reduces the possibility of the use of head crests as rudders. If a pterosaur were steeply banked in a turn, the head included, then turning the head strongly into the turn so that the head and its crest were acting as a rudder in the airstream would not only produce sideways thrust, but the tilt would also make the crest into a partial front elevator that would produce significant downward thrust, causing the pterosaur to go into some degree of dive. The downward force would be all the more disadvantageous because when a winged craft is banking in a turn, the reduction in direct vertical lift from the wings tends to cause the craft to lose altitude, costing critical maneuvering energy. On the other hand, if a crested pterosaur wished to make a descending turn, the head crest may have been useful.

Another viable set of potential rudders that has received little attention are the webbed feet. In birds that have them, foot webs can be and often are used for aerodynamic purposes, especially as air brakes when landing. If pterosaur legs were sprawled out to the sides to spread out the trailing portions of the inner brachiopatagium and the uropatagium during ordinary flight, the trailing feet were automatically held vertically so that the outer side of the foot was on the top edge and the inner side on the bottom edge. This is not the case with bats and birds, in which the trailing feet are more horizontal. The toe webs therefore could have acted as vertical air rudders. That could have been done either by swinging the lower leg inward or outward at the knee, and/or by swinging the foot outward at the ankle—the foot could not be flexed inward because the ankle would not allow it. The rudder effect could have been enhanced when needed by spreading out the toes to maximize the area of the webbing. Although in most pterosaurs the modest size of the feet would have kept the resulting rudder effect from being especially powerful, it would have helped produce a turn by pushing the rear of the pterosaur to, say, the right in a left turn, pointing the body in the direction the pterosaur wished to go—something that wing warping alone has trouble doing. Foot rudders could also have been used to help maintain a straight course via small, quick corrective movements, including helping to counter any accidental turning movements produced by head crests. The feet of some pterosaurs were too small to be effective rudders or air brakes. On the other hand, the presence of large, webbed feet in anurognathids despite their dry-land lifestyle supports the use of such appendages as auxiliary airfoils.

Now that we have dealt with roll and yaw, it is time to look at pitch control. The latter is linked to distribution of mass, specifically the center of gravity, relative to the axis of lift produced by the airfoils at any given moment. Those who design and operate fixed-wing aircraft must be very careful to ensure that the distribution of weight in the fuselage and of the engines, whether it be the permanent structure or the internal contents—humans, fuel, cargo, ammo—always produces a center of mass that correctly matches up with the axis of lift. Failure to

do so will result in the plane not being able to successfully take off if front heavy, or to stall on takeoff and crash if rear heavy. However, the two factors need not always be exactly aligned. In some cases the center of gravity is set somewhat ahead of the axis of lift so that the rear elevators have to be set to produce a downward force during level flight; doing this reduces the danger of stalling out, while improving sudden evasive dive performance in fighters. Biofliers, with their multijointed wings that are more flexible in their configuration, have even more room for slack.

With long tails aft balancing modest-sized heads fixed to short necks in the front, the fore-aft weight distribution looks well balanced relative to the wing base in rhamphorhynchoids. With big and sometimes enormous heads at the ends of sometimes long necks in front of the wing base, and hardly any tail behind, pterodactyloids appear much more front heavy, in a way not seen in similarly short-tailed bats and birds, big-headed toucans and hornbills excepted. Nor could the stiff, long necks of pterodactyloids that had them be pulled far back, as they can be in flexible-necked birds such as egrets and pelicans. Large-beaked pterosaurs were not as front heavy as they appear because their big bills were, like those of similar long-snouted birds, highly pneumatic—filled largely with air, they sported very low densities around a tenth that of normal tissues, about the same as extrastrength Styrofoam (Larramendi et al. 2021). Even so, large-headed pterodactyloids were more front heavy relative to the wings than were rhamphorhynchoids. Yet this was not a problem because the more forward weighted a pterosaur was, the more it could sweep its wings forward to shift the axis of lift far enough forward to match up with the forwardly placed center of mass. This was most especially the case in the very long-necked azhdarchids. How dramatic the difference in forward wing sweep would have been in more front-heavy pterosaurs compared to the rest is not firmly determinable and may have been visually subtle.

Because it is so easy for flying creatures to rapidly adjust the sweep of their wings at their many mobile arm joints, they can use that action to very quickly control pitch in order to climb, remain in level flight, or dive. To climb, sweep the wings forward the pertinent amount and the tail end will drop, so that the resulting upward pitch will cause a climb—temporary if no increase in power from increased flapping is applied, steady if it is not. Return the wing sweep to normal, and level flight resumes relative to the surrounding body of air if sufficient power is maintained. Sweep the wings back, and the nose pitches down and the flier descends regardless of whether flapping power is used or not. Wing sweep alteration is used by all biofliers—it is the primary or sole means of pitch control in aerial insects. In a high-speed dive, partial folding of the wings could reduce the frontal drag of the airfoils by up to half. Even so, no pterosaur had the ultrastreamlined body and wing form that would have allowed it to fast dive like a stooping falcon.

Another major means of pitch control is horizontal elevators. In conventional aircraft, movable panels provide such control.

These elevators can be placed in front of the wings, as in the earlier Wright Flyers and other planes over time, including some recent advanced designs. But in most, the elevators are either well behind the wings, or on the trailing edges of delta wings and flying wings. In the last situation, the aerodynamics can be dangerously tricky. The famed Northrop flying wing bombers—the piston-powered B-35 and its direct descendant, the B-49 jet—had so little distance between the center of mass and the elevators that they were too short coupled and at risk of tumbling end over end, which one of the B-49s did and crashed. It is computer stabilization that makes the Northrop B-2 and robotic stealth flying wings practical.

All bats and most birds have dedicated elevator control surfaces. The rare exceptions among birds are some Mesozoic examples, the fossils of which show very short tail feathers, too short for effective pitch control, so they had to rely on wing sweep alteration and thrust adjustment alone. Other birds use their tail fans as elevators in coordination with wing sweep alteration. Bats do much the same, except that their elevators consist of the combination of the rear inner membrane of the main wing and uropatagium attached to the legs. The same would have been true of pterosaurs with their uropatagia and the trailing portions of their inner brachiopatagia. Rhamphorhynchoids could have attained some pitch control with their long tails. Quickly swinging the tail up would have pitched the nose up into a climb, and a downward swipe of the tail would have helped point the nose down into a dive. Because the articulations of the forward vertebrae of the rod-tipped tail of *Pteranodon* tail appear to favor an up-and-down motion, it appears to have been part of an elevator complex. It might have supported an independent elevator surface or may have been integrated into the uropatagia.

The combination of immobile shoulder girdles; a fairly narrow-chorded main wing membrane that was supported by a single finger upon which stiff actinofibers were anchored, probably creating a shallow camber; the absence of boundary-layer airflow control interruptions on the wing tops; the absence of a long pteroid in some examples; and the presence of long tails in some indicate that pterosaurs as a whole were faster and less maneuverable than are mobile-shouldered, broad-chorded, multifingered, wing-warping bats in general. Yet pterosaurs were likely to have been somewhat slower than birds and more agile, especially when large pteroids were present and long tails were not.

Wing Power

In fixed-wing powered aircraft, the sources of lift and thrust are separate in that the rigid wings generate the lift while the engine's pistons or turbines produce the thrust. Most helicopters use the main, subhorizontal rotors to produce lift and thrust at the same time, although some have additional thrust-generating engines, and tilt rotors transform them from helicopters to winged machines pulled by vertical propellers. Attempts to produce human-carrying flapping machines have been made for engineering sport, but there are serious efforts to use flapping wings to create both lift and thrust in small drones. The latter are modeled on powered flying animals that are somewhat like helicopters in that the same airfoils produce both lift and thrust, but in their case the arm-borne airfoils are large horizontal wings that also act as vertical propellers to generate thrust.

Within the context of powered flight, using wings as propellers is marvelously energy efficient. Force equals mass times velocity squared, so velocity is the most important factor. That means that to generate a given amount of thrust, it costs less energy to slowly accelerate a large mass of air than to rapidly accelerate a small volume of air. World War II fighters pulled by big propellers accelerating large volumes of air, such as the P-51 Mustang, P-47 Thunderbolt, and P-38 Lightning, had very large flying ranges (when fitted with droppable external fuel tanks). Contemporary Me-262s, Meteors, and P-80 Shooting Stars, which were pushed along by turbines rapidly accelerating small volumes of air into the narrow intakes of their turbojets, had notoriously short ranges. Jet turbines have become fuel efficient by their transformation into turbofans that slowly accelerate the bulk of the air that passes through them—thus the plump engines suspended below the wings of late-generation airliners. Using entire wings to slowly move very large masses of air is about as fuel efficient as powered flight gets.

Bat flight can be extra energy efficient because the aerial mammals use their multiple, jointed, membrane-supporting fingers to semifold the wing inward during the upstroke such that it saves a third of the power. Because pterosaur archosaurs had a single stiff finger supporting an actinofiber-stiffened membrane, it is doubtful that they could fly as efficiently as bats. As for whether bats or birds are more energy-efficient fliers, there have been contradictory studies, with birds apparently having the advantage according to the most recent work, at least when comparing small examples of the two groups (Johansson et al. 2018). As one can see when watching small songbirds, they use an energy-saving, sort of bouncing flight in which they repeatedly cycle between a burst of lift and thrust-producing, power-demanding flapping followed by a period of no-cost streamlined ballistic travel with the wings fully folded. Bats are not able to do that. And because they were neither small enough nor able to tightly tuck up their wings, pterosaurs were not able to either. Although it does not save energy over a given distance traveled or a given period, some birds minimize flight muscle fatigue by repeatedly alternating flapping with gliding. That allows the flight muscles to rest somewhat before the next power burst, which pterosaurs of all sizes could do.

As efficient as wing propulsion is, much more so than walking the same distance, powered flight places high demands on the metabolic complexes of animals; it requires intense muscular exertion and corresponding energy expenditure per unit time, as well as possibly resulting in fatigue.

It is therefore not surprising that flight muscles make up almost a third of the total mass of some birds and bats, though

15 to 20 percent is more typical. But some birds with decent powered flight performance and acceptable flapping climb rates consist of only 6 percent flight muscle (Paul 2002). Ironically, the largest flight muscles are found in short-range birds such as fowl, including turkeys, chickens, partridges, and pheasants, which use their powerful, anaerobic-dominant and therefore whitish arm muscles to achieve very rapid, subvertical takeoffs to escape ground predators, and to travel short distances when flying fast is more suitable than walking or running the same distance. Some long-range flapping fliers such as ducks, geese, and swans actually have modest-sized flight muscles. These deep red, myoglobin-filled muscle fibers emphasize sustained aerobic power over long periods so they can power through the air for many hours. But with modest-sized flying muscles that do not have much anaerobic burst power, geese and swans have to use a laborious running takeoff followed by a gradual rate of climb to the desired cruising altitude. Pigeons are remarkable fliers because they can use their relatively big flight muscles to cruise at interstate speeds for long distances, and they are also able to take off vertically. They do the latter by wing clapping, which you can hear when they take off. Bringing the wings together directly over the body and then rapidly separating them at the beginning of the downstroke creates strong vortices that generate a dramatic increase in lift. The same aerodynamic trick is used by many insects such as bees—flying bees buzz because they are slapping their wings together at the top of each upstroke. Wing clapping is why bees and some other insects can fly with wings whose area would otherwise be too small to sustain flight using conventional aerodynamics. Whether any pterosaurs known or unknown were fast-takeoff wing clappers is not known, but anurognathids are possible candidates. Because some small bats can hover using wings that are not radically modified like those of the theropod hummingbirds, it is possible that some similar-sized pterosaurs did the same, perhaps unknown terrestrial examples. On the other hand, their multifingered wing membranes may allow bats to hover, and some flying mammals and dinosaurs hover in order to feed on the nectar of flowering plants, which were not available until the late Cretaceous, if then.

A caveat: the details of how the wings of small birds with their sophisticated feather airfoils, and of bats with their multifingered wings, work when producing thrust via flapping are very complicated and remain poorly understood. Studying the highly dynamic wing action of flying creatures is inherently taxing, and there has not been much money invested in the effort, although using bioflight as a source of ideas for improving the abilities of drones is changing that. Bird and bat flapping is so complex because it probably improves the efficacy of their flight in various ways. It would seem that the simpler, one-big-finger wings of pterosaurs would not have been as complex in action when flapping. If so, then they may indeed have been less efficient in terms of energy use and maneuvering, although the minimuscles within their membranes should have made up for some of their deficiency. Exactly how remains obscure

and, in view of our inability to study pterosaur wings in action, will always be so. At larger sizes the complexity differences would become less—the action of the steadily extended wing of a marine soaring albatross is not nearly as complicated as that of the flapping appendage of a robin, and probably not more multiplex than that of a wave-skimming ornithocheiroid. Ergo, there should have been less or no divergence in wing performance among the wings of the archosaurian titans.

Pterosaur Flight Repertoires

It is often said that animals that only glide—flying lizards, snakes, squirrels, and such—are not actual fliers because they do not power fly. This is silly. An albatross that is wave soaring for days on end over countless kilometers without flapping its wings is most certainly flying. As is a vulture using thermals. Same for a sailplane breaking a record in gliding height or distance. If something, animal or otherwise, has airfoils that allow it to at least glide, it is a flier. Flight comes in many forms, expressions, and variations.

The earliest-known pterosaurs were already fully developed fliers, and most were probably generalists with modest flight performance and correspondingly modest flight muscle mass. Many small and medium-sized pterosaurs, rhamphorhynchoids and pterodactyloids alike, appear to have been good performance generalists (Witton 2013; Venditti et al. 2020). But it is often hard to be sure about pterosaur flight performance estimations. Pterosaur aerodynamics are so distinctive from those of birds and bats, they often complicate figuring out what pterosaurs were doing in terms of aerial performance and habits. Most studies of pterosaur flight have focused on the giants, and the work has often been contaminated to uselessness by unrealistically low estimates of pterosaur masses (Witton 2013). Another vexing problem is that we do not know the actual wing plan profile for any pterosaur and probably never will—all wing restorations are speculative to a degree, and as a result, so are all flight performance estimates, which should be presented as a range of possibilities depending on varying possible wing areas and so on, rather than as firm conclusions. Also impossible to reliably restore are the details of the streamlining of the wing surface, such as the blending of the arm elements with the membranes. We do know that pterosaurs usually had wingspan to total mass ratios that either exceeded those of birds or were in the upper avian range, though a few were near or below the median. Pterosaurs were more like birds in that, if they had the narrow-chorded wing membranes restored herein, their wing area to total mass ratios were always well within the normal avian range for wing loading.

We also know that as pterosaurs evolved, so did flight specializations, notably in the anurognathids. Sporting well-developed wings and lacking tail stabilization, they should have been powerful flappers able to remain airborne for hours at a stretch, and fast and agile enough to capture flying insects on

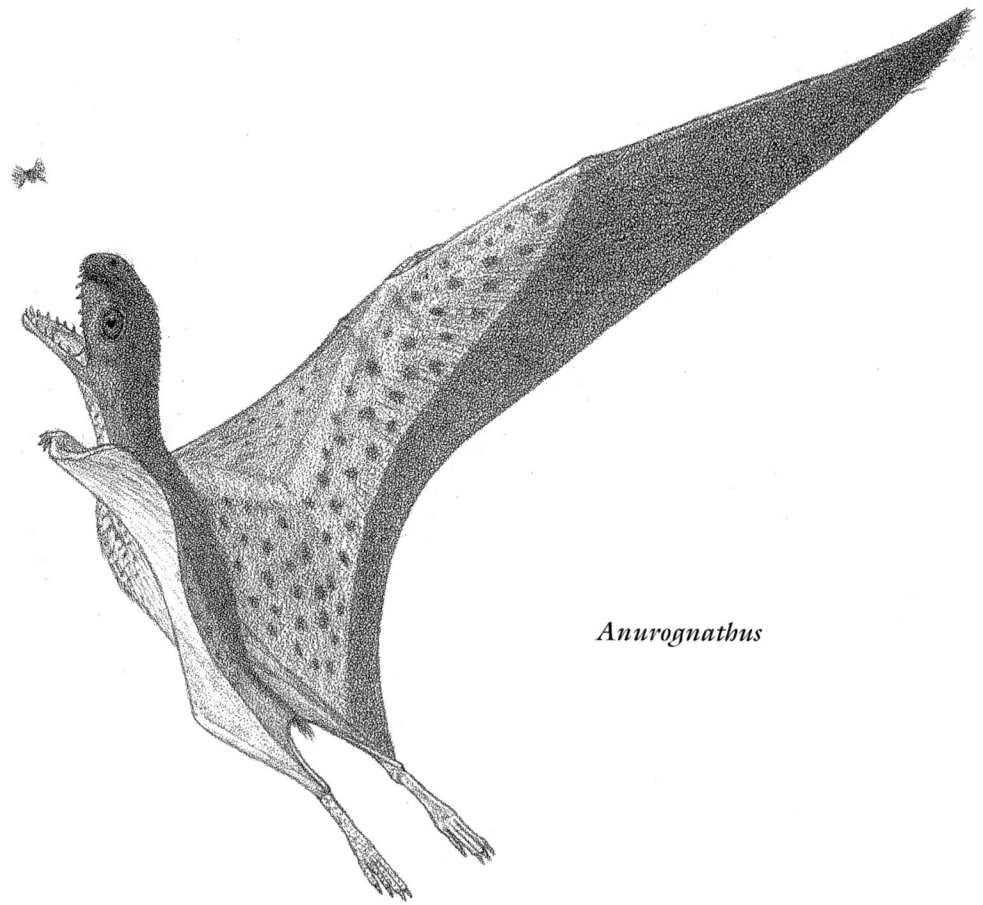

Anurognathus

the wing with ease (Witton 2013; Venditti et al. 2020). Whether anurognathids fully matched birds with comparable habits is problematic. Did anurognathids drink on the wing, like aerial insect-hunting and other birds and bats? This is possible and indeed probable.

For anurognathids, power flapping for hours was worth it despite the considerable energy costs because—like many bats and some anti-insect birds as well as dragonflies—by doing so they captured enough calories and other nutrients in the form of aerial insects to cover the energy expended in flight, plus maintain their bodies and reproduce. Some flappers even have enough energy left over from foraging that they engage in intense powered flight for no apparent reason. Flocking pigeons and starlings are prone to fly about in swirling, hard-turning flock formations for no apparent pressing reason—aerial predators are usually not present, so it may be a form of fitness testing, navigation orientation, or quite possibly social play. Mass crow flocks spend a portion of the evening irregularly flying from one location to another prior to roosting. So some fliers are able to acquire enough energy to burn some of it off in mass aerial relocation and acrobatics. Whether any pterosaurs did so is unknown.

It appears that the largest a bioflier can be and persistently power fly for many hours is about 20 kg (45 lb), the size of swans, which may be the largest birds to have evolved sustained power flying. Continuous cruising flight—energetically similar to walking at a very good clip or slow running—has to be powered aerobically, with the animal burning only as much oxygen at any given moment as its respiratory system can constantly provide without undue effort. But flapping flight can also power short anaerobic bursts, which are also used during fast running, including galloping. More than that and intense fatigue quickly sets in. Anaerobic power can also be used to get amazingly large animals off the ground and up to a substantial altitude, or over a few kilometers. Just how large a creature can be and still be at least a short-range power flier is not entirely clear; the biggest pterosaurs may well have been pushing the biological limits.

Also uncertain is how big a bioflier can be if it is mainly a glider. If a flier needs to spend a long time in the air on a daily basis for at least part of the year and is not hunting aerial prey, it would be to its selective benefit to minimize its flight costs by keeping powered flapping to a minimum. The way to do that is by gliding. By using gravity as a power source, a flier can travel a substantial distance at no more cost than is needed to hold the lift-generating wings in position. The last can be reduced to nearly nothing by an arm-locking mechanism in which the configuration of the bones, muscles, tendons, and ligaments allows the bioflier to make the wing semirigid. Gliding can cost dozens or more times less than power flying the same straight-line distance. If done in air that is not providing lift, the distance that can be traveled before the flight path coincides with the ground is limited. But if the aerialist has a sink rate that is less than that of the rising body of air it is in, then it will rise, which is passive soaring. Dynamic soaring does not rely on rising air but exploits the fact that when a strong wind blows, the air is slower close to a fairly flat surface such as a large body of water than it is a few meters higher up. When a flier flies into the wind, the faster air flowing over the top of the wing has lower pressure than the slower air below—the Bernoulli effect is achieved without the flier's wings producing it—and after rising as much as 15+ m (60 ft), the flier turns around and shallow dives to near the surface, repeating the procedure as often as needed. If this is done in a region of perpetual sufficient winds, it is possible for a soarer that can pick up food from the surface without landing to remain airborne with very little flapping for years. All forms of soaring are powered by a combination of gravity and the solar heating that produces rising air or wind.

Note that as energy efficient as soaring is, in practical terms it can be less so. That is because soaring usually requires not flying in a simple straight line from one place to another. So, unless they fly over suitable geography that coincidently matches the course they need to take, continental soarers traveling to a particular final spot may have to do so in a very erratic course that greatly increases the total aerial distance traveled and the time needed to make the journey. Even so, the overall reduction in energy expenditure via soaring during migration remains substantial in a number of birds that migrate primarily by soaring, storks and raptors being examples. But most migrating birds minimize travel time by power flying most or all of the way, including across oceans they cannot land on, geese and swans being large-bodied examples. If, on the other hand, a soaring animal is looking for sustenance, the erratic course of soaring can actually aid the search, as in flying scavengers and fishers.

Sources of rising air include solar-powered warm-air thermals, which are common over nonmountainous terrain with sufficient sunlight impacting the ground. Having exploited one thermal, the soarer can glide to the next—sometimes sighting in on the fluffy cumulus clouds that often mark a thermal's top—and so on from midmorning to late afternoon. When a general wind is blowing over ridges, hills, mountains, cliffs, and large waves in a direction not too parallel to the long axis of landforms, the resulting standing waves on the upslope side and over the top can provide a sustained updraft. In the case of hill and mountain ranges, this can be dozens to many hundreds of miles long, as long as suitable winds are in force. Direct sunlight is not required. Winds blowing steadily over large tracts of seas and oceans produce waves that act as miniridges—the resulting updrafts are all temporary, but new ones are constantly being formed. Dynamic soaring can be done over vast tracts of open seas and oceans day and night, though it does require that the

soarer fly in constant tight loops about 50 m (150 ft) across. When the air is still, waves themselves push on the air, their leading slopes forming small updrafts that fliers can exploit to reduce or eliminate the need for flapping while cruising barely above the rollers (Stokes and Lucas 2021). Such wave-slope soaring includes breakers along a coast, helping explain why birds often fly low along the shoreline.

Human-carrying soaring gliders range from hang gliders that weigh about half again as much as those who hang from them, to modest-performance slender-wing gliders with spans of 12–13 m (40–43 ft) that weigh 250 kg (550 lb) with the pilot, to ultra-high-performance sailplanes with 15 m (50 ft) wings that weigh over 500 kg (1,200 lb) fully loaded with pilot, possible passenger, and water ballast and can achieve remarkable distance/descent ratios of 45/1. Note that the highest-performing soaring gliders are not designed to be as lightweight as possible because their power source is gravity, and to glide requires that the glider weigh something—because a gas-filled airship weighs nothing, it cannot move through the air unless it is powered. Because gliding requires negative buoyancy in the air, if gliders are too light, then being insufficiently dense and wing loaded ironically leaves them with too low a mass relative to their drag to achieve the high gliding speed needed to produce the best overall performance. At the other extreme, being too heavy will make a glider sink too fast. So sailplanes are loaded to up to their optimal maximum design mass with water, which can constitute almost half the weight of the machine (Larramendi et al. 2021), or in some cases an auxiliary engine that is revved up only when needed. Some gliders join soaring birds in using temporary thermals as power sources. Ridge soaring is common and in the case of mountain ranges can allow extreme altitudes to be achieved, as well as tremendous distances. Dynamic soaring and wave soaring occur in a zone so close to the surface—birds occasionally incidentally dip a wing tip in the water, and the same was probably true of wave-skimming pterosaurs—that human-carrying machines do not soar this way, though there is an effort to apply this form of air transport to research and other drones.

The ultimate modern wave soarers are albatrosses; among birds over time they were the even more gigantic pelagornithids. With distance/descent glide ratios approaching 24/1, albatrosses and giant petrels combine dynamic soaring with wave-slope soaring. With their wings locked, the soaring energy expenditure is not much above the resting metabolism, and this is the most efficient form of nondrifting travel known among animals. That is all the more true because the near-zero cost of soaring occurs while the soarer is wandering about looking for food, rather than as part of a deliberate migration, in which not flying a straight course while not spending much time eating can pose a problem regarding energy in versus energy out. Although the oceanic soarers have very long wings, they are not lightly loaded. The slender wings are so narrow chorded that their area is not high compared to the mass of their large bodies. The

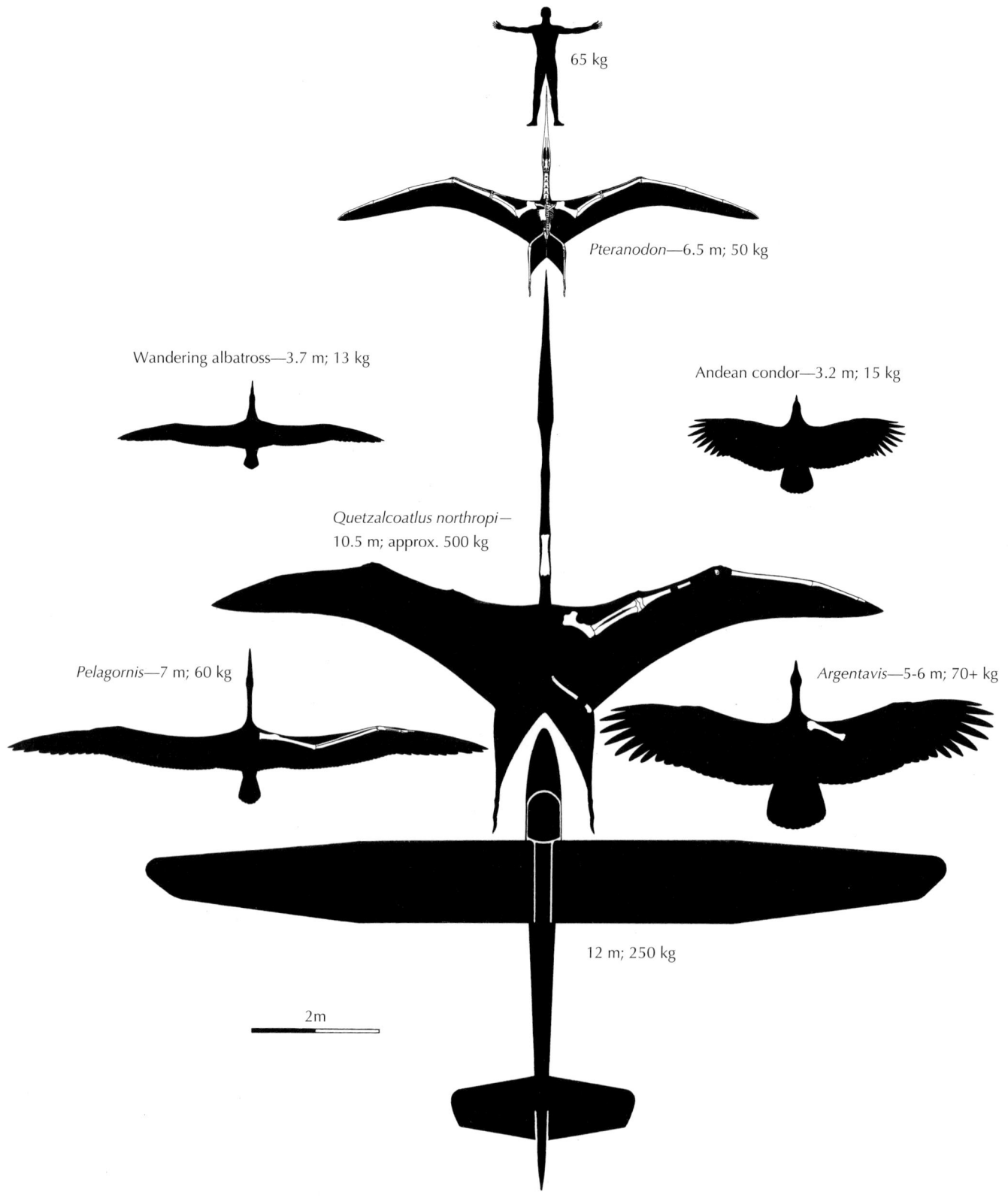

65 kg

Pteranodon—6.5 m; 50 kg

Wandering albatross—3.7 m; 13 kg

Andean condor—3.2 m; 15 kg

Quetzalcoatlus northropi—
10.5 m; approx. 500 kg

Pelagornis—7 m; 60 kg

Argentavis—5-6 m; 70+ kg

12 m; 250 kg

2m

Giant pterosaurs compared to large birds and sailplane

high wing loading is important to their ability to use gravity to power the high-speed downwind descents, often over 60 km/h (40 mph), needed to kinetically power the upwind climb in order to produce free lift via the surface wind–generated Bernoulli effect. Using near-surface soaring allows some albatrosses to stay at sea for up to five years at a stretch, flying 160,000 km (100,000 mi) in a year, not counting the dynamic soaring loops, at top speeds of 100 km/h (65 mph). The soaring performance of the often even more enormous pelagornithids was even more extraordinary in at least certain regards. Aside from the even greater dimensions, the faux-toothed birds had extreme wing-locking mechanisms that may have severely limited flapping. But critical details of pelagornithid soaring are obscure. Their mass-to-span ratio seems less than for albatrosses, and if their wings were correspondingly narrow chorded then they were probably not the high-velocity dynamic soarers that current albatrosses are. Less powerful oceanic winds prior to the exceptionally steep latitudinal global temperature gradient of the Late Cenozoic could have favored some form of slower soaring, perhaps putting more emphasis on wave soaring than on dynamic soaring.

The albatrosses and pelagornithids of pterosaurs were the great pterodactyloid ornithocheiroid ornithocheirids and pteranodonts of the oceans in the last two-thirds of the Cretaceous. The ornithocheiroid arm shows specializations for wing locking. Like those of modern maritime soarers, their inner arm bones are slender in front view, minimizing the frontal drag that is antagonistic to a very high distance/descent glide ratio. Calculations indicating that the marine pterosaurs were extremely lightly loaded because of probably excessively large wing membrane restorations and/or unrealistically low body mass estimates can be discounted. If they had narrow-chorded wings, then the relatively small-bodied ornithocheirids would have been a little less loaded than are albatrosses, suggesting flight dynamics more like those of pelagornithids. With a very low temperature gradient in the warm, nearly glacier-free Cretaceous, marine winds should have been mild by modern standards, so ornithocheirids look to have been slower soarers than albatrosses, and more prone to passive wave soaring than

dynamic soaring. On the other hand, *Pteranodon* had a relatively larger body than the ornithocheirids, and if its wing was narrow chorded it may have had more albatross-like wing loading, which suggests more dynamic soaring than that of ornithocheirids. It is possible that unknown ornithocheiroids were specialized for the deep oceans, especially the Pacific, which was even larger then.

Of particular interest are the pteranodontids' closest relatives, the nyctosaurids. Ossification of some arm tendons suggests wing locking approaching or reaching the pelagornithid level. With the longest wings relative to their exceptionally small bodies, they were very probably the most lightly wing-loaded pterosaurs. In addition to showing that they were the pterosaurs least well adapted for ground locomotion, the loss of the small fingers means that nyctosaurs had the most aerodynamically streamlined wings among the group. The extreme pterosaurs were most similar to the most extreme gliding birds, the frigate birds, which in turn are the most pterosaur-like of birds. Frigate birds are the most lightly loaded known flying dinosaurs, and their basic appearance is the most reminiscent of the pterosaurs, with their long, slender, strongly kinked wings. Frigate birds are extremely aerial, spending little time anywhere but the air. But they are not deep-oceanic birds; they remain near islands that are their home bases.

In whatever way oceanic pterosaurs soared over open water, when along cliff-lined coastlines they would have taken advantage of the updrafts generated by the steep terrain near the waterline to move about, and to climb up to high nesting sites if they were using such, with minimal effort.

Ornithocheiroids were the most evolved soaring pterosaurs, but they were far from the first. The lifestyle seems to have appeared in rhamphorhynchoids as early as the Triassic—specifically in the slender-inner-winged caviramians, and in the campylognathoidids, whose very long wings suggest soaring (Witton 2013), perhaps of a gull or frigate bird type. Soaring appears to have evolved in very early birds too, its energy efficiency having strong natural selective value. Being typically nocturnal, few bats soar, but some diurnal flying foxes use thermals and slopes to do so.

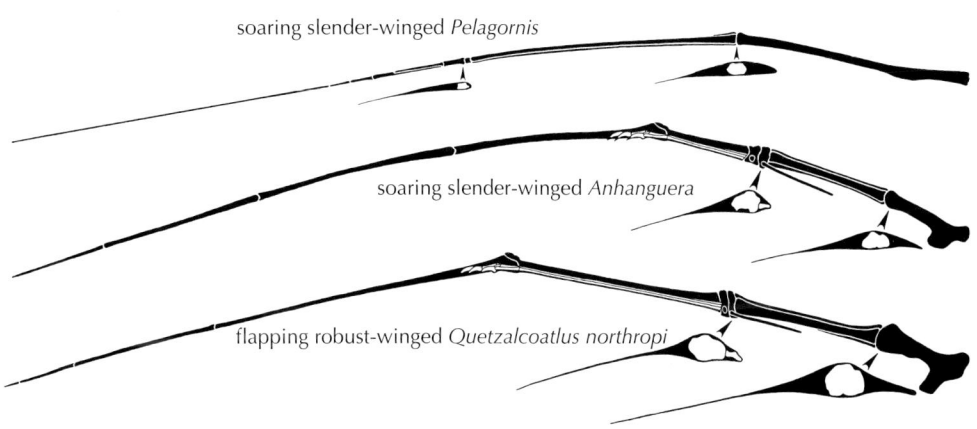

soaring slender-winged *Pelagornis*

soaring slender-winged *Anhanguera*

flapping robust-winged *Quetzalcoatlus northropi*

Soaring versus flapping wing profiles compared

The flight type of the pterosaur group that included the most colossal fliers, the continental azhdarchids, has been the focus of much research and conflicting results. Because these were nonmarine pterosaurs, they were not very low-altitude soarers like the ornithocheiroids. The big question is whether they were passive soarers that used thermals and ridges, like vultures, condors, and the extinct and sometimes gigantic *Teratornis*, or powered fliers that rarely or never soared, like turkeys and bustards. Many have presumed the first was most true (Habib 2010; Witton 2013), on the theory that because big modern birds are energy-saving soarers, all the big pterosaurs should have been as well. But some of the heaviest living flying birds are actually nonsoarers that rely on flapping flight, and some of them are short-range burst fliers that use an intense spurt of anaerobic power to get into the air quickly and then rapidly tire and reland. The big azhdarchids were probably too large to power fly for long distances.

Azhdarchids were shorter winged relative to their mass than most pterosaurs, being the only members of the group whose span-to-mass ratios fall close to the avian median, so they were quite unlike the low-altitude ornithocheiroid soarers. Presumably azhdarchid wing chords were fairly broad by pterosaur standards; the wing area to mass ratio appears to be a little below the avian average but still well within norms. Azhdarchid wings were correspondingly so much more heavily loaded than those of soaring continental birds that the glide ratio should have been too steep to allow these superpterosaurs to be passive soarers (Goto et al. 2020; Venditti et al. 2020). Also antithetical to soaring azhdarchids is the robustness of the inner wing elements, astonishingly so in the giant examples. This could hardly have been more different from the drag-minimizing frontal slenderization of the inner wing of even the biggest soaring ornithocheiroids, as well as avian soarers, which helped provide the streamlining needed to slip through the air and thereby maximize gliding efficiency. Especially notable is the azhdarchid pectoral crest, an extraprominent arcing structure indicative of what should have been the most powerful pectoralis musculature among pterosaurs, or for that matter any fliers. Also bulky were bulging Popeye-like elbows and wrists that would have interfered with the airstream atop and beneath those sections of the wings.

The azhdarchid combination of short wings; oversized wing musculature, which, as powerful as it was, probably could not propel long flights; and the thick, drag-inducing inner wing that the massive musculature and its robust supporting and strengthened bones required to exist indicates that the land dwellers' wings were adapted to maximize flapping thrust power production over static wing gliding, and therefore these were short-range burst fliers (Paul 1991, 2002). If so, then the flight muscle cells should have been configured to maximize quick, high-intensity anaerobic power over a few moments, which would have restricted powered flight range, rather than longer-term but less-intense aerobic power, which could have sustained somewhat longer trips. This is adaptively logical in that as big as the azhdarchids were, they lived in a world infested with even larger and faster-running predaceous theropod dinosaurs, so they needed to be able to get their up-to-half-ton, bear- and horse-sized bodies into the air very fast when threatened. That the azhdarchids could not then flee tremendously far would not be important relative to the critical aerial escape mechanism. And the ability to spontaneously travel a few kilometers in a given trip in the air at substantially less total energy cost, at a much faster rate, and with no danger of being attacked than when walking would also have been a major advantage. The initial flapping to altitude could have been followed by a distance-lengthening glide, further extended by fatigue-reducing cycling between flapping and gliding before landing if desired—a technique used by some albatrosses when sufficient winds are not available. An advantage of short wings is that they would have facilitated low flight through narrow spaces between tall vegetation in the landscapes azhdarchids dwelled in, similar to the short, broad wings characteristic of forest birds. The common presumption that thick-armed azhdarchids soared to cover long distances appears to be no more applicable than it is to the big ground-foraging turkeys and bustards, which never soar.

About half of all bird species migrate, and the possibility that some terrestrial pterosaurs migrated is considerable. The very dark polar winters shut down plant productivity, and even during the warmer portions of the Mesozoic they were chilly enough for snow. During global climatic cooling, high-latitude winters were outright severe, with blizzards and extended frosts during parts of the Mesozoic. The pressure to fly equatorward during the fall and then return toward the poles for the late spring and summer when the sun was above the horizon for longer, even continuously, would have been compelling. However, with the global north-south temperature gradient much less in the Mesozoic than in the modern world, the pressure for species to migrate, and the distances traveled, may have been generally much less than is common today. Fliers can migrate long distances because flying costs only about two-thirds as much as walking the same distance, even less if soaring is exploited. If pterosaur flight was less efficient than avian flight, it could have decreased their migratory range, as may be true of bats vis-à-vis birds (Johansson et al. 2018). Migration also makes it possible to avoid geographical barriers and hazards, such as dangerous river crossings and the big predators that afflict terrestrial travelers. By making a few short flights amounting to some tens of kilometers a day, and feeding most of the day, short-flight azhdarchids could have traveled in the low thousands of kilometers in the spring and fall. Thermal and ridge soaring could have facilitated longer trips if any land pterosaurs practiced such energy-efficient flight.

It is tempting to assume that as pterosaurs evolved, their flight became progressively more sophisticated. That was true in broad terms as various pterosaur clades expanded into previously unfilled aerial niches over Mesozoic time (Witton

2013; Venditti et al. 2020). But evolution is not intentionally progressive; it is adaptive to the specific life circumstances of a breeding population at a given time. Many modern birds have limited flight performance compared to birds all the way back in the Mesozoic, to the point of being flightless in many cases. Among pterosaurs the flight abilities of Early Jurassic *Dimorphodon* appear to have been less impressive than those of known earlier Late Triassic eopterosaurs, indicating a decrease in aerial performance. Nor do the flight abilities of Early Cretaceous filter-feeding *Pterodaustro*, which lays claim to the lowest wingspan/total mass ratio yet known among pterosaurs, appear particularly impressive, although the ratio was still well up in the avian range.

Takeoffs and Landings

Like airplanes, pterosaurs and other aerial beasts are highly configured to fly, but as per the principle that—unless it goes into deep space—what goes up must come down, all aerial objects must at some time take off and land. As a result, fliers need to have the means to do so, and to get around to at least some extent when not airborne. Birds and bats in high places can just drop into the air and glide or begin flapping. Or they can start the flight with a leap. If a breeze is available, launching into the wind is preferable, unless doing so is not compatible with the direction a bioflier wishes to travel between trees, or when departing a cliff. These factors would also have been applicable to pterosaurs of all sizes when launching from high locations.

More difficult is taking off from flat ground, against the gravity well of the entire planet. A few flying birds and many bats are not well adapted to do so. Even flapping wings can be hard pressed to produce enough lift at zero wind velocity to readily get a body into the air; an initial booster is needed to be adept at it. In biofliers the booster is a push-off from the limbs. These are the normally powerful hindlegs in bipedal birds, but the weak legs of some of the highest-performing aerialists such as swifts and frigate birds are not well adapted for taking off from flat ground.

Among bats, the group most adept at ground locomotion, the terrestrial prey-stalking vampires, need to be able to lift up fast. Mammalian quadrupedal walkers and fliers logically rely on a push-off more from their powerful arms than from their comparatively weak legs.

Because most pterosaurs were strong-armed walking and flying quadrupeds with relatively weak legs, they are likely to have used their forelimbs as the main push-off mechanism during takeoff, as do bats (Habib 2010; Witton 2013). This adaptation appears to have been taken to an extreme in the azhdarchids. Their exceptionally robust-boned and powerfully muscled inner arms were well adapted for combining a rapid leap off the ground followed by a rapid flapping flight to climb away from danger. A downside of this system was that the resulting thick inner wing—because it was well muscled and strong boned—interfered with high-performance soaring.

Although pterosaurs were probably practitioners of arm-assisted takeoffs, the inner arms of some rhamphorhynchoids

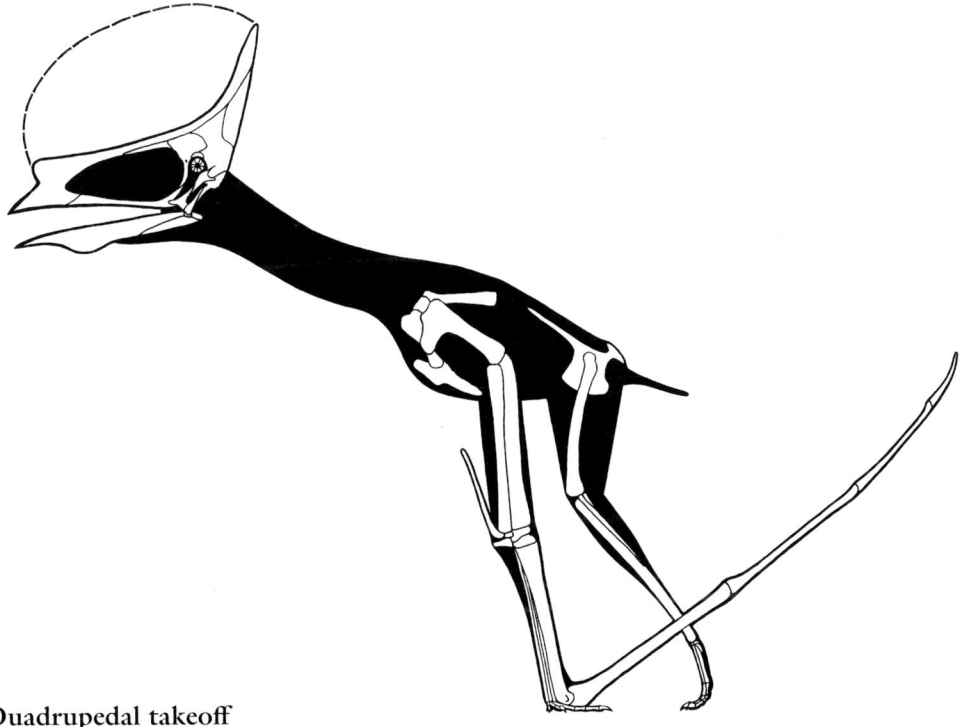

Quadrupedal takeoff

appear rather short for this purpose. It is possible that some pterosaurs on occasion would push off with the hindlimbs if enough wind were available, or run into the wind, especially in the case of big pterodactyloids. The best candidates for bipedal takeoff in pterosaurs are caviramians; their humerus was exceptionally slender and only a little stronger or more powerfully muscled than the femur.

It has been plausibly argued that water-loving ornithocheiroids used the arm push-off to get off the surface of the water, in part because most of them had better-developed inner arms than expected in soarers—the nyctosaurs being an exception—and because the ability to get back in the air after a spashdown, intended or not, was important. But the idea of oceanic ornithocheiroids spending much time floating on deep-water waves fishing like pelicans and albatrosses is problematic because they would have been sitting ducks—or in this case sitting ornithocheiroids—for being picked off by the host of big marine predators lurking under warm Cretaceous salt waters. Floating albatrosses can seize prey on the surface with reasonable safety because they typically inhabit cool to cold waters where active large predators are not numerous. Warmer-water pelicans minimize the danger of attack from below by seizing prey from the surface of shallow waters (which also hinders the prey fish diving away). Because they were creatures of the Mesozoic tropics and subtropics—there is evidence that ornithocheiroids avoided chillier high-latitude waters—floating would have rendered the pterodactyloids vulnerable to the many species of deadly fish and reptiles that packed those seas. Even worse, at any given time much of the sea surface has little in the way of easy prey, so open-ocean birds search for spots where the surface is being churned up by frenzied schools and bait balls of fish trying desperately to escape underwater predators attacking from below—predators that would be as happy to consume floating pterosaurs as they would fish. Surface floating is all the more dangerous because it is very difficult for eyes barely above the water to see and thereby trigger escape from submerged predators, while the latter can readily see their targets silhouetted against the sky above and beyond. At the same time, float feeding involves the significant energy expense of powered takeoffs.

Their high wing loading prevents big albatrosses from flying slowly enough to readily pick up aquatic prey on the wing, and their short head-neck length further limits their ability to quickly reach out for a tasty fish on pitching waves from the air. The longer beaks and perhaps slower flight of pelagornithids should have allowed them to dip feed—their ability to relaunch from the water being limited, as it is for frigate birds. Slower airspeeds combined with beaks up to over a meter long should have allowed giant marine pterodactyloids to feed on the wing. Concentrations of fish driven to the surface by their underwater marauders would have been easy pickings, including inert body parts left behind by the swimming killers. Ornithocheiroids would have done best by slowly wheeling on their great wings over the reptiles and sharks in the sea as they went about their

business, taking advantage of the submarine hunters driving small fish and cephalopods en masse to the surface to be snatched up by their sword-length beaks, all the while staying safe from the dangerous undersea predators. This was not a particularly hard life; weighing about as much as an adult female human, a gigantic yet energy-efficient soaring *Pteranodon* had to eat only a dozen or more foot-long fish a day, even if feeding its nestlings back onshore. As for the strongly muscled inner arms of the marine pterodactyloids, they may indicate that they did more flapping than their avian wave-soaring analogues.

Having been in the air from just a few moments to months, flying pterosaurs needed to land. Doing so requires a flier to slow down dramatically, sufficiently to avoiding a crash landing. One way to do so when landing on a high location is to bleed off speed by approaching it from below, in which case the final landing speed can be close to zero, with the free fingers perhaps making the first contact with the branch or rock edge, as in bats. This gravity-assisted landing option is not available when landing on the ground or water. In that case, the objective is to land as slowly as possible, even while traveling down the gravity well. One way to do that is via a controlled stall, approaching the landing zone with the leading edge of the wings strongly tilted upward to make their undersides into drag-producing air traps, while keeping the more rapid airflow over the top of the wings stable enough to still generate enough lift to prevent the stall from occurring too early and causing a sudden hard crash. Increasing the wing camber to maximize both lift and drag is of great advantage when landing. Jet airliners use leading-edge wing slats to help do so, and pterosaurs may have increased the fore-aft arc of the inner wing by depressing the propatagium. To produce extra drag, webbed feet could be deployed as auxiliary air brakes as discussed earlier, which is convenient in that putting the legs down vertically pre-positions them to act as initial landing gear during touchdown. This is similar to some aircraft, in which the nearly vertically deployed wing flaps are entirely for generating drag during landing while producing no extra lift, as in those of the Spitfire fighter. Extending the free fingers could have added a little braking drag. As the feet get close to the ground, the wings' angle of attack is further increased to the point that the top airflow becomes too turbulent to generate lift, and the deliberately stalled animal drops the last very short distance to the ground.

Depending on various factors, including wind speed, the landing could be fairly static, with the feet contacting the ground and no further movement occurring, or it could include deceleration running for a short distance, as appears to be recorded by a pterosaur trackway (Mazin et al. 2009). In that example, the two feet made contact in parallel fairly close to one another, the toes dragged a little as the body bounced a bit, the feet made a second contact followed by the hands slightly out to the side, then another step with hands and forefeet, and then the pterosaur walked off. In a water landing, the webbed feet were used as water skis to help cushion the initial contact, as in waterbirds.

Pteranodon feeding on fish and belemnite schools driven to the surface by the mosasaur *Platecarpus* (those and front pterosaur to same scale)

Respiration and Circulation

The hearts of turtles, lizards, and snakes are three-chambered organs incapable of generating high blood pressures. Crocodilian hearts are incipiently four chambered but are still low pressure. Reptile lungs, although large, are internally simple structures with limited ability to absorb oxygen and exhale carbon dioxide. Although they have a dead end, the lungs of some lizards and crocodilians may have unidirectional airflow. Lizard and snake lungs are operated by straightforward rib action. Crocodilian lung ventilation is more sophisticated. Muscles attached to the pelvis pull on the liver, which spans the full height and breadth of the rib cage, to expand the lungs. This action is facilitated by an unusually smooth ceiling on the rib cage that allows the liver to easily glide back and forth. The presence of a rib-free lumbar region immediately ahead of the pelvis, and,

basal archosaur

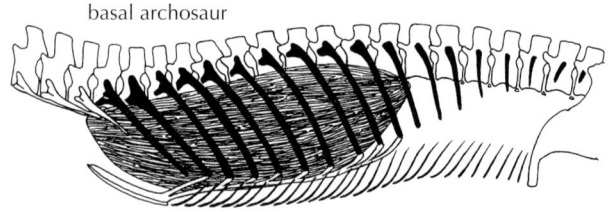

crocodilian with liver pump operated by pelvic muscles

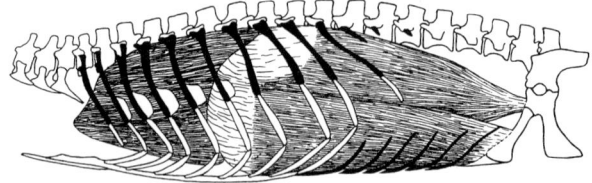

pterosaur with air sacs operated by pelvic muscles

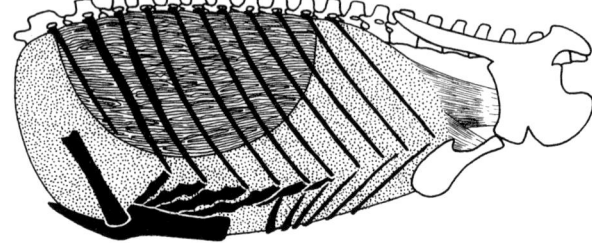

bird with air sacs operated by ribs and sternum

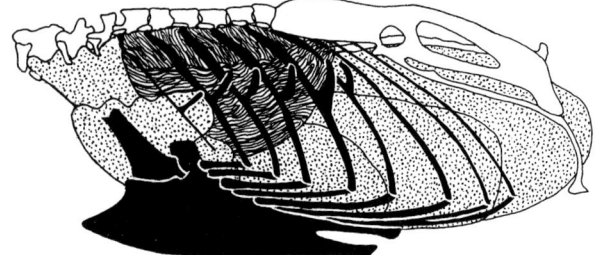

Respiratory complexes of archosaurs

at least in advanced crocodilians, a mobile pubis in the pelvis—very unusual in tetrapods—enhances the action of the muscles attached to it.

Birds and mammals have fully developed four-chambered, double-pump hearts able to propel blood in large volumes at high pressure. Mammals retain fairly large dead-end lungs, but they are internally very intricate, greatly expanding the gas exchange surface area. The lungs are operated by a combination of rib action and the vertical, muscular diaphragm. The presence of the diaphragm is indicated by the existence of a well-developed, rib-free lumbar region, preceded by a steeply plunging border of the rib cage, on which the vertical diaphragm is stretched.

It is widely agreed that all dinosaurs very probably had fully four-chambered, high-capacity, high-pressure hearts (Paul 2002, 2012, 2016). Their respiratory complexes appear to have been much more diverse. Most researchers also agree that the theropod dinosaurs, especially the avepods ancestral to birds, evolved increasingly birdlike respiratory complexes. Birds have the most complex and efficient respiratory system of any vertebrate. Because the lungs are rather small, the chest ribs that encase them are fairly short, but the lungs are internally intricate so they have a very large gas-exchange area. The lungs are also rather stiff and set deeply into the strongly corrugated ceiling of the rib cage. The lungs do not dead end; instead, they are connected to a large complex of air sacs whose flexibility and especially volume greatly exceed those of the lungs. Some of the air sacs invade the pneumatic vertebrae and other bones, but the largest sacs line the sides of the trunk; in most birds the latter air sacs extend all the way back to the pelvis, but in some, especially in flightless examples, they are limited to the rib cage. The chest and abdominal sacs are operated in part by the ribs; the belly ribs tend to be extralong in birds with well-developed abdominal air sacs. All the trunk ribs are highly mobile because they attach to the trunk vertebrae via well-developed hinge articulations. The hinging is oriented at an angle that compels the ribs to swing outward as they swing forward, inflating the air sacs within the rib cage, and then the sacs deflate as the ribs swing backward and inward. In most birds the movement of the ribs is enhanced by ossified uncinate processes that form a series along the side of the rib cage. Each uncinate process acts as a lever for the muscles that operate the rib the process is attached to. In most birds the big sternal plate also helps ventilate the air sacs. The sternum is attached to the ribs via ossified sternal ribs that allow the plate to act as a bellows on the ventral air sacs, the sternum dropping to further inflate abdominal air sacs as the chest ribs swing forward. In those birds with short sterna, the flightless ratites, and in active juveniles, the sternum is a less important part of the ventilation system. Some Mesozoic birds with short sterna retained the dinosaurian gastralia between the pelvis and sternum, and these probably helped ventilate the abdominal air sacs.

The system is set up in such a manner that most of the fresh inhaled air does not pass through the gas-exchange portion of the lungs but instead goes first to the air sacs, from where it is injected through the lungs in one direction on its way out. Because this unidirectional airflow eliminates the stale air that remains in dead-end lungs at the end of each breath and allows the blood and airflow to work in opposite, countercurrent directions that maximize gas exchange, the system is very efficient. Some birds can sustain cruising flapping flight at altitudes as high as Mount Everest, at 8,500 m (28,000 ft), and equaling those of jet airliners; more energy-efficient soarers can reach over 11,000 m (37,000 ft).

There is no reason to doubt that pterosaur hearts were fully four-chambered, high-pressure organs (Paul 2002, 2012; Witton 2013). Because the ancestors of pterosaurs are not documented, the evolution of their respiration cannot be detailed. What we do know is that their breathing apparatus appears to have paralleled or converged with, and combined features of, those of both birds and crocodilians, along with adaptations of their own (Paul 2002; Witton 2013). The birdlike attributes included the well-developed pneumatic complex in which abdominal air sacs helped ventilate the lungs, presumably with considerable unidirectional air flow. The small volume of the pterosaur body suggests that the lungs must have been small, as in birds, and possibly rigid. As for how the air sacs were ventilated, in basal pterosaurs the main ribs were fairly mobile, and they were joined to the sternum via mobile sternal ribs. This should have allowed the back end of the sternum to change the volume of the air sacs near that location, as in birds. But unlike in the latter, the extra ossified projections that improved the leverage of the respiratory muscles were not uncinate processes on the ribs; they were the sternocostapophysis projections on the sternal ribs. In derived pterodactyloids the ribs started fusing with the vertebrae, limiting and finally eliminating the ability of the ribs to operate the air sacs.

That is where the crocodilian aspect of pterosaur breathing comes in. The pterosaur pubis was not mobile as it is in crocodilians, but the unusual prepubis at the end of the pubis was movable. It probably acted like the crocodilian pubis to help change the volume of the abdomen (Witton 2013). In doing so, it would have ventilated the air sacs in that region in all pterosaurs, aided by the gastralia as in avepod theropods, including basal birds. Being predominantly like that of birds, the air sac–ventilated respiration of pterosaurs should have been almost or perhaps as efficient at oxygenating their blood.

But pterosaurs may have had an additional breathing apparatus, one that is not found in birds and is present in bats: the wings. In bats the ultrathin wing membrane makes up 85 percent of the animal's surface area and, perfused with blood vessels, may add a tenth or more to oxygen intake. In pterosaurs the top of the wing was probably sealed off by the actinofibers, but the dense layer of blood vessels under the membrane appears well suited for acting as auxiliary lungs.

Sporting avian-like respiration and bat-like wing lungs, pterosaurs should have been able to respire well enough to fly at extreme altitudes. Whether they could fly as high as birds we cannot know, but they may not have had to in view of the scarcity or absence of Mesozoic mountain ranges as tall as some are today.

Mammalian red blood cells lack a nucleus, which increases their gas-carrying capability. The red blood cells of reptiles, crocodilians, and birds retain a nucleus, so those of pterosaurs should have as well.

Feeding Apparatus and Digestive Tracts

The beaks and conical teeth of most pterosaurs were adapted for grabbing and holding on to food items, which they would then have had to swallow whole. The simple spiky teeth could not reduce items, and their jaws were often too weakly built and muscled to do much food processing. A recent analysis has used tooth microwear patterns to help sort out what kinds of foods specific pterosaurs were chowing down on (Bestwick et al. 2020), and another focuses on the power of the jaws based on reconstructions of the bite muscles of various derived pterodactyloids (Pêgas et al. 2021; includes prior examinations of isotope ratios in teeth). These studies often affirm what is already thought, and in other cases the research may settle controversies—the results are integrated into the group and species descriptions herein. Tooth microwear has not yet demonstrated dramatic shifts in feeding style from juveniles to adults. It is possible that food items could have been held in the throat pouch, when present, for a time if there was not yet room in a full belly. The very large size of fish found in some body cavities, up to 60 percent of trunk length, shows that such pterosaurs could swallow enormous prey intact and let their digestive tract acids and enzymes take care of it (Witton 2013). Swallowing such big items would have required that the esophagus be very distensible, as it is in birds that swallow large prey whole. As well as fish, saltwater pterosaurs dined on unshelled cephalopods (Hoffman et al. 2020). A few early rhamphorhynchoid pterosaurs with multicusped teeth could chew prey and plant material to some extent; why this seemingly useful ability was quickly lost in the group is a mystery. Among Cretaceous pterodactyloids, the strong-jawed and exceptionally powerfully muscled dsungaripterids and perhaps chaoyangopterids could crush hard-shelled creatures. Tapejarids may have been able to do some food processing with their mouths, especially those with deeper skulls and more powerful jaw muscles. The teeth of istiodactylid predator scavengers were like animal traps and appear to have been able to cut up items before they were gulped down. Istiodactylid hyoids are similar to those of scavenging birds, which adds to the evidence that those pterosaurs fed on carcasses (Jiang et al. 2020). Filter-feeding pterosaurs used their combs of slender teeth to ingest very small items in very large numbers.

Because the abdominal regions of all pterosaurs were quite small, and because some of that internal space was occupied by air sacs lining the walls, pterosaur digestive tracts were always low in volume and short. Archosaurs are prone to having gizzards in about the middle of the digestive tract, after the stomach and before the intestines, and a wide variety, including many birds, deliberately seek out and swallow gizzard stones, or gastroliths, to help process food. The absence of such grit preserved with many pterosaur specimens indicates that such stones were not widely employed by members of the group. But they have been found within the flamingo-like filter-feeding *Pterodaustro*, the front teeth of which look as if they were strengthened for gathering up small gravel, which was then used to help grind up the small hard-shelled creatures it fed on.

The big fish found inside fossil pterosaurs could have powered the creatures for a day or more, so some pterosaurs did not necessarily have to constantly work hard to obtain sufficient sustenance. Any plant matter that pterosaurs consumed should have been relatively easy-to-digest components such as fruits and seeds, their digestive systems being too small to effectively process leaves. Fruit seeds have been found in the abdominal cavity of a tapejarid pterodactyloid.

Pellets that may have been retched up by pterosaurs suggest that they periodically unloaded indigestible items such as bones, as do some birds and cats; that the pellets contain bones indicates that pterosaur stomach acids were not extremely strong, as is true of birds. Unlike dinosaur fossil feces, or coprolites, which are fairly common, pterosaur examples are few. One had just been voided from a deceased individual. Pterosaur coprolites might be rare because their feces may have usually been too liquid and soft to readily fossilize, as is true of those of their avian relations. It is not surprising that the few other examples of pterosaur poop appear to be those of filter-feeding ctenochasmatid pterodactyloids found in association with their trackways; the feces are laden with a high density of hard bits of the small creatures they consumed (Qvarnström et al. 2019).

Pterosaurs as Food

For Mesozoic eaters of flesh, pterosaur necks, legs, and especially flight muscles would have been appealing meals, and a number of their skeletons show evidence of being bitten by large fish, sharks, marine reptiles, and dinosaurs (Witton 2013). Whether these particular specimens record scavenging versus predation is not determinable, but it is likely that both were involved. It is also likely that large predaceous and omnivorous pterosaurs picked up and put down smaller pterosaur species and juveniles, in some cases perhaps of their own species. Pterosaurs were very likely to have stolen food from other pterosaurs, again of their own species, when the opportunity arose—some birds such as sea eagles and frigate birds are frequent food stealers. Also on the menu of Mesozoic predators would have been pterosaur eggs

and hatchlings, which is one reason many pterosaurs nested in isolated locations where there were few or no egg stealers.

Senses

The usually large eyes and well-developed optical lobes characteristic of pterosaurs indicate that vision was their primary sensory system, as it is in all birds (Witton 2013). Bats are never blind; their vision ranges from poor to very well developed in some fruit bats. The poorly developed color vision of most mammals is a result of the nocturnal habits of early mammals, which reduced vision in the group to the degree that eyesight is often not the most important of the senses—the high-quality color vision of primates, humans included, is a mammalian anomaly. Reptiles and birds have full-color vision extending into the ultraviolet range, so pterosaurs almost certainly did as well. Reptile vision is usually as good as or better than that of mammals, and birds tend to have very high-resolution vision both because their eyes tend to be larger than those of reptiles and mammals of similar body size and because they have higher densities of light-detecting cones and rods than mammals. The cones and rods are also spread at high density over a larger area of the retina than in mammals, in which high-density light cells are more concentrated at the fovea (so our sharp field of vision covers just a few degrees). Some birds have a secondary fovea. Day-loving raptors can see about three times better than humans, and their sharp field of vision is much more extensive, so birds do not have to point their eye at an object as precisely as mammals to focus on it. Birds can also focus over larger ranges, 20 diopters compared to 13 diopters in young adult humans. The vision of the bigger-eyed pterosaurs, particularly the flying insect–pursuing anurognathids, may have rivaled this level of performance.

The pterosaurs' big eyes have been cited as evidence for both day- and night-dominant lifestyles. It is the structure of the retina and pupil (unknowable for pterosaurs) that determines the type of light sensitivity, but some researchers have used the differing configurations of the sclerotic rings to try to determine what some pterosaur eyes were optimized for. The results indicate that filter-feeding pterodactyloids were largely nocturnal, like some water-straining birds, and other pterosaurs had different light level preferences (Schmitz and Motani 2011), but these conclusions have been disputed. Not needing keen vision to see their food, some filter feeders look as if they had the smallest eyes among pterosaurs.

Birds' eyes are usually so large relative to the head that they are nearly fixed in the skull, so looking at specific items requires turning the entire head, although the larger retinal area of focus reduces this need. The same was likely to have been true of smaller-headed pterosaurs. Pterosaurs with larger heads should have had more mobile eyeballs that could scan for objects without rotating the entire head. Most pterosaurs' eyes faced to the sides, maximizing the area of visual coverage at the expense of the view directly ahead, so they lacked the ahead, stereo vision present in some birds. The notable exceptions were the swallow-like anurognathids, whose broad heads caused the orbits to face partly forward, presumably to facilitate tracking and snapping up flying insects on the wing. Even so, a limited field of visual overlap, perhaps including some three-dimensional binocular processing that would have aided the targeting and manipulation of food items, may have been present in some pterosaurs. Another anurognathid-specific optical feature among pterosaurs was eyes that faced partly upward; this appears to have been an adaptation for spotting high-flying bugs silhouetted against the bright sky and ambushing them with a climbing attack from below.

Most birds have a poorly developed sense of smell, the result of the lack of utility of this sense for flying animals, as well the lack of space in heads whose snouts have been reduced to save weight. Exceptions include some vultures, which use smell to detect rotting carcasses hidden by deep vegetation, and grub-hunting kiwis. The large snouts of the big-headed pterosaurs should have provided them with the space needed for well-developed olfactory tissues, but the small olfactory portions of their brains suggest these fliers did not in fact have such. That this was true of the azhdarchids works against their having been specialized scavengers.

Mammals have exceptional hearing, in part because of the presence of large, often movable outer ear pinnae that catch and direct sounds into the ear opening, and especially because of the intricate middle ear made up of three tiny elements that evolved from jawbones. In some mammals hearing is the most important sense, bats and cetaceans being the premier examples. Reptiles and birds lack sound-catching fleshy outer ears, and there is only one middle ear bone. The combination of outer and complex inner ears means that mammals can pick up sounds at low volume. Birds partly compensate by having more auditory sensory cells per unit length of the cochlea, so sharpness of hearing and discrimination of frequencies are broadly similar in birds and mammals. Where mammalian hearing is markedly superior is in high-frequency sound detection. In many reptiles and birds, the auditory range is just 1–5 kHz; owls are exceptional in being able to pick up from 250 Hz to 12 kHz, and geckos go as high as 10 kHz. In comparison, humans can hear 20 kHz, dogs up to 60 kHz, and bats the exceptional 100 kHz, which is integral to the echolocation systems many of them use. At the other end of the sound spectrum, some birds can detect very low frequencies: 25 Hz in cassowaries, which use this ability to communicate over long distances, and just 2 Hz in pigeons, which may detect approaching storms. It has been suggested that cassowaries use their big, pneumatic head crests to detect low-frequency sounds, but pigeons register even lower-frequency sounds without a large organ.

In the absence of fleshy outer and complex inner ears, pterosaur hearing was in the reptilian-avian class, and pterosaurs could not detect very high frequencies. Nor were the auditory lobes of pterosaur brains especially enlarged. Even so, hearing should have been important for detection of prey and especially of predators when on the ground, and for communication when grounded and in flight, in all species.

New research indicates that some pterosaurs had sensitive snout tips featuring sensory microorgans in tiny foramina that helped them sense and gather food, including when probing for items beneath surfaces (Martill et al. 2020).

Vocalization

No reptile has truly sophisticated vocal abilities, which are best developed in crocodilians. Some mammals do, with insect-hunting bats having developed their voices into a form of highly sophisticated aerial sonar. A number of birds have limited vocal performance, but many have evolved a varied and often very sophisticated vocal repertoire not seen among vertebrates other than humans. Songbirds sing, and a number of birds are excellent mimics, to the point that some can imitate artificial sounds such as bells and sirens, and parrots and a few other avians can produce understandable humanlike speech. Some birds, such as swans, possess elongated tracheal loops in the chest that are used to produce high-volume vocalizations. Cassowaries call one another over long ranges with very low-frequency sounds. Pterosaurs could not have had vocal abilities exceeding those present in their fellow archosaurs, and sophisticated echolocation can be ruled out. It is rather doubtful that any pterosaurs had vocal abilities matching the more sophisticated examples seen in birds—singing and mimicking pterosaurs are unlikely—but such cannot be entirely ruled out. The same is true of the extralong trachea beyond that present with long necks, there not being much space available in the pterosaurs' small bodies. The very long trachea of long-necked pterosaurs would have been well suited for generating deep, low-frequency calls. It is possible that some small pterosaurs along the lines of anurognathids or something similar had a low-grade echolocation ability like some swifts, but as in the latter this would have been for avoiding collisions with large objects, not for targeting small insects. Vocalization is done primarily through an open mouth rather than through nasal passages, so complex nasal passages in the large snouts of pterosaurs that had them may have acted as supplementary resonating chambers.

Disease, Pathologies, and Injuries

Pterosaurs lived in a world filled with diseases and other dangers to their health and well-being. The infectious disease problem was accentuated by the global greenhouse effect, which maximized the tropical conditions that favor disease organisms, especially bacteria and parasites. Biting insects able to spread assorted diseases were abundant during the Mesozoic; specimens have been found in amber and fine-grained lake sediments. Reptile and bird immune systems operate somewhat differently from those of mammals; in birds the lymphatic system is particularly important. Presumably the same was true of their pterosaur relations.

Of particular medical interest is the pterosaur wing membrane, the thin tissue of which was more vulnerable to aerially disabling injury than the feather airfoils of birds. In order to minimize the potentially lethal consequences of not being able to fly because of a torn brachiopatagium, the dense layer of blood vessels in bat membranes promotes remarkably rapid healing of the airfoil. The similarly dense vascularization of the underside of the pterosaur brachiopatagium is compatible with their having a comparable ability to recover from wing injury.

Pterosaur skeletons often preserve numerous pathologies (Witton 2013). Bone depressions, pitting, and overgrowths record internal diseases and disorders, parasites, possible cancers, and physical injuries. The very thin walls of some pterosaurs' bones would have been vulnerable to impact injuries. A common affliction of old individuals was arthritis in the wrists and finger base joints, from a lifetime of ground, takeoff, and flight activities that overstressed the joints. Some partly fractured bones show evidence of healing, and sometimes the tip of the wing finger was lost. Completely fractured limb bones were probably lethal to these delicate fliers. That pterosaurs were strongly quadrupedal on the ground may have made it harder for them to recover from injuries to their already delicate membranous arm wings. A bird might be able to move about on the ground well enough on its two legs to escape predation and get some nourishment until it recovers from a wing injury, but a pterosaur with a damaged arm could have been immobilized both on the ground and in the air.

BEHAVIOR

Brains, Nerves, and Intelligence

Because they calculate what to do, brains are biocomputers that conduct analog-digital mass parallel processing over neural networks. In general, the larger—especially relative to body mass—and more complex brains are, the more capable they are. But this does not mean that small, simple brains have abjectly low performance. Fish and lizards can retain new information and learn new tasks. Many fish live in organized groups. Crocodilians care for their nests and young. Octopi are notorious schemers and escape artists. Social insects with tiny neural systems live in organized collections that rear their young, enslave other insects, and even build large, complex architectural structures. It is often proposed that the demands of maneuvering a body at high speeds in the three-dimensional environment of the air require very sophisticated neural control systems. But this concept should not be taken too far. The ground and associated objects often present a complicated topography, and coordinating multijointed limbs to traverse such a landscape can be a great neurological challenge, especially running on long, erect limbs at high speed. Some flying insects are as aerially adept as birds and bats.

Even when fossilized, pterosaur brain cavities (endocasts) are rarely preserved intact because the delicate structures are usually crushed. When those that are available are compared to body mass, the brains of both rhamphorhynchoids and pterodactyloids appear to have been between the reptilian maximum on the one hand and the bird-bat minimum on the other (Witton 2013). The same was true of the brains of protobirds. Pterosaur brain structure was quite birdlike and correspondingly complex, somewhat more so in pterodactyloids than in rhamphorhynchoids, showing that some neurological evolution occurred in the group through the Mesozoic. Sustained vertebrate flight appears to have been associated with and perhaps required brains larger and more intricate than those of reptiles. Pterosaur brains were not as large as those of birds but had similar structure (Witton 2013), and they were larger and more complex than those of reptiles, implying that pterosaur intelligence was correspondingly intermediate. Although small-brained animals can exhibit complex behavior, it is likely that pterosaur behavior was somewhat more stereotypical than that of flying dinosaurs and mammals.

Did pterosaurs enjoy flying, as birds often seem to, and as human pilots do? Because flight was a critical component of most pterosaur lives, there would have been strong selective pressure to make flying a desirable activity. That could be achieved by making flight feel pleasant, as with other vital functions. When not pushing limits to the point of overtaxing, pterosaur flight was likely to have been agreeable, perhaps sometimes fun, especially if aerial play was involved.

The great dimensions of some pterodactyloids posed a potential problem in terms of the time required for electrochemical impulses to travel along the nerves. In the longest-necked and longest-winged pterosaurs, a nerve input from the wing tip to the brain, and the responding command back to the outer arm muscles that operated the outer wing out to the tip, would have had to travel as much as 8 m (25 ft). This would have been a disadvantage when quick input and response were critical during flight. Synaptic gaps where chemical reactions transmit information slow down impulses, so this problem could have been minimized by growing individual continuous nerve cords as long as possible.

Pterodaustro in flight

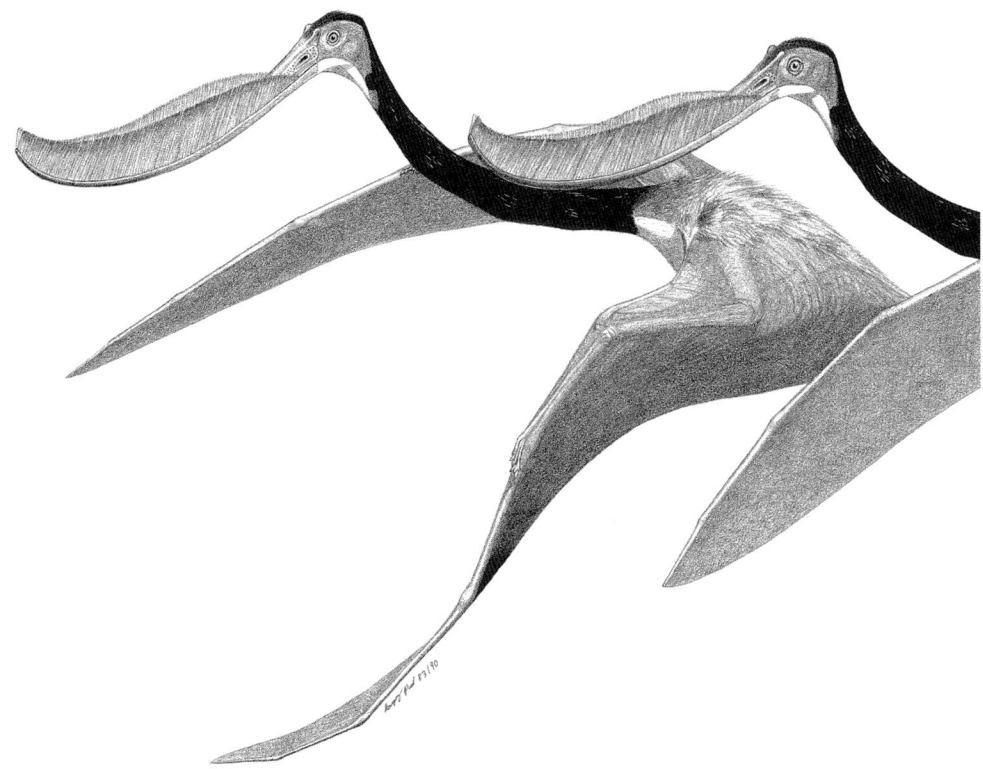

Social Activities

Modern reptiles do not form organized groups. Birds and mammals often do, but many do not—most big cats, for instance, are solitary, but lions are highly social. Some deer form herds, but not all. Many fish swim in schools, and many do not.

There is limited evidence as to whether pterosaurs formed flocks. Their trackways are too scarce to provide potential information. The best evidence for some form of flocking is the hundreds of *Pterodaustro* specimens that have been found in one location. It is not clear whether this abundance of individuals in a particular place records true, organized flocks, or mass aggregations resulting from ganging up to filter feed on abundant invertebrates in shallow lake waters, or breeding colonies on the shoreline. Terrestrial pterosaurs that foraged in vegetation and on the ground may have moved in flocks, probably small, or perhaps not. Also good candidates for flocking were shoreline pterosaurs, which may have foraged together on beaches or gone out in force in search of concentrations of sea prey. An intriguing question is whether those pterosaurs that began to fly shortly after hatching flew in flocks segregated by age, or in groups of all ages—the danger of being cannibalized by larger adults would tend to favor the first option in at least some species, but some trackways appear to show pterosaurs of differing sizes and possibly ages congregating together (Li et al. 2021). Also not known is whether pterosaurs ever flew in mixed-species flocks, as some birds sometimes do. The wave-soaring lifestyle of marine ornithocheiroids would have been as antithetical to flocking as is that of albatrosses.

Reproduction

It has been suggested that some pterosaur species included robust and gracile morphs that represent the two sexes (Bennett 2001; Witton 2013). Although the sexual dimorphism hypothesis is highly plausible, it is difficult to fully confirm because it is possible that the two forms represent different species, or ontogenetic stages. Males are often more robust than females, but there are exceptions. Female raptors are usually larger than males, for instance, and the same is true of some whales. It is when the robust morph bears a large head crest, which is lacking on the smaller morph from the same sediments, that the evidence that the former is the male of the species is stronger. If this is true, then there was quite a large size disparity between the sexes in some pterosaurs, such as *Pteranodon*. In such cases males both dominated females and possibly protected a harem of females from other male competitors. In contrast, there seems to be no size difference between the sexes of *Nyctosaurus*, which may have nested at the same locations as *Pteranodon*.

There is little doubt that a major if not the sole evolutionary force driving the evolution of pterosaur head crests was display—even if secondary functions occurred—largely or entirely for reproductive purposes. The crests aided in identifying members of a species to other members of that species and helped ensure success in sexual competition. Females may have used their crests to signal males that they were suitable and fertile mates. Males used them both to intimidate male rivals and to attract females for mating. Healthy animals in their reproductive prime are generally able to dedicate more resources to growing superior-quality displays that can be used to give females the impression that a given male is prime reproductive material. The reproductive advantage provided by an impressive crest caused it to be of overall advantage to males notwithstanding the energy needed to grow and maintain the structure and to overcome the aerodynamic drag it created. Pterosaurs are likely to have engaged in ritual displays and vocalizations during competition and in courtship that have been lost to time. Because the aerodynamically streamlined crests were laterally flattened, their owners had to present their broad surfaces sideways to the viewer in some manner, perhaps by turning the head to the side when facing one another, or by overlapping heads.

Why do so many if not all pterosaurs sport cranial crests, while their closest flying analogues, birds, usually shun them, especially the hard variety? Many birds use fluffy feathers to form a head display, which in some cases can be smoothed down to improve streamlining during flight. Nor is it likely a coincidence that some flying birds that do have hard crowns, such as hornbills and pelicans, have large heads like the similarly crested pterosaurs. Bird heads are prone to be relatively small, which probably has a lot to do with many being smooth topped, although a fair number of galliform birds, including chickens and turkeys, and duck relations have soft and bony combs and casques. An advantage of display crests is that they can help keep reproductive competition fairly peaceful, which minimizes the risk of injury or worse. However, the genetic compulsion to reproduce propels the males of many species to engage in more dangerous combat, and such may have occurred among pterosaurs using their sharp beaks and teeth.

Reptiles along with some birds and mammals, including humans, achieve sexual maturity before reaching adult size, but most mammals and extant birds do not. The time at which a species becomes sexually mature depends in part on the rate of growth. If it is slow, then it is likely the animal will need to start spawning while it is still growing, because chances are good that slow growers that wait many years to reach full size before reproducing will be killed before they can carry out that genetically critical life task. If growth is fast, then it might be acceptable to wait the relatively short time until final mass is reached. There is a trade-off in that reproducing while growing can get reproduction underway sooner than otherwise, but reproduction is expensive, and if it starts during growth it will slow the latter down, especially for females. Another factor pertinent to pterosaurs is the timing of the advent of powered flight during their growth. Frequent flight is so energy expensive that it slows down growth if it starts before adult size is reached, as

occurs in megapode fowl that begin to fly early in life, unlike in other birds, or bats. There appears to have been a slowdown in the growth of pterosaurs during their ontogeny (Witton 2013); whether this resulted from the onset of reproduction and/or flight is currently an open question. Head crests did not become fully developed until late in growth in at least some pterosaurs, favoring the beginning of reproduction with maturity.

In reptiles and birds, the penis or paired penes (if either are present) and the testes are internal, which was also the case in pterosaurs. Most birds lack a penis, and whether any pterosaur shared this characteristic is unknown but plausible, especially because of their small bodies and hips. Copulation was presumably a quick process, as it is in modern flying vertebrates.

Most animals have twin ovaries that produce eggs in pairs, and the pair of eggs associated with one pterosaur specimen indicates this was true of pterosaurs (Wang et al. 2015). This is interesting in that birds have just one ovary, and this has been thought to be a means of weight reduction among fliers. Many or all dinosaurs, including the living examples, lay highly calcified, hard-shelled eggs in most or all cases, but those of pterosaurs were more typically reptilian in having softer, thinner, more parchment-like shells (Witton 2013). Also reptilian is that the eggs were fairly small relative to the mothers who formed and deposited them, which had the advantage of saving weight.

There are two basic reproductive strategies: r-strategy and K-strategy. K-strategists are slow breeders that produce few young, while r-strategists produce large numbers of offspring rapidly. One advantage of rapid reproduction is that it compensates for high loss of juveniles from genetic defects, accidents, incompetence during early independence, diseases, and especially predation. Producing large numbers of young also allows a species to quickly expand its populations when conditions are suitable, so r-strategists are "weed species" able to rapidly colonize new territories or to promptly recover their population after it has crashed for some reason. A disadvantage of rapid reproduction is that it requires lots of energy to produce so many offspring, so animals that do not suffer high levels of juvenile predation are typically slow-breeding K-strategists.

The handling of offspring after hatching or birth ranges from one extreme of abandoning eggs, usually laid in large numbers, immediately after depositing them properly, to the other extreme of lavishing attention on the young, usually only a few, including feeding, until they are fully grown. Juveniles that receive no or limited parental attention and quickly leave the nest, if that is where they started out, are precocial, being well developed for locomotion and self-feeding after hatching or birth. Juveniles that get a lot of parental care tend to be altricial; they are initially too poorly developed to move about, much less feed themselves.

Megapode fowl are unusual among birds in that they carefully incubate their eggs in mound nests that generate heat via fermenting vegetation until they hatch, and then they pay no attention to the precocial nestlings, which quickly leave the nest. Some birds, such as the living big flightless ratites, are r-strategists, producing large numbers of eggs that hatch into precocial chicks that quickly leave the nest and feed themselves under the guidance and care of adults, a practice also seen among ducks and other anatids. Many birds and bats lean more toward the K-strategy side in that they lay just a few or in some cases one egg a year and lavish extensive parental attention and feeding on their helpless altricial young.

It is not entirely clear how fast pterosaurs bred. With paired ovaries producing small eggs, they had the potential to be r-strategists. But because they were usually small-bodied fliers, it is questionable whether pterosaurs laid more than a few eggs per year. And a limited breeding season, perhaps occurring only every few years as in albatrosses, may have limited the pace of reproduction in the gigantic marine pterodactyloids. Like the big oceanic birds, the ornithocheiroids would have taken advantage of their seagoing soaring capability to breed on remote islands that had the tremendous advantage of being free of nonaerial predators of eggs, chicks, and adults—although flying predators

in the form of birds like the toothed ichthyornids, or small or juvenile pterosaurs cannot be ruled out. The relative lack of danger to offspring would have further reduced pressure on the giant pterosaurs to breed rapidly, in which case they were K-strategists. On the other hand, the land-based pterodactyloids such as the azhdarchids had the large bodies and time needed to generate and broadcast large numbers of eggs, so they would have been under strong pressure to be r-strategists in order to compensate for the heavy losses of eggs and chicks to the many predators in their habitats, all the more so if they did not care for their young, or were even cannibals.

It is probable that some or most pterosaurs laid their eggs in solitary nests, or a few nests in the same location, perhaps to take advantage of suitable soil conditions. There is some provisional fossil evidence of pterosaur nesting colonies, and it is nearly certain that the island-utilizing ornithocheiroids nested in colonies, perhaps enormous in scale and perhaps consisting of multiple species.

A *Nyctosaurus* nesting colony

The thin, flexible shells of pterosaur eggs would have prevented them from being exposed and incubated by their parents, so they had to have been buried and warmed by ground heat. The thin shells would have favored their being laid in moist soils so that they would not desiccate. It is possible that some pterosaurs built vegetation-fermenting mound nests like those of megapodes. No living known tetrapod that buries its eggs later provides the hatchlings with intense parental care. However, crocodilians guard their nests, megapode fowl carefully monitor and maintain the proper temperature of their mound nests, and there is evidence that some mound-nesting dinosaurs cared for their young after they hatched but were still in the nest. So it is possible but not certain that some or all pterosaurs did something similar at least until the eggs hatched, and perhaps beyond. Additional evidence that baby pterosaurs were often not taken care of much or at all is provided by the numerous examples of very small juvenile flaplings of the size expected of hatchlings, with wings spanning less than 150 mm (under 0.5 ft), as well as increasingly mature flying juveniles being known from the same deposits as their fully grown parents. That could happen only if the baby rhamphorhynchoids and pterodactyloids took to the air immediately upon emerging from their eggs, like megapode chicks (Witton 2013; Hone et al. 2020).

There are reasons to conclude that not all pterosaurs were fully precocial and nonparental. A number of species are known from a substantial number of large juveniles and adults, but no hatchling-size or small juvenile specimens are known. This pattern indicates that the hatchlings remained in or near the nest for an extended period before becoming aerial and independent, which in turn suggests they were fed by their parents. The altricial development of the embryos in the eggs of some pterosaurs is compatible with this scenario. This arrangement appears to have been taken to its greatest expression by the ocean-soaring ornithocheiroids. In those cases, the smallest examples found in marine sediments are already a quarter to half or more adult size (Witton 2013; Bennett 2017). This indicates that the chicks were fed and cared for by their parents over many weeks or months on their breeding islands until they were sufficiently mature to fend for themselves over deep waters.

The candidates for a ratite or duck-goose-swan–like pattern of adults leading their chicks, perhaps still flightless, perhaps not, across the landscape in search of food for the young to eat are the more terrestrial pterodactyloids—dsungaripterids, istiodactylids, and azhdarchoids. Alternatively, the young of some or all of these vagile pterosaurs were independent, perhaps moving about in their own age-cohort flocks.

GROWTH

All land reptiles grow slowly (Paul 2002, 2012). This is true even of giant tortoises and big, energetic (by reptilian standards) monitors. Some marsupials and large primates, including humans, grow at the same rate as or only a little faster than the fastest-growing land reptiles. Other mammals, including other marsupials and a number of placentals, grow at a modest pace. Still others grow very rapidly; horses are fully grown in less than two years. No extant bird, including the ostrich, takes more than a year to grow up, but some of the very recently extinct and gigantic island ratites appear to have taken a few years to complete growth. The secret to fast growth appears to be having an aerobic capacity high enough to allow the growing juvenile, or its adult food provider, to gather the large amounts of food needed to sustain rapid growth. But, as noted above, starting to reproduce or power fly while in the growth phase slows down the rate of growth, enough so that even high-metabolic-rate juveniles that are reproducing and/or flying can have reptile-like rates of growth.

At the microscopic scale, the bone matrix is influenced by the speed of growth. Because most pterosaur bones are badly crushed, they are difficult to examine, but the three-dimensional preserved examples of small and medium-sized rhamphorhynchoids and pterodactyloids that have been examined indicate that growth was initially moderately fast and not as swift as in modern birds, but then slowed as perhaps reproduction or flight began (Witton 2013; Hone et al. 2020). Growth was therefore stretched over years. If the giant marine pterodactyloids raised their nestlings until their wings spanned a few meters, then the growth rate should have been similar to that of albatrosses, whose chicks fledge at about the same size, in their case adult size. In that situation, an ornithocheiroid chick may have taken half a year or more to fledge. Over that time, it should have required about 100 kg (220 lb) of food—perhaps stored in its caretaker's big throat pouch—0.5 kg (1.1 lb) or less a day, albeit less than that early on and then more as the growing chick approached fledgling mass.

As animals grow, some change form only a modest amount, such as turtles, while others undergo radical changes, as in humans and birds. Hatchling pterosaurs were prone to having large but short-beaked heads without crests, low tooth counts, short necks, and somewhat short wings. Crests, when present, often began to show up in midgrowth but in some examples became notable only at nearly full size. How long pterosaurs lived is not known because not enough bones have been examined to obtain sufficient statistics. They may have lived about as long as birds of similar size, a few years on up, with the gigantic forms being the longest lived. Suffering low rates of predation and disease, albatrosses can live over half a century. It is possible that the giant marine ornithocheiroids also lived for many decades. Living much more dangerous lives on predator- and disease-infested land, azhdarchids probably enjoyed shorter lives, perhaps of 20 to 30 years, similar to those of the large predaceous avepod dinosaurs whose habitats they shared.

ENERGETICS

Vertebrates can utilize two forms of power production. Aerobic metabolism involves the direct use of oxygen taken in via the lungs to power muscles and other functions. This system has the advantage of producing power indefinitely without intense fatigue, but it is limited in its maximum power output. An animal that is walking or flying at a modest speed for a long distance, for instance, is exercising aerobically. The other scheme, anaerobic metabolism, involves chemical reactions that do not immediately require oxygen. This system has the advantage of being able to generate about ten times more power per unit of tissue and time. But it cannot be sustained for an extended period, and it produces toxins that can lead to serious illness and even death if sustained at too high a rate for too long. Anaerobic power production also builds up an oxygen debt that has to be paid back during a period of recovery. An animal that is running or flying full tilt is exercising anaerobically.

Most fish and all amphibians and reptiles have low resting metabolic rates and low aerobic exercise capacity. They are therefore bradyenergetic, and even the most energetic reptiles, including the most aerobically capable monitor lizards, are unable to sustain truly high levels of activity for extended periods. Many bradyenergetic animals are, however, able to achieve very high levels of anaerobic burst activity, such as when a monitor lizard or crocodilian suddenly dashes toward and captures prey. Because bradyenergetic animals do not have high metabolic rates, they depend largely on external heat sources, primarily the ambient temperature and the sun, for their body heat, so they are ectothermic. As a consequence, bradyenergetic animals tend to experience large fluctuations in body temperature, rendering them heterothermic. The temperature at which reptiles normally operate varies widely depending on their normal habitat. Some are adapted to function optimally at modest temperatures of 12°C (52°F). Those living in hot climates are optimized to function at temperatures of 38°C (100°F) or higher, so it is incorrect to generalize reptiles as "cold blooded." In general, the higher the body temperature, the more active an animal can be, but even warm reptiles have very limited long-term activity potential. Because their aerobic power systems are weak, reptiles cannot generate high blood pressures, so they cannot pump blood high above heart level, which limits their height (Paul 2002, 2012, 2017b; Perry et al. 2009).

Most mammals and birds have high resting metabolic rates and high aerobic exercise capacity. They are therefore tachyenergetic to some degree and are able to sustain high levels of activity for extended periods. The ability to better exploit oxygen for power over time is one of the advantages, and perhaps the chief advantage, of being tachyenergetic. Tachyenergetic animals also use anaerobic power to briefly achieve the highest levels of athletic performance, but they do not need to rely on this as much as reptiles and can recover more quickly. Because tachyenergetic animals have high metabolic rates, they produce most of their body heat internally and so are endothermic, specifically tachyendothermic. As a consequence, tachyenergetic creatures can achieve more stable body temperatures. Some, like humans, are fully homeothermic, always maintaining a nearly constant body temperature when healthy. Many birds, and mammals including bats, however, allow their body temperatures to fluctuate to varying degrees on a daily and/or seasonal basis. So they are heterothermic, albeit in a more controlled manner than reptiles because they can go into high-temperature mode by ramping up internal heat production, which bradyaerobic reptiles cannot do. The ability to keep the body at or near its optimal temperature is another big plus of having a high metabolic rate. Normal body temperatures range from 30°C to 44°C (86°F to 105°F), with birds always at least 37°C (98.6°F), but because these temperatures are sometimes less than those of active reptiles, it is not entirely correct to characterize tachyendotherms as "warm blooded." High levels of energy production are also necessary to do the cardiac work that creates the high blood pressures needed to be a tall animal (Paul 2002, 2012, 2017b; Perry et al. 2009). Typically, mammals and birds have resting metabolic rates and aerobic exercise capacity about ten times higher than those of reptiles, and differences in total daily to yearly energy budgets are even higher.

However, there is substantial variation from these norms in tachyenergetic animals. Among birds, the smaller examples tend to be hyperenergetic even when accounting for their size—metabolic rates tend to decrease per unit mass as total weight increases—as are many small mammals. Other flying birds, some of the raptors in particular, have relatively modest resting metabolic rates, somewhat below the avian-mammalian average for a given body mass. Interestingly, despite spending much of their time in energy-efficient soaring, marine birds tend to have higher-than-typical metabolic rates, with frigate birds being an exception. Birds that hunt flying insects at night tend to have resting energy production on the low side. Bats also show a great deal of metabolic variation. Again, those that are nocturnal hunters of flying insects have resting metabolic rates below the mammalian norm, while the rest are at the mammalian norm, and none are as hyperenergetic as other small mammals and birds can be. Some slow-moving mammals like monotremes, tenrecs, mole rats, and sirenians have metabolic rates not all that much higher than those of reptiles.

Bradyenergetic reptiles enjoy the advantage of energy efficiency, allowing them to survive and thrive on limited resources, and it takes a long time to starve them. Tachyenergetic animals can sustain much higher levels of activity, which can be used to acquire even more energy, which can then be dedicated to the key factor in evolutionary success: reproduction. Having lower sustained maximal power production severely limits what reptiles can do compared to tachyenergetic endotherms. Tachyendothermy has allowed mammals and birds to become the dominant large land animals from the tropics to the poles, on the land and in the air. But reptiles remain very numerous and successful in the tropics, and to a lesser extent in temperate zones.

As diverse as the energy systems of vertebrates are, there appear to be things that they cannot do. All insects have low, reptile-like resting metabolic rates. When flying, larger insects use oxygen at very high rates similar to those of birds and bats. Insects can therefore achieve extremely high maximal/minimal aerobic metabolic ratios of 100/1, allowing them to be both energy efficient and aerobically capable, or tachyaerobic, as well as endothermic. Insects can do this because they have a dispersed system of tracheae that oxygenate their muscles. No vertebrate has both an extremely high aerobic capacity and a very low resting metabolism, probably because the centralized respiratory-circulatory system requires that the internal organs be large relative to total mass and work fairly hard even when resting in tachyenergetic vertebrates. In most birds and mammals, the aerobic maximum/minimum is ten to one, but in some it can be three dozen fold. An insect-like metabolic arrangement should not, therefore, be applied to the flying pterosaurs.

When flying, insects over 1 g (0.36 oz) have to be tachyaerobic to have enough oxygen-fueled power to sustain flapping flight, and they have to be endothermic so they can maintain the high body temperature needed to achieve optimal power production, and to operate their neurological systems at the peak level of performance that flying demands. The same is true of powered flying vertebrates, whether they are sustaining flight by flapping or even low-energy soaring, because that requires them to be on top of their mental game. Brains above the size of reptiles are always associated with elevated metabolisms because they need a well-developed oxygen supply. High aerobic capacity requires large locomotory muscles to utilize all that oxygen, and large, highly aerobic muscles require a high-capacity circulatory-respiratory system to supply the oxygen-hungry muscles. Reptiles tend to have less in the way of limb muscles. Erect gaits are always associated with high aerobic capacity because they favor the high walking speeds that only tachyaerobes can sustain. Among nonaquatic vertebrates, all that have extensive external insulation are tachyenergetic endotherms; the insulation helps them retain the body heat they generate. Not all endotherms are insulated. Usually these are large, low-latitude mammals that use bulk to retain heat, humans among them, but interestingly, the small tropical naked bat lacks substantial fur. Conversely, it is maladaptive for an ectotherm to be insulated because that serves only to prevent it from taking in the heat it needs from the surrounding environment.

Because all the above indicators of high aerobic endothermy apply to the flying, big-arm-muscled, fuzzy, large-brained, sometimes tall-statured pterosaurs, the long-ago debate over whether pterosaur energetics were largely reptilian or avian-mammalian in nature has ended in favor of a broad consensus that their power production and thermoregulation were logically closer to those of flying birds and mammals (Paul 2002; Witton 2013). And because the amount of anatomical diversity in pterosaurs was modest, the variability in their energetics should have been modest, too—similar to the situation in flying birds and bats, and less similar to the situation in the widely variable dinosaurs and among mammals as a whole. It is not likely that any pterosaurs were as hyperenergetic as birds, especially the smaller avians. If the anurognathids were nighttime insect interceptors, then they are good but not certain candidates for having rather low resting metabolisms, like nightjars and nocturnal insectivorous bats. Conversely, marine pterosaurs may have possessed higher-than-average metabolisms, or some or all may have been in the lower range of frigate birds.

One horsepower (hp) is the work that can be aerobically sustained without undue fatigue by a large workhorse over a period of some hours, such as turning a wheel pump or pulling a plow. When going all out anaerobically for a brief period, as in pulling a heavy sled in a country fair competition or winning the Triple Crown, a horse can do about 15 hp. Human athletes can put out 0.33 hp indefinitely, and about 2.5 briefly, over twice that of a nonathlete. When taking off, the biggest half-ton azhdarchids would have been anaerobically powered by about 12 hp, which is a little higher than the engine power of the smallest ultralights. The 2 hp that the flight muscles of a 100+ kg (220+ lb) superpterosaur could aerobically sustain was probably not enough to stay in the air via continuous flapping.

Producing so much power risks overheating, especially when the external climate is hot, and a pterosaur's expansive wing tops would have picked up lots of solar infrared, which would not have helped matters. The pterosaur's primary cooling organ via convection and radiation should have been the wings with their tremendous surface area, especially the shaded undersides with their high density of blood vessels. Big, bare-surfaced, blood vessel–filled pterosaur beaks and crests appear well suited for radiating heat out into the environment and would have had the advantage of being exposed even when their owners were not flying. Birds use their beaks, and crests when they have them, for thermoregulation, but most do not have such large structures, so they are not critical for that purpose. The great variation in the size and presence of pterosaur beaks and crests indicates that they were not critical to this function in that group either, which makes sense since they had their wing membranes for thermal control. Radiating heat out into the environment works only when the air temperature is lower than that of the body; if it is higher, then blowing air over a surface actually causes a heating effect like a blast furnace. In that case, keeping heat out while using evaporative cooling becomes critical. Body temperature can be allowed to rise a few degrees, but to help keep it from becoming lethal and damaging the brain, pterosaurs may have used some form of gular fluttering like birds.

As far as the weather, being high-energy animals had big climatic advantages for pterosaurs. Not dependent on direct solar heating—though pterosaurs may have basked when it was a favorable option, as do birds and mammals—pterosaurs could operate around the clock in the air and on the ground in variable conditions, including cloudy, rainy, and even snowy weather if pterosaurs at higher latitudes or elevations were caught by a storm. Pterosaurs would have had more of a problem than birds with retaining heat when it was chilly. When the latter are not flying and it is quite cool or cold, the tightly folded and thick wing feathers provide extra insulation. For pterosaurs, their gangly wings stuck out into the air even when folded, and the exposed membranes provided no insulation and would have been something of a heat sink even though blood flow to the sheets would have been minimized to reduce heat loss. A limited ability to cope with severe cold combined with the associated seasonal shortage of food probably compelled pterosaurs to migrate away from areas that experienced harsh winters, returning during the warmer seasons.

Until the 1960s it was assumed that high metabolic rates and/or endothermy were an atypical specialization among animals, being limited to mammals and birds, perhaps some therapsid ancestors of mammals, and flying pterosaurs. The hypothesis was that being tachyenergetic and endothermic is too energy expensive and inefficient for most creatures, and this combination evolved only in special circumstances such as the presence of live birth and lactation, or powered flight. Energy efficiency should be the preferred status of animals because it reduces their need to gather food in the first place. More recently, it has been realized that varying forms of tachyenergy definitely are or probably were present in large flying insects, tuna, lamnid sharks, some basal Paleozoic reptiles, some marine turtles and the ocean-going plesiosaurs, ichthyosaurs, mosasaurs, brooding pythons, basal archosaurs, basal crocodilians, pterosaurs, all dinosaurs including birds, some pelycosaurs, therapsids, and mammals. Energy-expensive elevated metabolic rates and body temperatures appear to be a widespread adaptation that has evolved multiple times in animals of the water, land, and air. This should not be surprising in that being highly energetic allows animals to do things that bradyenergetic ectotherms cannot do, and DNA selection acts to exploit available lifestyles that allow reproductive success without caring whether it is done energy efficiently or not. Whatever works, works. So, many animals do live on low, energy-efficient budgets, while others use more energy to acquire yet more energy that can be dedicated to reproducing the species.

A long-term debate asks what specifically it is that leads animals to be tachyenergetic and endothermic. One hypothesis proposes that it is habitat expansion, given that animals able to keep their bodies warm when it is cold outside are better or exclusively able to live in chilly conditions, whether near the poles, at high altitudes, in deep waters, or during frosty nights. Another hypothesis proffers that only tachyenergetic animals with high aerobic capacity can achieve high levels of sustained activity regardless of ambient temperature, whether at sea level in the tropical daylight or at the poles during winter nights, and that ability is critical to maintaining high energy. Certainly the first hypothesis is true, but it is also true that all the many animal groups that feature high energy budgets and warmer-than-ambient body temperatures also thrive in warm and even hot climes where they beat out the bradyenergetic creatures in activity levels. So both hypotheses are operative, and which is more so depends on the biocircumstances.

GIGANTISM

In the 1800s, famed discoverer of giant pteranodonts Edward Cope proposed what has become known as Cope's Rule, the frequent tendency of at least some portion of animal groups to evolve gigantism, as in the often colossal dinosaurs as well as birds, sea reptiles, and ancient mammals he also uncovered. Being really big has numerous advantages, including greater ability to resist being prey and/or more successfully preying on other animals; greater energy efficiency per unit mass in metabolism as well as in cost of locomotion; resistance to fast starvation—hummingbirds have to eat as much as they weigh each day, which is why they are so frenetic, while swans are much calmer because they down only a quarter of their own mass daily, and ostriches even less; internal thermal stability; and the ability to produce more and/or larger offspring. Downsides include greater sudden vulnerability to extinction over the long term because of population disruption during an environmental crisis; smaller adult populations as a result of larger individual food budgets and slower K-strategy reproduction if it is via live birth; and exclusion from many habitats suitable for small animals, such as arboreal niches, as well as from difficult terrain. The latter would have been especially pertinent regarding the big continental pterodactyloids—their stilt-like, delicate-boned limbs, and their need to always be able to get airborne at a moment's notice, meant they were limited to living largely on reasonably open flats; similar-sized dinosaurs could inhabit much steeper, rougher ground and dense vegetation. The high respiratory capacity of the tachyenergetic pterosaurs, and of birds, was probably critical for their ability to become big (Paul 2002; Witton 2013). The energy cost of being big should not be exaggerated; even the biggest azhdarchids, which were basically pterosaurian bears with wings, needed to gulp down only three or four housecat-sized animals a day to keep their half-ton forms fit and flying.

Both flying and flightless birds have benefited sufficiently from being big to evolve gigantic species. Bats have not done so as much; the largest flying foxes have wings less than 1.7 m (6 ft) across and weigh only 1.5 kg (3+ lb).

Giant pterosaurs compared with large birds and mammals

a *Quetzalcoatlus*—WS 10.5 m; approx. 500 kg

b *Quetzalcoatlus?*—WS 5.2 m; 65 kg

c *Pteranodon*—WS 6.5 m; 50 kg

d *Anhanguera?*—WS 9.5 m; approx. 100 kge

e *Tapejara*—WS 8 m; approx. 200 kg

f *Pelagornis*—WS 7 m; 60 kg

g Wandering albatross—WS 3.7 m; 13 kg

h *Argentavis*—WS 5-6 m; 70+ kg

i *Teratornis*—WS 3.3 m; 15 kg

j Mute swan—WS 2.4 m; 20 kg

k Great bustard—WS 2.5 m; 18 kg

l *Dinornis*—400 kg

m *Dromornis*—400 kg

n Ostrich—110 kg

o Giraffe—1,500 kg

p Golden-crowned flying fox—WS—1.7 m; 1.5 kg

q *Meganeura*—0.55-0.65 g

Although they evolved from small archosaurs, and many were small, pterosaurs—particularly the Cretaceous pterodactyloids—are famous for their tendency to develop gigantic forms even as they maintained the ability to become airborne with bodies as heavy as horses, bears, and moas. However, the size of the biggest azhdarchids should not be exaggerated. Some illustrations are not based on rigorous skeletal plans and give the impression that the likes of *Quetzalcoatlus* boasted a body as hefty as a giraffe, but bulls of the ultratall ungulates weigh three to five times as much.

The big questions about pterosaurian gigantism include why no Triassic/Jurassic rhamphorhynchoid became really large, why some Cretaceous pterodactyloids did, and why some of the latter greatly exceeded the size of the largest flying birds to the point that some pterosaurian fliers weighed around half a ton more than any living big bird, including the flightless examples. It is often said that ostriches cannot fly because they are too big, but the largest ornithocheirids and azhdarchids were not only heavier than ostriches; some of the latter matched the size of and maybe even outmassed the biggest extinct flightless birds—the moas of New Zealand, elephant birds of Madagascar, and thunderbirds of Australia. Suggestions that the bigness of superpterosaurs meant they could not fly are not compelling. Truly flightless azhdarchids should have had severely atrophied outer wings—the inner arms had to remain functional for ground movement—similar to the reduced arms of flight-free birds. Azhdarchid wings had the normal span and area relative to mass as those of flying birds, providing strong evidence that they could get into the air.

Why a few of the late-appearing rhamphorhynchoids became fairly large but none became truly gigantic can be plausibly but not firmly attributed to their retention of a stable rather than dynamic form of flight based on the presence of a long tail—it may not be a coincidence that no long-tailed Mesozoic birds became gigantic either. Another factor may have been the usual retention of a short inner arm and hand, which limited the span of the inner wing, which was strongly constructed and levered for takeoffs as well as for climbing and level flight. That no Triassic or Jurassic pterosaur became big can therefore be blamed on the failure of the rhamphorhynchoids to do so, whatever it was that hindered them from doing so.

Conversely, the more dynamic flight abilities of short-tailed pterodactyloids and their greater inner wingspans may have given pterodactyloids the potential to become gigantic. It is not surprising that the ornithocheiroids evolved into gigantic soarers of the Cretaceous seas because they preceded the marine birds that would do the same in the Cenozoic, so it is clearly a viable lifestyle. The evolution of the big azhdarchid walking stalkers may also have its later avian equivalent—the giant *Teratornis*, which had wings spanning up to 5–6 m (16–19 ft) and weighed as much as a human. Although superficially vulture/condor-like in overall form, it may also have been adapted to hunting small game.

The pelagornithids were tremendous fliers with wings up to 7 m (23 ft) across, and they weighed over 50 kg (170+ lb), but they were edged out by the relatively smaller albatrosses with nearly 3.7 m (12 ft) wingspans and weights of a dozen or so kilograms (30 lb) that survive today, possibly because stiffer ocean winds of the cooling Late Cenozoic favored the latter.

A structural factor that may limit the size of birds is their lack of bone support for their entire wingspan. Making up about a third of the total span, the outer primaries of the biggest teratornes and pelagornithids were themselves amazing, oversized structures up to and over 1 m long (3–4 ft)—imagine having one of those in your feather collection. Growing longer feathers, which have to be replaced via molting on a yearly basis, may not be practical for physiological and structural strength reasons. Having a stiffer bone supporting the entire leading edge of the wing, pterosaurs were not similarly limited regarding wingspan and loading. On the other hand, superbirds might be able to get around the span problem by making more of their wing length be bone, but perhaps that would make the outer primaries too small a portion of the total span to be efficient. Some researchers claim that the pterosaur wing membrane was more efficient at generating lift for a given span than those of birds, but this appears problematic. More plausible is that using the strongest set of limbs, the arms, to take off allowed the superpterosaurs to push off more easily than birds, which have to take off with their relatively weak hindlegs (Habib 2008; Witton 2013). If that is true, then the quadrupedal gait that usually hindered their competition with birds was an advantage in this particular regard.

A possible nonstructural reason that birds have not become as big as the biggest pterosaurs is the lack of an available niche that would drive the same expansion to wings of 10 m or more (33+ ft) and masses exceeding those of ostriches during the Cenozoic, with its non-Mesozoic geography and climate, and its domination by mammals rather than dinosaurs.

One factor likely to have favored Mesozoic giants of marine skies is that the many shallow seaways of the Cretaceous were being fed large loads of nutrients from the many rivers and streams that emptied directly into them. That may have pumped up the productivity of the small oceans, favoring high densities of the small sea creatures that the ornithocheiroids fed on, boosting their size to extreme levels. It is possible that the general retreat of the seaways at the end of the Mesozoic limited the size of pteranodonts, and the few, last known marine pterosaurs found so far are not as large as earlier forms when the seaways were more extensive, but more remains from the end of the Mesozoic are needed to be sure.

Unclear is why Late Cretaceous birds did not become gigantic alongside or in competition with pterodactyloids. Possibly only neornithines have had the bioevolutionary potential to acquire wings spanning many meters, and that group did not become numerous until the Cenozoic.

With masses from 350 g (0.75 lb) to around half a ton, the variation in pterosaurs was about fifteen hundredfold. That is a lot; ranging from 2 g to 1.6 kg (0.07 oz to 3.5 lb), bat mass variation is just eight hundredfold. On the other hand, birds vary from 2 g to over a third of a ton for a range of around two hundred thousandfold, although the mass range of fliers is thirty-five thousandfold. All dinosaurs, birds included, differ in mass by a factor of fifty million, the upper end being around a hundred tons. If birds are excluded, the smallest dinosaurs were under a quarter kilogram (less than 9 lb), and size differed less than half a millionfold. Mammals takes the size differential cake, with a maximum/minimum ratio of two hundred million, the biggest beast being two hundred tons.

It is not surprising that the mass range of animal groups that can fly is much less than that of groups whose bigger members are not airborne. It is plausible but not certain that the greatest pterosaurs maxed out the size potential of flying animals. Both Mesozoic pterosaurian and Cenozoic avian ocean soarers, each present for millions of years, shared similar wingspans, suggesting that the aerodynamics of power gliding so close to the surface of the waves, including in storms, make bigger wing spans impractical in that environment. It is possible that a nonstructural aerodynamic problem involving the lifestyle of gigantic biofliers, such as food availability relative to the costs of dedicating so much of an animal's body mass to big, heavily muscled wings that burn a lot of energy, is the critical limiting factor. Or limitations in bone, muscle, and tendon strength, and in metabolic and power systems may constrain nonmarine flying vertebrates to the amazing dimensions and bulk of the continental azhdarchids. But what these limits are is not possible to estimate—if marine mammals were extinct, who would calculate that some of the tachyenergetic mammals, with their high consumption of oxygen, could hold their breath long enough to dive thousands of feet into high-pressure depths? And the supermassive azhdarchids seem to have appeared just before the final extinction, so it cannot be ruled out that they were on the way to becoming even larger had they had more time to test the structural, energetic, and habitat limitations on the size of flying giants. There used to be calculations that firmly established that *Pteranodon* was as big as a flier can be, but these have been disproven by bigger pterosaurs. Likewise, if dramatically bigger superpterosaurs with well-developed wings are found, then any calculations that the biggest currently known azhdarchids were the biggest possible will be moot, and another round of debate over whether the new monster pterosaurs are the largest animals of the air will begin anew.

MESOZOIC OXYGEN

Oxygen was absent from the atmosphere for much of the history of the planet, until the photosynthesis of single-celled plants built up enough of it to overwhelm the processes that tend to bind it to various elements such as iron. Until recently, it was assumed that oxygen levels then became stable and made up about a fifth of the air for the last few hundred million years. It has been proposed that oxygen levels have instead fluctuated strongly over time. The methods used to estimate past oxygen levels suggest that they reached a uniquely high level of about a third of the atmosphere during the Late Paleozoic, when the great coal forests were forming and, because of the high oxygen levels, often burning. It is notable that this was when many insects achieved enormous dimensions by the standards of the group, including the dragonfly relatives with wings over 0.5 m (2 ft) across. Because insects bring oxygen into their bodies via a dispersed set of tracheae, the size of their bodies may be tied to oxygen levels.

Soon afterward, oxygen levels may have plunged precipitously, sinking to a little over half the current level by the Triassic and Jurassic as the rhamphorhynchoids and then early pterodactyloids appeared and began to evolve. In this case, oxygen availability at sea level would have been as poor as it is at high altitudes today. Making matters worse was the high level of carbon dioxide. Although it was not high enough to be directly lethal, the combination of low oxygen and high carbon dioxide would have posed a serious respiratory challenge.

Reptiles subjected to low-oxygen conditions become more sluggish and grow more slowly, whereas some birds can fly higher than Mount Everest. If oxygen was scarce in the Mesozoic, then the ability of pterosaurs to achieve high levels of sustained activity and rapid growth is all the more remarkable and is evidence that they evolved systems able to efficiently take in and utilize oxygen at high levels while coping with excess carbon dioxide. In this context, the evolution and success of pterosaurs in the Late Triassic and Jurassic may have resulted from the development of efficient air sac–driven respiratory systems, which would have allowed them to breathe as easily at low altitudes as birds do today at high altitudes.

During the Cretaceous, oxygen levels are estimated to have crept upward toward modern levels, although they never reached the current concentration in the Mesozoic. This rise of oxygen may have allowed the pterodactyloids to be the first pterosaurs to finally evolve great size.

But there is a problem. A different method of estimating oxygen levels also calculates that there was a big dip in oxygen at the beginning of the Mesozoic but then indicates that it soared to the present level early in the Triassic before edging up higher, perhaps much higher, in the later Mesozoic. If so, then most of the above discussion is moot, and pterosaurs of all sizes would have been able to easily exploit oxygen to power their active lives. Figuring out the actual oxygen content of the atmosphere in pterosaur times remains an important challenge.

PTEROSAUR SAFARI

Assume that a practical means of time travel has been invented, and, *The Princeton Field Guide to Pterosaurs* in hand, you are ready to take a trip to the Mesozoic to see the pterosaurs' world. What would such an expedition be like? Here we ignore the classic paradoxical issue that plagues the very concept of time travel: What would happen if a time traveler to the pterosaur era did something that changed the course of events to such a degree that humans never evolved to travel back in time and disrupt the timeline?

One difficulty that might arise could be the lack of modern levels of oxygen and extreme greenhouse levels of carbon dioxide (which can be toxic for unprepared animals), especially if the expedition traveled to the Triassic or Jurassic. Acclimation could be necessary, and even then, supplemental oxygen might be needed at least occasionally. Work at high altitudes would be especially difficult. Another problem would be the high levels of heat chronically present in most pterosaur habitats. Relief would be found at high latitudes, as well as on mountains.

If the safari were to go to one of the classic Mesozoic habitats that included gigantic dinosaurs, the biggest problem would be the sheer safety of the expedition members. The bureaucratic protocols developed for a Mesozoic expedition would emphasize safety to minimize the chances of losing any participants. Modern safaris in Africa require the presence of a guard armed with a rifle when visitors are not in vehicles in case of an attack by big cats, hippos, buffalos, rhinos, or elephants. Similar weaponry is often needed in tiger country, in areas with large populations of grizzlies, and Arctic areas inhabited by polar bears. The potential danger would be even greater in the presence of flesh-eating dinosaurs as big as rhinos and elephants, which would be able to run down a potentially out-of-breath human unused to the low oxygen. It is possible that theropods would not recognize humans as prey, but it is at least as likely that they would, and the latter would have to be assumed. Aside from the desire to not kill members of the indigenous fauna, rifles, even automatic rapid-fire weapons, might not be able to reliably bring down a 5-ton allosauroid or tyrannosaur, and heavier weapons would be impractical to carry. Nor would the danger come from just the predators. A herd of whale-sized sauropods would pose a serious danger from trampling or impact from tails, especially if they were spooked by humans and either attacked them as a possible threat or stampeded in their direction. Sauropods would certainly be more dangerous than elephants, whose high level of intelligence allows them to better handle situations involving humans. The horned ceratopsids, even less intelligent than rhinos, and probably with the attitudes of oversized pigs, would pose another major danger. Even medium-sized dinosaurs could pose significant risks. An attack by sickle-clawed dromaeosaurs, for instance, could result in serious casualties. So could assault by a squadron of parrot-beaked peccary-like protoceratopsids. As for the pterosaurs, the giant azhdarchids are not creatures one would want to get particularly close to.

But there would be another danger that would be as small as it is big: microbes. Expedition members would be at risk of picking up exotic Mesozoic disease organisms they would not be immune to, and, at least as bad, there would be the danger of contaminating the ancient environment with a host of Late Cenozoic viruses, bacteria, and parasites that could seriously disrupt Jurassic and Cretaceous life.

Because of the combined menaces small and big, time-traveling pterosaur watchers would probably be banned from directly interacting with the ancient habitats. Instead of walking about under the Mesozoic sun breathing fresh air, binoculars in hand, they would always have to wear microbe-proof biohazard suits. The vehicles and habitats would likewise need to be sealed against microbes getting in or out. Dwelling in pterosaur habitats would be a lot like living on the moon or Mars—a very artificial experience in which the paleonauts were significantly detached from the fascinating world around them, always respiring pretreated air. An advantage of being in biosuits would be temperature control, which would eliminate dealing with the extreme heat prevalent in much of the Mesozoic. Travel by foot would probably be largely precluded in habitats that included big theropods, sauropods, and ceratopsids. Expedition members would have to move about when on the ground in vehicles sufficiently large and strong to be immune to attacks by colossal dinosaurs. Movement away from the vehicles would be possible only when drones showed that the area was safe. Even in places lacking giant dinosaurs, there would be the peril of a biosuit being breached by an assault by a smaller dinosaur—any such penetration from any cause would require some level of medical care, quarantine at least. Defensive weapons might be necessary, although pepper-spray guns might suffice. Yet another danger in some Cretaceous habitats would be elephant-sized crocodilians that might snap up and gulp down whole a still-living human unwary enough to go near or in the waters where pterosaurs were present.

A safe way to observe pterosaurs would be remotely via drones that could track the winged archosaurs when on the ground, and even better follow them in the air. Piloted ultralights would work too. Also relatively safe would be visiting the isolated islands that marine pterosaurs roosted and nested on—although getting too close to adults defending nests would be problematic. Safe too would be observation of the wave-soaring ornithocheiroids from vessels and perhaps light planes, although they too might need to be sealed against microbes. The waters below would be filled with big marine reptiles and sharks, but people in a boat would be in no greater danger than they are today.

Assume that some or all pterosaurs were not killed off by the K/Pg impact and continued into the Cenozoic, or that the impact did not occur and pterodactyloids were not liquidated 66 million years ago. What would the evolution of animals of the air have been like in that case?

Certainly pterodactyloid pterosaurs would have continued for millions of years, perhaps tens of millions, and possibly even to today. How long they would have lasted and how well they would have done would have depended in part on their diversity at the time. If only a handful of species made it to circa 60 million years ago, then the group would have been at high risk of extinction. The more species flapping about, the more resilient they would have been. A major pterosaur problem would have been increasing competition from the fast-evolving birds.

The pelagornithids began to appear in the Paleogene and started to become large around 40 million years ago. Albatrosses appeared late in the Neogene. It is possible that had pteranodonts not gone extinct, the big marine birds might never have appeared because the ocean-soaring lifestyle would already have been occupied by the superpterodactyloids. Or, oceanic birds may have evolved and then outcompeted the membrane-winged archosaurs, or both may have thrived in parallel, with giant pterosaurs and birds soaring past one another over the ocean waves and sharing island nesting sites. These possibilities are hard to sort out in part because of the low diversity and populations of such big animals. In the latter case, the diversity of marine birds would probably have been less than it is. The pteranodonts would probably not have changed all that much, having largely perfected the adaptive design of seagoing soarers long ago.

On the continents, the azhdarchids would have been competing with land dinosaurs if they too survived. In that case, the big land pterosaurs could have remained in force for a short time, or up to today. If instead terradinosaurs suffered severe losses or went extinct, the superpterosaurs would have had a chance to become a major factor in the global fauna, with some perhaps becoming flightless, and some even more gigantic. But mammals would probably have become dominant in a dinosaur-free world and generated strong competition for the also-furry pterosaurs, which the latter may or may not have survived.

In the unlikely case that there were aerial insect-hunting pterosaurs at the end of the Mesozoic that continued to flitter about, they could have hindered to some degree the comparable swallows, swifts, and nightjars, been outcompeted by such birds, or coexisted with at least some of them. What can be confidently proposed is that, even if some small pterosaurs were present, birds would have continued to dominate the niches of small-bodied flying archosaurs. It can also be predicted with good confidence that echolocating bats would have appeared over 50 million years ago and taken possession of the night skies. Because of their small size, the mammalian fliers are likely to have evolved even if dinosaurs and/or pterodactyloids had not gone extinct before then. Also likely to have evolved are the bigger fruit-consuming bats. And if pterosaurs did make it to modern times, it probably would not have had much impact on the spectacular evolution of the dinosaurs of the air, from big running ratites to swimming pelicans, and from ducks to assorted raptors to pigeons to parrots to corvids to countless songbirds—birds would rule the daytime skies.

PTEROSAUR CONSERVATION, KEEPING, AND CONSUMPTION

If we take the above scenario to its extreme, assume that pterosaurs managed to survive to modern times. Also imagine that we humans or something very like us evolved and produced a technological civilization similar to the one we have. How would pterosaurs be doing?

The overall situation for modern pterosaurs would probably be grim. For tens of millions of years, large mammals remained stable in average size, only to see a sharp decline as prehistoric game-hunting humans became the leading factor in the extinction of a large portion of the megafauna that roamed much of the earth toward the end of the last glacial period. Within recent historical times humans extirpated the colossal moas of New Zealand and the elephant birds of Madagascar, and matters continue to be bad for most wildlife on land and even above and in the oceans in the age of agriculture and civilization. Very probably, terrestrial pterosaurs would have made good eating, like most birds, with most of the meat coming from the shoulders and

arms, and next the legs. Marine pterosaurs might have been more problematic if they were fishy and/or oily, although some accounts indicate that saltwater birds are quite edible. Also edible should have been the nutritious soft-shelled eggs of the archosaurs of the air. Big land pterosaurs would have been vulnerable to thrown rocks, flung bolas, tossed spears, shot arrows, and fired bullets, and their possibly abandoned eggs would have made excellent free breakfasts. The ensuing enormous loss of habitat to farming, logging, extraction, urbanization, pollution, and the host of human activities associated with agriculture, civilization, and industry would have been similarly devastating. It is open to question whether any large continental or island-bound pterodactyloids would have survived to today, and those that did would have been the subject of intense conservation efforts to prevent their final termination, with debates raging about the ethics of keeping some of them in zoos. Marine pterosaurs would likely have been doing somewhat better, but like the big albatrosses they would have been under serious stress from disruption of nesting sites, overfishing of their prey species, and being caught by fishing gear, and conservation efforts would have resulted. At the same time, big sea pterosaurs, like big marine birds, would have benefited from human ocean-going activities, using the air wake of powered ships to enhance gliding, and consuming the offal from such ships, as well as the fish driven to the surface by fishing nets.

Without a host of small pterosaur species flittering about, wild pterosaurs would have been rare encounters for humans—birds would have been our common archosaurian flying friends.

The place to most easily see big pterosaurs would have been zoos, in what might have been called pteroaviaries. Note that the ready means of preventing large captive birds from flying away from open enclosures—wing clipping—would not have been available for the primary feather-free pterosaurs; perhaps strapping weights to their wings would have sufficed to keep them ground bound. It would likely have been ill advised for keepers to enter enclosures containing the giant pterosaurs because of the high possibility of being attacked. As animal rights ethics evolved, imprisoning pterosaurs for human viewing and other issues would have become increasingly controversial.

WHERE PTEROSAURS ARE FOUND

Because pterosaurs are in fact long gone and time travel violates the known nature of the universe, we have to be satisfied with finding the remains they left behind. With the possible exception of very high altitudes, pterosaurs lived in all places on all continents and flew over vast expanses of ocean, so where they are found is determined by the existence of conditions suitable for preserving their bones and other traces, eggs and footprints especially, as well as by conditions suitable for finding and excavating the fossils. For example, if a pterosaur habitat lacked the conditions that preserved fossils, then that fauna has been totally lost. The same is true of deposits containing their fossils that have eroded away. Or, if the fossils of a given pterosaur fauna are currently buried so deep that they are beyond reach, then they are not available for scrutiny.

All but a very small percentage of carcasses are destroyed soon after death. Many are consumed by predators and scavengers, and others rot or are weathered away. Even so, the number of animals that have lived over time is immense. Because at any given time from the Late Triassic to the end of the Mesozoic many millions of pterosaurs were probably alive, amounting to many billions over the 170 million years they existed, most of them juveniles and small adults, the number of pterosaur fossils that still exist on the planet is large, probably numbering in the many millions of individual specimens.

Of these, only a tiny fraction of 1 percent have been found in or near the very small portion of pterosaur-bearing formations that are exposed on the surface where the fossils can be accessed, or in the deeper quarries that allow additional remains to be reached. Even so, the number of pterosaur fossils that have been scientifically documented to at least some degree is considerable. The question is where to find more of them.

Much of the surface of the planet at any given time is undergoing erosion. This is especially true of highlands. In erosional areas, sediments that could preserve bones and other traces are not laid down, so highland faunas are rarely found in the geological record. Fossilization can occur in areas in which sediments are being deposited quickly enough and in large enough quantities to bury animal remains before they are destroyed.

Land areas undergoing deposition tend to be lowlands downstream of uplifting highlands that provide abundant sediment loads to be carried in or into streams, rivers, lakes, or lagoons that settle out to form beds of silt or sand. Therefore, large-scale formation of terrestrial fossils occurs only in regions experiencing major tectonic activity. Depositional lowlands can be broad valleys or large basins of varying sizes amid highlands, or coastal regions. As a result, most known nonmarine pterosaur habitats were flatlands, with little in the way of local topography. In some cases, the eroding neighboring highlands were visible in the distance from the locations where fossilization was occurring; this was especially true in ancient rift valleys and along the margins of large basins. Prime conditions for pterosaur fossils include the fine-grained, low-oxygen lake beds and lagoon bottoms best suited for preserving the pterosaurs' delicate skeletons, sometimes in a complete state, even including soft tissues. Ashfalls can aid lake-bed fossilization by first killing the fliers and then covering them quickly. Coarser floodplain deposits laid down under more tumultuous conditions are not so suitable for good fossilization of pterosaur bones, but they do preserve important partial remains and some skeletons of pterosaur giants. The preservation problems of such land deposits are the reason no nearly complete azhdarchid superpterosaur skeleton has yet been found. In deserts, windblown dunes or slumped dunes wetted by rains can preserve bones and trackways, but pterosaur remains rarely survive these processes. Vast sediments laid at the calm bottoms of interior seaways and continental shelves offered good conditions for preserving the remains of ocean-soaring pterosaurs that sank down onto their surfaces, which is why a few nearly complete *Pteranodon* skeletons are known.

Of the pterosaur remains and traces that are within potential reach, many are found by individuals, often but not always paleozoologists, searching for their remains on and just under the surface of suitable Mesozoic sediments, typically but not always in dry regions with little vegetation to hinder the visual search. Also productive are eroding cliffs, especially along coastlines where erosion is especially rapid. Finding pterosaurs under such circumstances has changed little since the 1800s. It normally consists of walking slowly while stooped over, often under a baking sun, often afflicted by flying insects, and in the case of coastal cliffs sometimes in chilly, wet, and windy circumstances, looking for telltale traces. Broken pieces of bone on the surface may indicate that a bone or skeleton is eroding out. One hopes that tracing the broken pieces upslope will soon lead to bones that are still in place. In recent years GPS and digital technologies have greatly aided in determining and mapping the position of such fossils.

But a major location of pterosaur fossils is quarries dug into the fine-grained lake and lagoon deposits that best preserved them, called Konservat-Lagerstätten for the abundance of their high-quality fossils. These quarries may exist explicitly for collecting fossils, like those being excavated by local residents and paleontologists in the Jehol beds of northeastern China. Commercial quarries extract the high-quality stone for various purposes and find fossils on the side. Most famously, the Solnhofen lithographic slate quarries of Bavaria have also produced *Archaeopteryx* specimens.

Because pterosaur paleozoology is not a high-priority science backed by large financial budgets, and because the number of people searching for and excavating pterosaurs in the world in a given year is only in the dozens—which is markedly more than in the past—the number of pterosaur skeletons that now reside in museums is still only in the many hundreds, most from the ancient lake beds of northeastern China.

In the lab, fine tools are often used to meticulously eliminate some or all of the sediment from the bones and any other remains. Acids can also be used to remove dissolvable sediments such as limestones without mechanical means. Slab mounts, in which the flattened skeleton is left half encased in the flat, hard plate that was first deposited as mud at the bottom of a lake, lagoon, or seabed, are common for pterosaur remains. Most pterosaur bones are left intact, and only their surface form is documented unless they happen to be broken open. Increasingly, certain bones are opened and sectioned to reveal their internal structure for various purposes: to examine bone histology and microstructure, to count growth rings, to search for traces of soft tissues, and to sample bone isotopes and proteins. It is becoming the norm to conduct CT scans on skulls and complex bones to determine the three-dimensional structure without invasive preparation, as well as to reduce costs. These can be published as conventional hard copies and in digital form. There is increasing reluctance to put original bones in mounted skeletons in display halls because delicate fossils are better preserved and more readily studied when properly stored. Instead, the bones are molded, and lightweight casts are used for display skeletons.

Landowners who allow paleontologists onto their land sometimes get a new species found on their property named after them, informally or formally. So do volunteers who find new pterosaurs. Who knows, you may be the next lucky amateur.

PTEROSAUR COUNT

The ten dozen or so named and valid pterosaurs represent only a tiny fraction of those that existed for 170 million years. In a few locations and formations, the pterosaur faunas of those particular times and places are well documented, such as the Solnhofen and related deposits of Late Jurassic southern Germany, and the Middle Jurassic to Early Cretaceous Jehol beds of northeastern China. But most species were never fossilized, many of those that were lie in sediments too deep to be sampled, and many sediments that are accessible have yet to be worked. Birds, the living flying archosaurs closest to pterosaurs, currently number nearly 10,000 species. If pterosaurs were as species numerous over a portion of their span, and a species lasted on average a couple of million years, then there could have been over half a million pterosaur species. But birds are much more anatomically diverse than pterosaurs, ranging from flightless ostriches to hovering hummingbirds to a plethora of songbirds many of which are minor variants of the same form, to waterfowl and seabirds and so forth. Bats add up to around 1,000 species. But like most birds, bats become independent only at maturity, while pterosaurs apparently were prone to go out on their own right out of the egg. If the youngsters filled the lifestyle roles of small adults, that should have reduced the number of species to perhaps a few hundred at the height of group diversity in the Late Jurassic and Early Cretaceous. Rhamphorhynchoid diversity would, however, have been low early in their evolution in the Late Triassic and Early Jurassic, while the diversity of pterodactyloids appears not to have been high in the Late Cretaceous. If an average of a couple of hundred pterosaur species were alive at a given time, then there may have been over 15,000 species total during the Mesozoic, four-fifths of which were pterodactyloids, and of all those pterosaurs we know but a fraction of 1 percent.

Late Triassic (Rhaetian–Norian–Carnian)

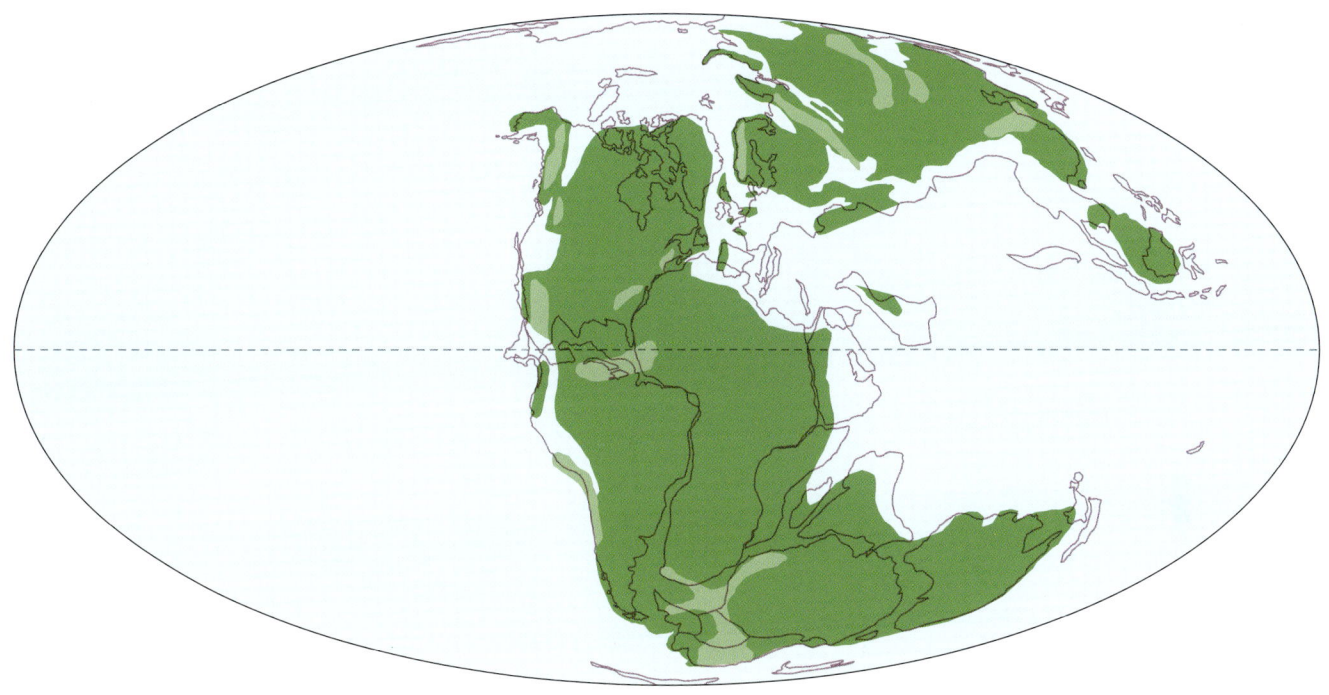

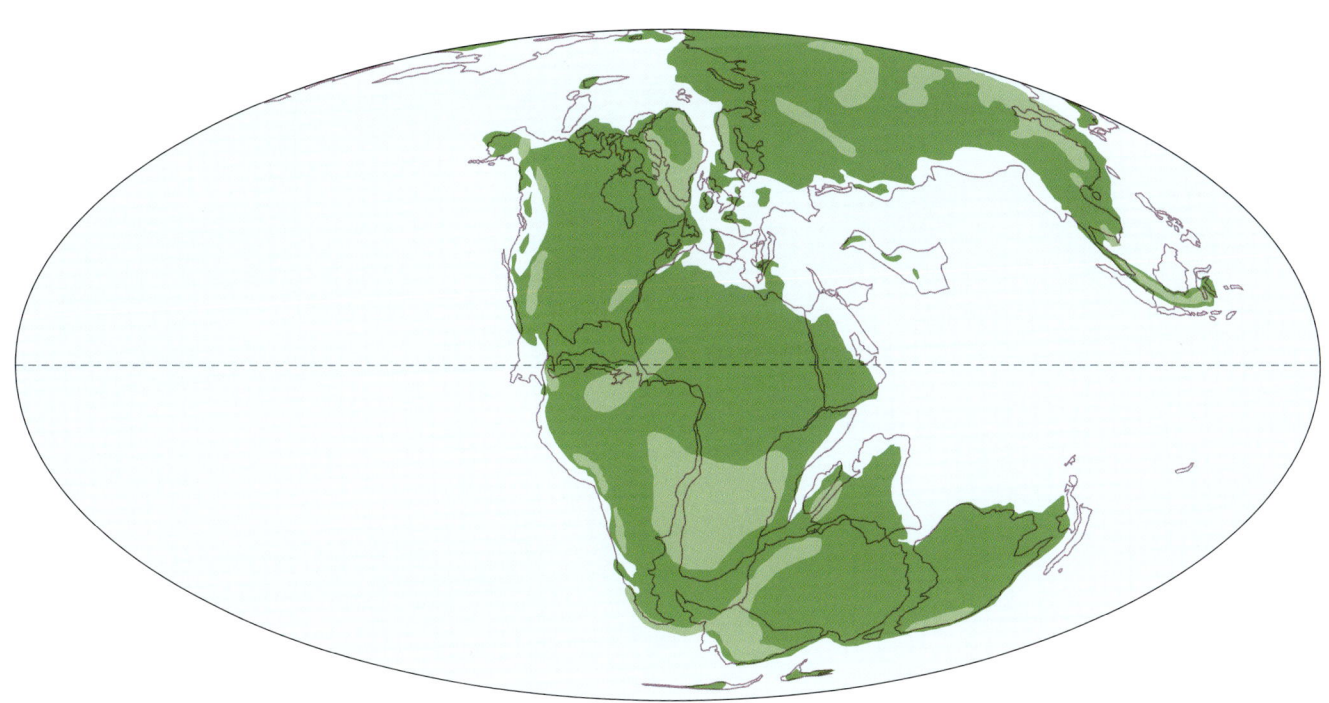

Early Jurassic (Sinemurian)

Middle Jurassic (Callovian)

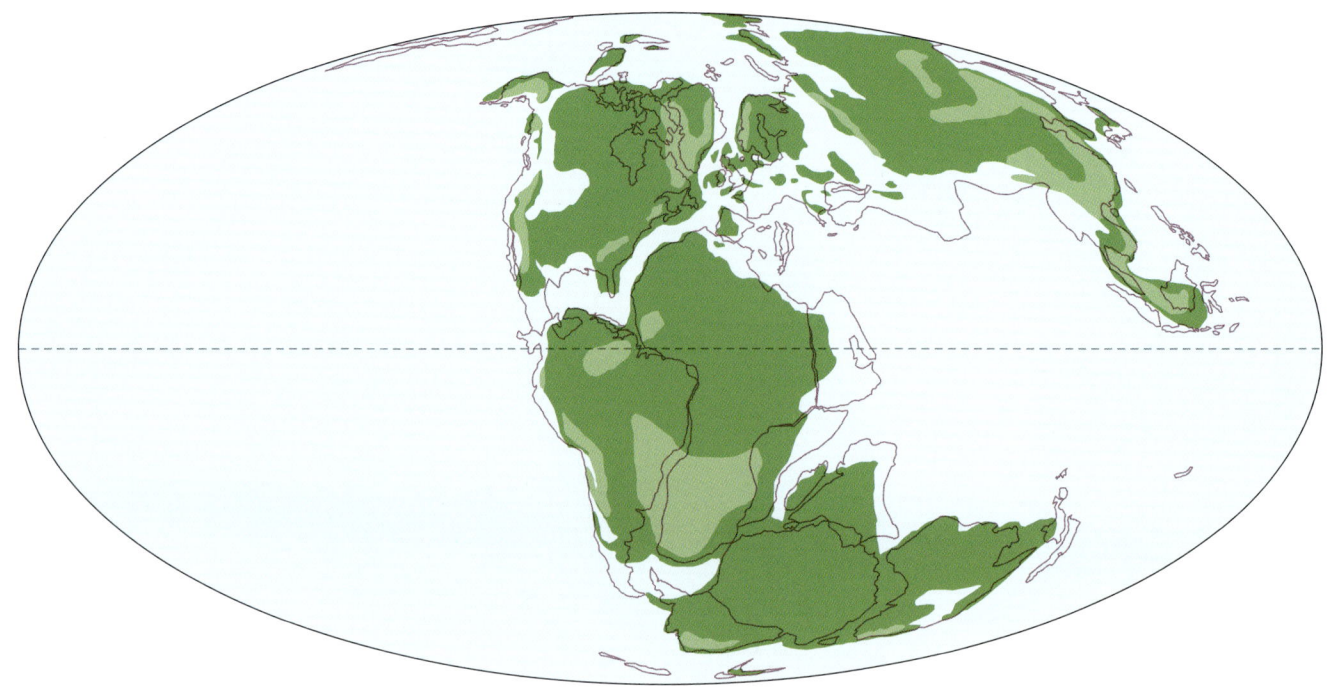

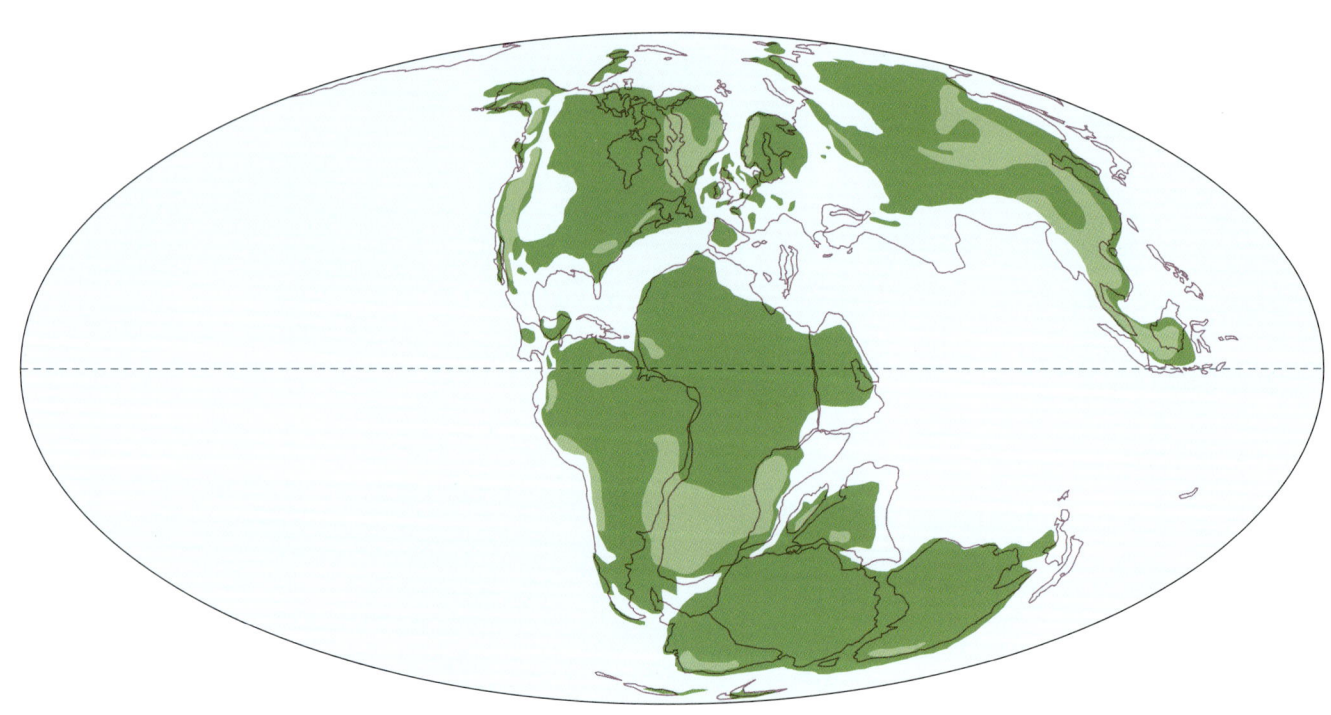

Late Jurassic (Kimmeridgian)

Early Cretaceous (Valanginian–Berriasian)

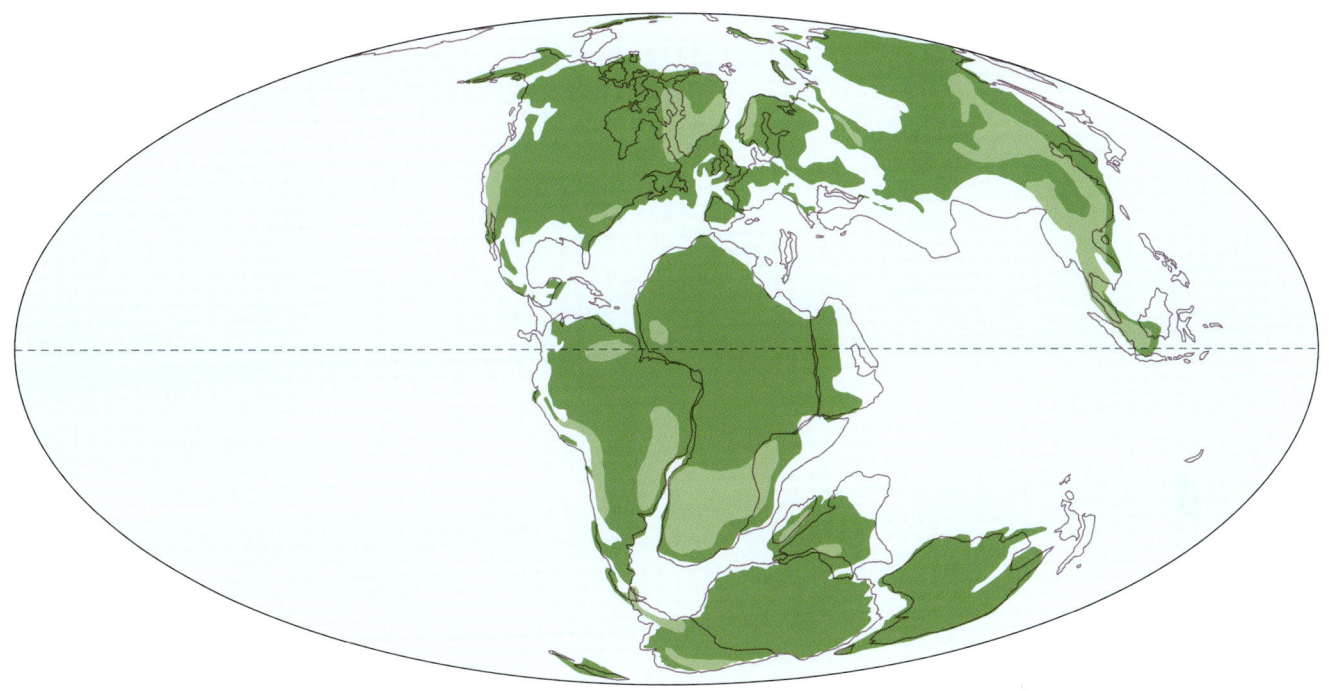

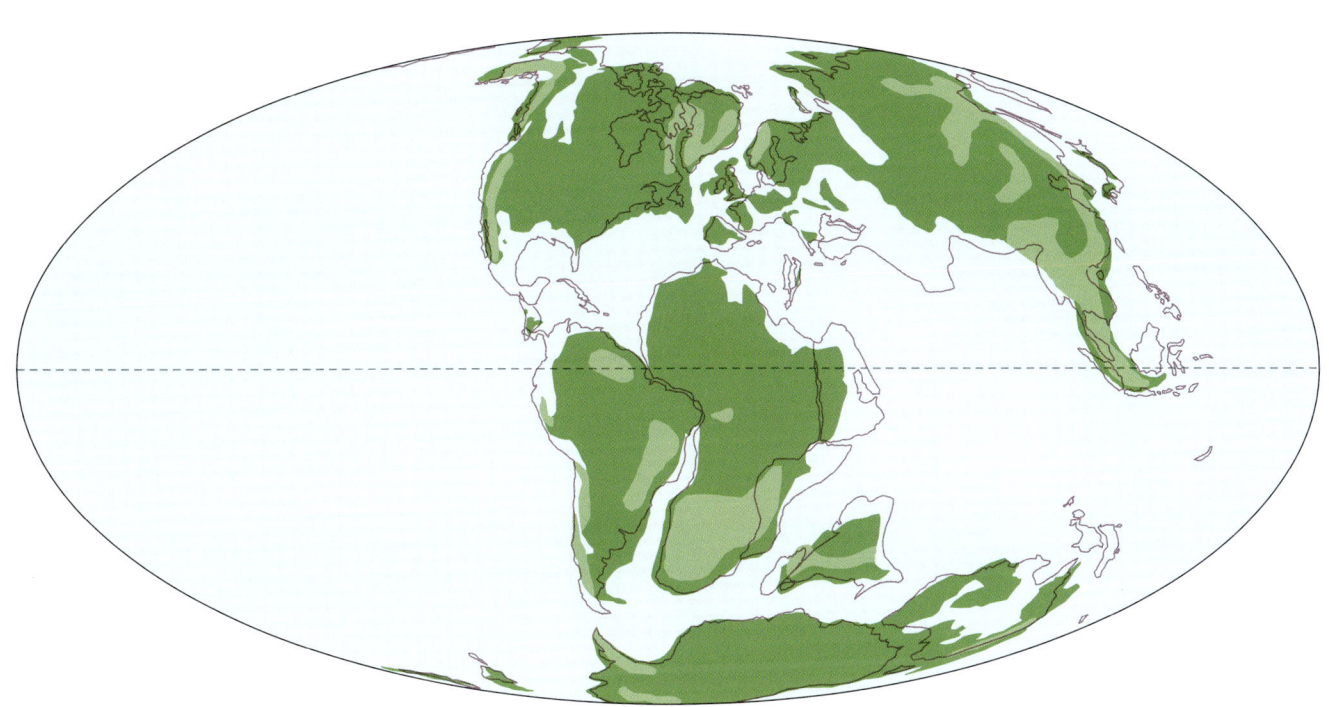

Early Cretaceous (Aptian)

Late Cretaceous (Coniacian)

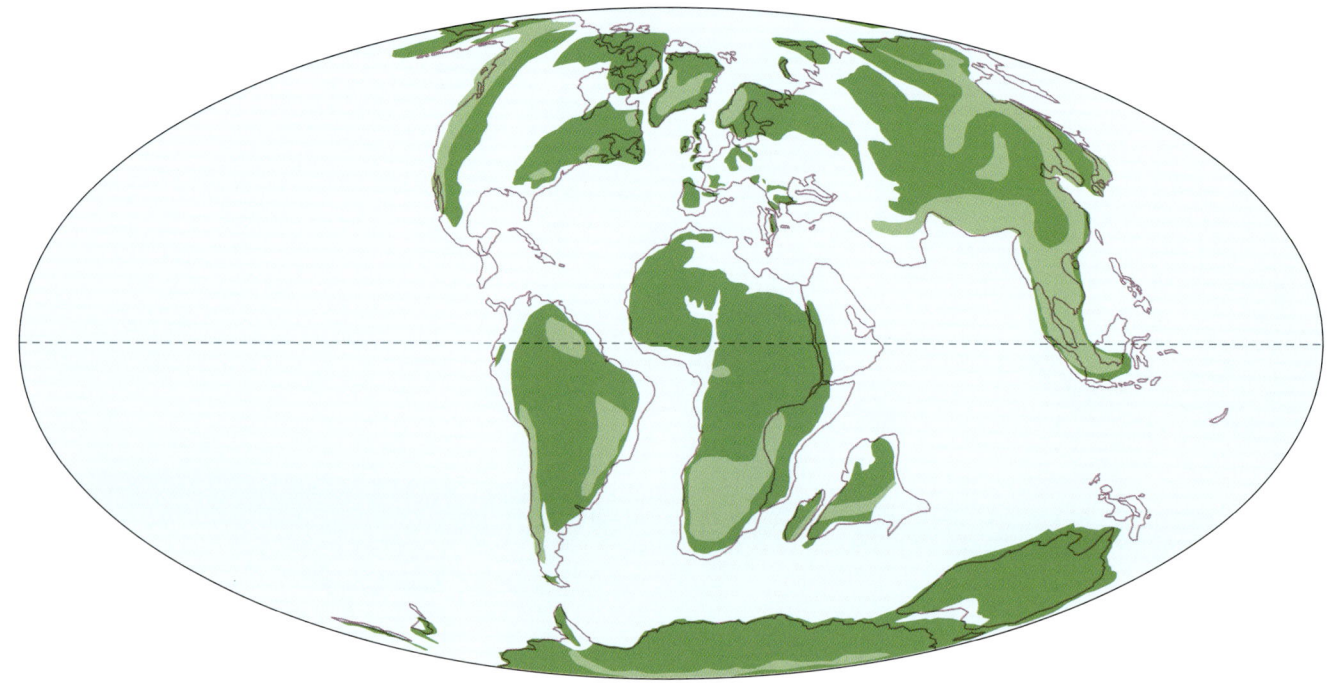

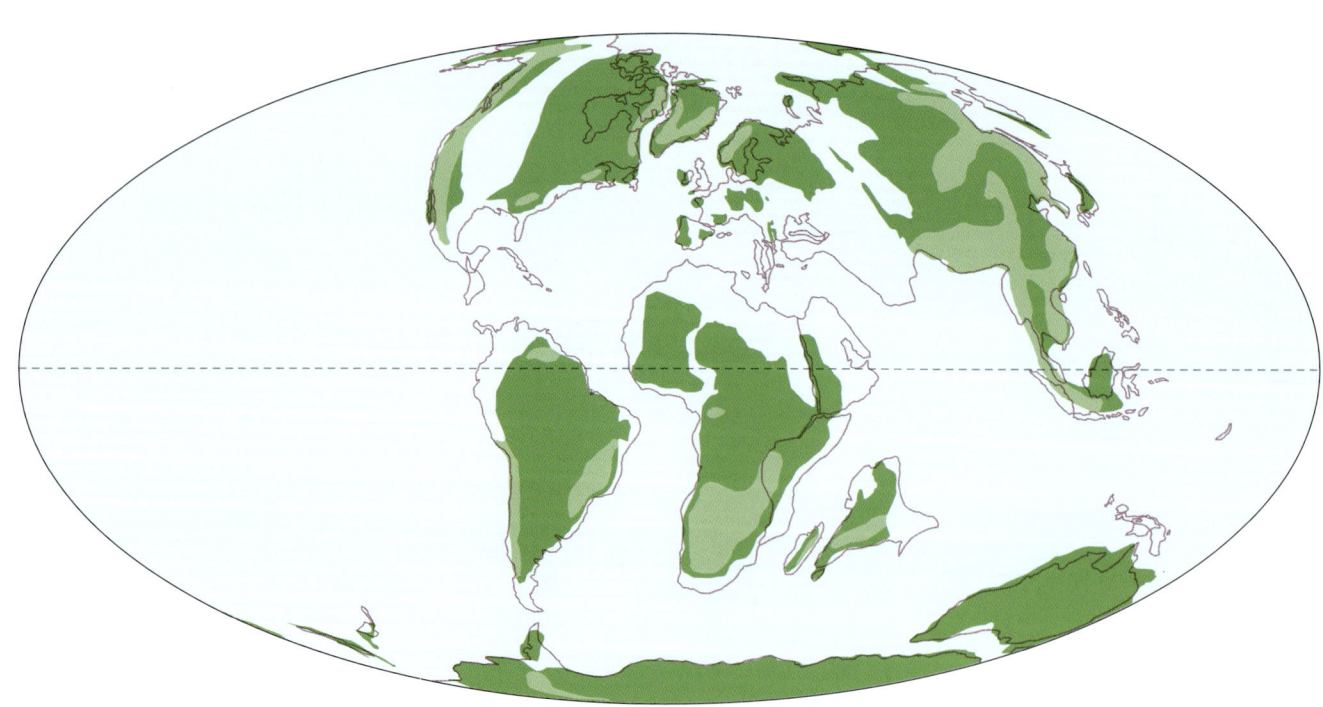

Late Cretaceous (Campanian)

USING THE GROUP AND SPECIES DESCRIPTIONS

Over twenty dozen pterosaur species have been named, but a significant portion are invalid. Some are based on inadequate remains, such as one or a few bones, that are taxonomically indeterminate, although a number must represent real species when they are distinct from known species or are all that is known from formations that lack significant other remains. Because so many pterosaur species are based on at least fairly complete remains found in the fine-grained Konservat-Lagerstätten deposits, the ratio of valid to unsubstantiated species is much lower than for the remains of dinosaurs, which are less often complete. Other pterosaur species are junior synonyms for species that have already been named from the same geological formations. *Pteranodon ingens*, for instance, turned out to be the same species as the previously named *Pteranodon longiceps*, also from the upper Niobrara Formation, which had been named shortly before, so the former species name is no longer used. In many cases small skeletons were thought to be the adults of small-bodied species because it was presumed that, like most birds and bats, pterosaurs did not fly until mature. The realization that many pterosaurs were fliers soon after hatching prompted the reassignment of many small specimens to species that include bigger adults. *Sinopterus dongi* is widely seen as consisting of a few named species, also from the Jiufotang Formation, that are really its youngsters. Some eight specimens from the Romualdo Formation of assorted sizes and featuring different beak tip crests may all belong to *Anhanguera blittersdorfi*, or at least to that genus, or only some may, or perhaps none. Because juvenile pterosaurs were competing in a given habitat space, this guide usually presumes that smaller specimens are juveniles of a species known from adults from the same level of the formation, unless compelling evidence indicates otherwise. This guide includes those species that are generally considered valid and are based on sufficient remains. A few exceptions are allowed when a species based on a single bone or little more is important in indicating the existence of a distinctive type or group of pterosaurs at a certain time and place.

The species descriptions are listed hierarchically, starting with major groups and working down the ranking levels to genera and species. Because many researchers have abandoned the traditional Linnaean system of classes, orders, suborders, and families, there is no longer a standard arrangement for the pterosaurs—many pterosaur genera are no longer placed in official families—so none are used here. In general the taxa are arranged phylogenetically, with more-derived groups nested within more-basal clades. There are a number of problems. It is more difficult for the general reader to follow the various groupings without the traditional scheme. Worse, recent phylogenetic studies are producing very different results, especially for the pterodactyloids, so there is no consensus on the detailed relationships of the groups, and some species are being placed in different clades by different analyses. This is not surprising because the incompleteness of the fossil record hinders a better understanding of fossil relations, the great majority of pterosaur species that lived are not known, many of those that are known are documented by incomplete remains, and it is not possible to examine pterosaur relationships with genetic analysis. Lacking a well-founded consensus to follow, I have used a degree of personal choice and judgment in arranging the groups, and the species within groups. Some of these placements reflect my considered opinion, but most are arbitrary choices among a large array of competing research results. Most of the phylogeny and taxonomy offered here is not a formal proposal. Disputes and alternatives concerning the placement of pterosaur groups and species are often but not always mentioned.

Under the listing for each pterosaur group, the overall geographic distribution and geological time span of its members are noted. This is followed by the anatomical characteristics that apply to the group in general, which are not repeated for each species in the group. The anatomical features usually center on what is recorded in the bones, but other body parts are covered when they have been preserved. The anatomical details are for purposes of general characterization and identification, reflecting as much as possible what a pterosaur watcher might use; they are not the result of extensive technical phylogenetic diagnoses. The type of habitat the group favored is briefly listed; this varies from specific in some types, to very generalized in others, to not known in a few. Also outlined are the restored habits that probably characterized the group as a whole, including effects related to growth when that is an important factor. The reliability of these conclusions varies greatly. There is, for example, no doubt that pteranodonts with long, toothless beaks that are consistently found in marine sediments consumed fish and the like rather than plants. But the diet of the similarly gigantic azhdarchids has been more controversial, and what many pterosaurs ate remains very obscure.

The naming of pterosaur genera and species is often problematic. In part this is because what were considered distinct genera and/or species at one time are now proposed to be juveniles or different sexes of another genus or species, or vice versa, by at least some, while others may disagree. Many species are based on fragmentary remains that are problematic. In other cases there is disagreement on how much difference between species constitutes different genera, which are also based on phylogenetic relationships, and stratigraphic levels also play a role. *Pteranodon* is seen by some as constituting a number of genera, the others being *Geosternbergia* and *Dawndraco*. Because the species are all similar, this work opts to put them all in three time-separated species of *Pteranodon*, with some moderately large skeletons seen as immatures or females, but the choice is by no means certain. Disputes and alternatives concerning the designation of pterosaur genera and species are often but not always mentioned. Following most species names is a reference. Some cite the paper in which the genus and/or species was first named, and sometimes more recent papers are included that provide more detailed information on the fossils. None of the references in this book are older than a few decades, but even so, the information may be open to dispute. No non-English papers are cited, but some references are not readily accessible.

The entry for each species first lists the dimensions and estimated mass of adults; those for juveniles are not listed. For all species for which adults are apparently known, wingspan (WS) is measured or estimated as the span when flying and is measured between perpendiculars with the elbow, wrist, and finger base flexed as they were during flight; fully straightening out the wing increases the span by about 5 percent. For those species for which a sufficiently complete profile-skeletal restoration is available, the total length (TL) is given for the combined skull and skeleton, measured along the length of the skull from the tip of the keratin beak and the curve of the neck to record the length if the animal were straightened out, rather than between perpendiculars as posed. The values presented are general figures for the size of the largest known adults of the species and do not always apply to the estimated values for specific specimens whose skeletons are restored. Because the number of specimens for a particular species is a small fraction of those that lived, the largest individuals are not on hand to be measured; "world record" specimens can be a third or more heavier than is typical. The sizes of species known only from immature specimens are not estimated. All values are, of course, approximate, and their quality varies depending on the completeness of the remains for a given species; tail length in rhamphorhynchoids is highly variable in ways that are not consistent, so if much or all of the tail is missing, it is not possible to reliably estimate total length. Because wing dimensions are more consistent, the spans can always be estimated for adults. If the species is known from sufficiently complete remains, the dimensions and mass are based on the profile-skeletal restoration. Profile-skeletals can be used to estimate the volume of pterosaurs, which can then be used to calculate the mass, with the portion of the volume that was occupied by lungs and air sacs taken into account. As explained above, pterosaur specific gravity densities ranged from about 0.95 in the small rhamphorhynchoids, to around 0.9 in more-pneumatic small pterodactyloids, to as low as about 0.75 in big-winged giant pterodactyloids (Larramendi et al. 2021). Also needing to be considered are the wing membranes, which, although only a small fraction of a millimeter in thickness, were so extensive that they added almost a tenth to the total mass even assuming fairly narrow wing chords. When remains are too incomplete to directly estimate dimensions and mass, these are extrapolated from relatives and are considerably more approximate. Of particular interest are the azhdarchids. Only a medium-sized species is complete enough to make a direct volume estimate. If the resulting mass of that form is scaled up isometrically to the

gigantic examples such as *Quetzalcoatlus northropi*, cubing the starter value produces a mass exceeding 600 kg (1,400 lb), a value than cannot be ruled out. But in birds, and apparently in pterosaurs, mass relative to wingspan scales to less than a power of 3, closer to 2.5. In that case the gigantic *Quetzalcoatlus* would have been around 450 kg (1,000 lb). The mass estimates herein are higher than those calculated in the deeper past, but they are often somewhat less than those produced by recent efforts. Mass estimates were not attempted for a number of incomplete species whose dimensions and body form are insufficiently known. Both metric and English measurements are included. All original calculations are metric; they are often imprecise, and conversions from metric to English are often further rounded off.

The next line outlines the fossil remains, whether they are skull or skeletal material or both, that can be assigned to the species with reasonable confidence to date; the number of specimens varies from one to many dozens. The accuracy of the list ranges from exact to a generalization. The latter sometimes results from recent reassignment of specimens from one species to another, leaving the precise inventory uncertain. Profile-skeletal and/or skull restorations have been rendered for those species that are known from sufficiently complete remains and were available to execute a reconstruction as the book was being produced—the pace of discovery is so fast that some new finds could not be included.

This book includes by far the most extensive pterosaur skull-skeletal library yet published—the adult and juvenile specimens that have been restored can be found via https://press.princeton.edu/books/hardcover/9780691180175/the-princeton-field-guide-to-pterosaurs, as well as their estimated wingspans and masses. The restorations show the bones, but not restored cartilage, as solid white set within solid black profiles that include the restored muscles, tendons, digit pads, keratin sheaths, and other nonbony tissues, some of which have been preserved in a few cases, such as throat pouches and nuchal ligaments above long necks. Missing portions of bone head crests are tentatively restored as solid black, and preserved soft tissue crests are outlined with solid lines; in some cases potential soft tissue crests are outlined with dashed lines. It is highly probable that many or even all pterosaurs had soft crests or partial crests that were rarely preserved—the famous sword crest of *Pteranodon longiceps* may have supported a larger nonbony structure—but because fossil data are lacking, they are not included in the profiles. In some cases, only the skull is available for illustration. In many cases the skull and skeletal restorations are of adults. But because many pterosaurs appear to have become abundant—perhaps *the* most abundant active flying members of the aerial fauna—while they were still juveniles, often just out of the eggs, a number of juveniles have been illustrated, sometimes as a same-scale growth series. In some cases the skeletons of the smallest juveniles are reproduced both to scale to the much larger adults, and at a larger scale to allow the details to be seen. The skeletons are posed in a common basic posture in order to facilitate cross comparisons. Of late it has become common to pose pterosaur skeletons in a quadrupedal takeoff pose with the wings pushing off the ground. In visual terms this is an awkward, viewer-perplexing posture that obscures the shape of the most important element in the fliers—the wings—and the proportions of the wings relative to the rest of the animal cannot be readily discerned. In this volume the wings are posed in a more viewer-friendly, vertical plan view at the top of the upstroke.

The accuracy of the restorations ranges from very good for those that are known from extensive remains and for which a detailed description and/or good photographs of the skeleton are on hand, down to approximate if much of the species remains are missing or have not been well illustrated. A number of skeletons and skulls show only those bones that are known, which ranges from a large fraction to nearly complete, whereas others have been filled out to represent a complete skeleton, and in others only major sections, such as a skull, hand, and so forth are not drawn in. Reliable information about exactly which bones have and have not been preserved is often not available, so the widely used term "rigorous" for incomplete skeletal restorations is best avoided in favor of "known bone only"—a skeletal is truly rigorous if it is executed with as much scientific accuracy as possible with the data on hand. A few samples of top-view profile-skeletals of basic pterosaur types have been included, and top views of skulls are included when sufficient information is available. Some representative examples of shaded skull restorations have been included with some of the major groups. The same has been done with a sample of muscle studies, whose detailed nature is no less or more realistic than are the particulars found in full-life restorations, which, if anything, involve additional layers of speculation.

The color profiles are based on the fully or nearly adult profile-skeletal restorations or skulls of those species deemed of sufficient quality for a full-life profile restoration. The more atypical a species is, the more likely a life profile was produced based on a less complete set of remains. In some cases only the skull is preserved well enough to warrant a life restoration to the exclusion of the overall body. The colors and patterns are entirely speculative except in one case where the color banding pattern, but not the actual colors, of a head crest was preserved. Preserved pigment capsules used to restore the colors of fossil dinosaurs have not yet been used to assay a number of pterosaurs. The color patterns for a given species are similar in all the life restorations in this book. The pose chosen for the profile-skeletals automatically exposes the underside of the wing closest to the viewer; in some cases the life appearance of the underwing is restored, and usually the topside is illustrated with the cutoff line just above the main trunk. Because it is not entirely certain how wing membranes attached to the legs and tails, there is variation in this factor among the illustrations. Those who wish to use the skeletal, muscle, and life restorations herein as the basis for commercial and other public projects are reminded to first contact the copyright illustrator.

The particular anatomical characteristics that distinguish the species are listed, and these too are for purposes of general identification for putative pterosaur watchers, not technical species diagnoses. These differ in extent depending on the degree of uniformity versus diversity present in a given group as well as the completeness of the available fossil remains. In some cases the features of the species are not different enough from those of the group to warrant additional description. In other cases not enough is known to make a separate description possible.

Listed next is the formal geological period and, when available, the stage from which the species is known. As discussed earlier, the age of a given species is known with a precision of within a million years or less in some cases, or as poorly as an entire period in others—for example, it can only be said that the well-known *Dsungaripterus weii* lived sometime within the 45 million years of the Early Cretaceous, while *Pteranodon longiceps* was flying above the warm waves of the Coniacian/Santonian boundary circa 86 million years ago, and the titanic *Quetzalcoatlus northropi* was striding about when it was liquidated by the K/Pg crisis 66 million years ago. The reader can refer to the scale on the timeline chart to determine the age, or age range, of the species in years (see pp. 94-5). Most species exist for a few hundred thousand years to a couple of million years before either being replaced by a descendant species or going entirely extinct. In some cases it is not entirely clear whether a species was present in just one time stage or crossed the time boundary into the next one. In those cases the listing is "and/or," as in late Santonian and/or early Campanian.

Next the geographic location and the geological formation from which the species is known so far are listed. The paleo-maps of coastlines at the end of this section (see pp. 87-90) can be used to geographically place a species in a world of drifting continents and fast-shifting seaways, with the proviso that no set of maps is extensive enough to show the exact configuration of the ancient lands when each species was extant. I have tended to be conservative in listing the presence of a specific species only in those places and levels where sufficiently complete remains are present. Some pterosaur species are known from only a single location, whereas others have been found in an area spread over one or more formations. In some cases formations have yet to be named, even in areas that are well studied. Many formations were formed over a span that was longer than that of some or all of the species that lived within them, so when possible the listing of just the overall formation a given species is from is avoided. For example, *Pteranodon* is known from the Niobrara and Pierre Shale Formations. In the species entries, *P. sternbergi* is listed as coming from the middle of the Niobrara, *P. longiceps* from its upper levels, and *P. mayesi* from the lower Pierre Shale. Noted next are the basic characteristics of the pterosaur's habitat in terms of rainfall and vegetation as well as temperature when it is not generally tropical or subtropical year round. Environmental information ranges from well studied in heavily researched formations to none in others. If the habits of the species are thought to include attributes not seen in the group as a whole, then they are outlined, including changes with growth when pertinent. Listed last are special notes about the species, when called for. In many cases other pterosaurs that the species shared its habitat with are listed. Possible ancestor-descendant relationships with close older or younger relatives are sometimes noted, but these are always tentative. This section also notes alternative hypotheses and controversies that apply to the species.

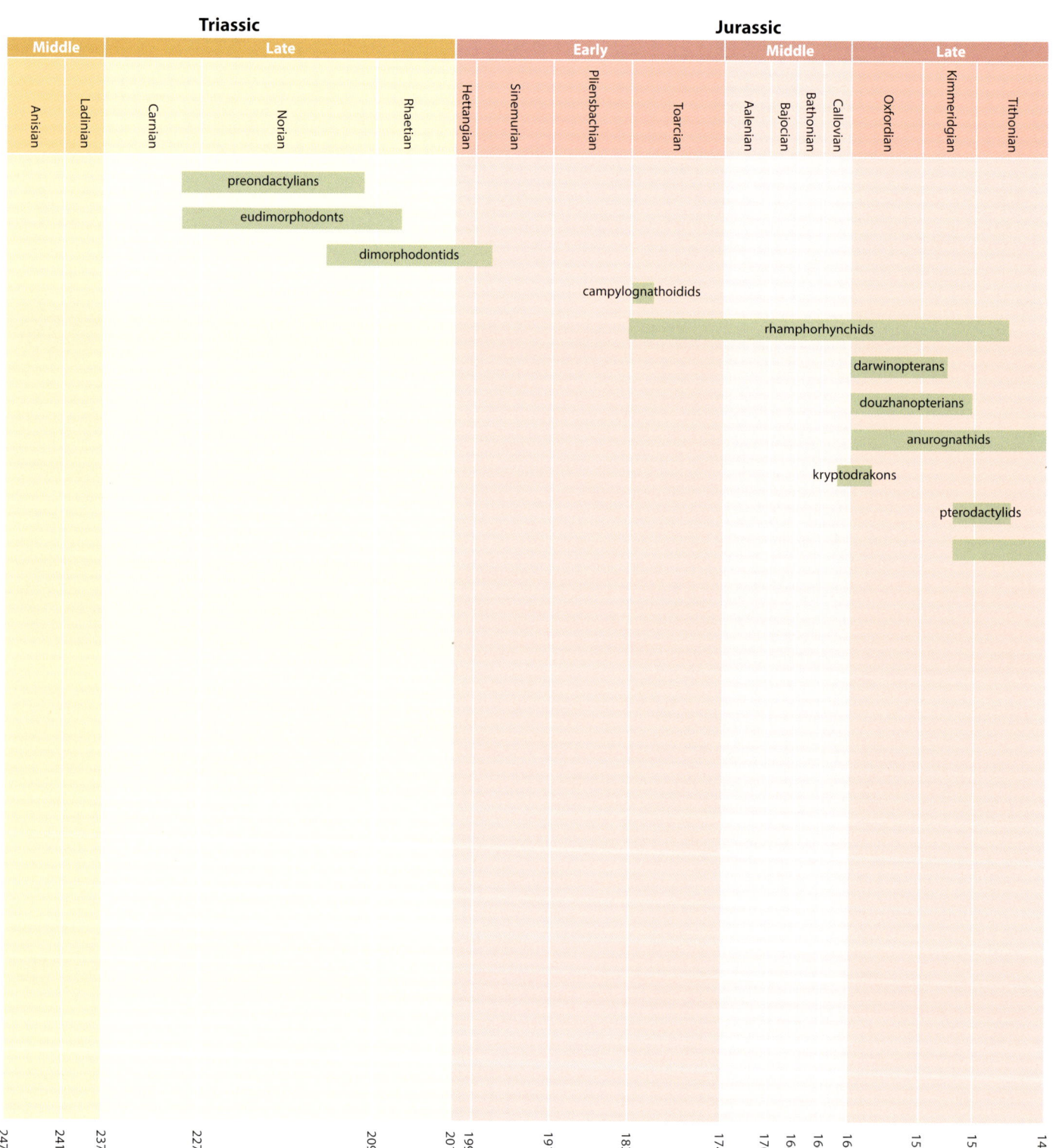

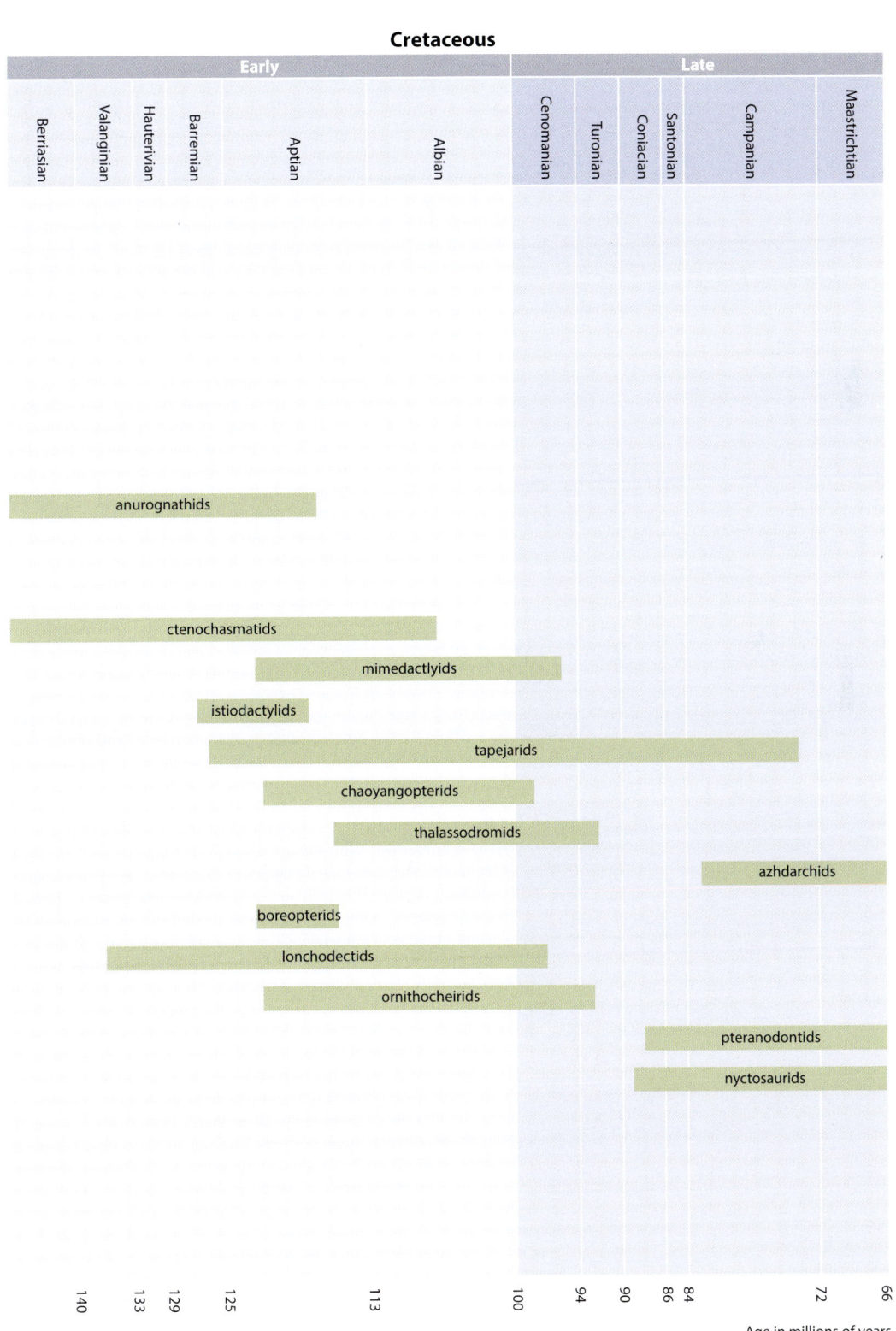

Cretaceous

Early						Late					
Berriasian	Valanginian	Hauterivian	Barremian	Aptian	Albian	Cenomanian	Turonian	Coniacian	Santonian	Campanian	Maastrichtian

anurognathids

ctenochasmatids

mimedactlyids

istiodactylids

tapejarids

chaoyangopterids

thalassodromids

azhdarchids

boreopterids

lonchodectids

ornithocheirids

pteranodontids

nyctosaurids

140   133   129   125   113   100   94   90   86   84   72   66

Age in millions of years

Quetzalcoatlus and juvenile tyrannosaur

PTEROSAURS

SMALL TO GIGANTIC FLYING ARCHOSAURS OF THE LATE TRIASSIC TO THE END OF THE MESOZOIC ON ALL CONTINENTS AND OCEANS

ANATOMICAL CHARACTERISTICS Moderately variable. Head generally large, sometimes extremely so, usually bears midline crest consisting of soft and/or bony tissue, nasal openings almost always large and/or placed aft, upper temporal openings face partly sideways. Neck not extremely flexible. Trunk short, teardrop shaped, gastralia present, sternal ribs ossified and often bear small sternocostapophyses. Tail slender. No clavicles or furcula, front portion of sternum a narrow keel, main body a large plate, somewhat convex ventrally, scapulas usually strap shaped, narrow coracoids articulate with sternum near midline at keel-plate juncture, shoulder joints face strongly sideways and innermost arms subhorizontal when walking, humeri not highly elongated, pectoral crests prominent, radius-ulna slender and tightly appressed to one another, wrists consist of large, interlocking carpal blocks with pteroid pointing inward and small preaxial pointing forward, hands digitigrade, metacarpals 1–3, very slender if present, metacarpal 4 robust and ends with pulley-folding articulation for its finger, fingers 1–3 short, slender, flexible, and large clawed if present, finger 4 fairly robust, stiffened, and extremely long, with four or three elements. Sacrum long, pelvis with long, forward-projecting ilia, short backward- and upward-projecting ilia, short vertical pubes tipped with mobile, forward-projecting prepubes, ischia short and platelike, semierect or erect leg posture achieved by round femora head at end of partly inward-directed stalk articulating with cup-shaped hip socket facing laterally to varying degrees, femora long, slender, and gently bowed, slender tibias longer than femora, fibulas reduced, simple hinge-jointed ankles, feet plantigrade, four metatarsals subequal, fairly long and slender, not tightly bound together, toes subequal, moderately long, flexible, clawed. All largely or entirely quadrupedal, forelimbs dominant in that arms more strongly built and muscled than legs. Trackways show hands always at least as far apart from midline as feet or farther, gauge moderate to narrow. Arms and legs support membrane airfoils, propatagium leading edge supported by pteroid, outer main membrane stiffened by actinofibers, parts of head, neck, body, inner arms, and legs insulated with pycnofibers, foot scales form a nonoverlapping mosaic pattern.

REPRODUCTION AND ONTOGENY Probably all laid soft-shelled eggs in pairs buried in the ground. Growth rates moderate, many if not all began to fly well before adulthood.

HABITS AND HABITATS All capable of flapping flight, otherwise highly variable. From strongly continental to highly oceanic, from burst-powered fliers to extreme soarers. None strongly herbivorous.

SMALL TO MEDIUM-SIZED PTEROSAURS OF THE LATE TRIASSIC TO THE EARLY CRETACEOUS ON ALL CONTINENTS

ANATOMICAL CHARACTERISTICS Somewhat variable. Head not extremely large, teeth always present. Neck never elongated. Tail almost always long and stiffened by very long, ossified tendons. Scapular blades fairly long, subhorizontal, do not articulate with neural spines. Main metacarpals short. Fifth toes well developed, clawless, hooked support for uropatagium. Possibly more bipedal than pterodactyloids.

HABITAT Moderately variable. From strongly continental to shoreline, from modest-range to very high-performance flappers, some soarers. Scarcity of trackways may indicate that relatively little time was spent on the ground.

HABITS Variable. Burst-powered fliers to soarers. Fishers, predators, insectivores.

NOTES Survival of group into the Late Cretaceous via anurognathids uncertain.

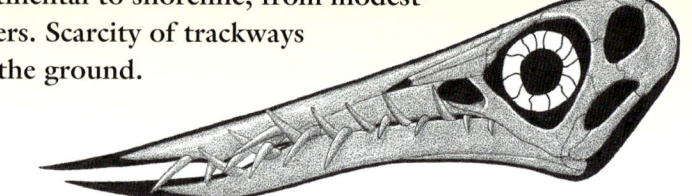

Rhamphorynchus shaded skull

EOPTEROSAURS

SMALL RHAMPHORHYNCHOIDS OF THE LATE TRIASSIC OF EUROPE AND GREENLAND

ANATOMICAL CHARACTERISTICS Head variable, modest in size, teeth mainly subvertical except at front end of jaws, which project forward, throat pouch long. Sternocostapophysis projections on sternal ribs absent in at least some examples. Pteroids short.

HABITS Flight performance from relatively limited to well developed. Fishers, possible small-game predators, and fruit and other soft vegetation omnivores.

NOTES Earliest and most basal known pterosaurs. Absence from at least some other continents probably reflects lack of sufficient sampling.

PREONDACTYLIANS

SMALL EOPTEROSAURS OF THE LATE TRIASSIC OF EUROPE

ANATOMICAL CHARACTERISTICS Uniform. Head aside from crest fairly low, subtriangular, nasal opening very large, lower jaw shallow, teeth along first three-quarters of upper jaw, first half of lower, teeth single cusped, very variable in size. Tail lacks long, ossified tendons. Neck length moderate. Wingspan short relative to mass, inner wings and legs fairly long, feet large.

HABITS Flight performance relatively limited.

NOTES These are the only long-tailed pterosaurs known to lack long, ossified tail tendons.

Preondactylus buffarinii (Dalla Vecchia 1998)
0.46 m (1.5 ft) WS, 0.045 kg (0.1 lb)

FOSSIL REMAINS Nearly complete skull and skeleton.

ANATOMICAL CHARACTERISTICS No bony crest present, teeth single cusped. Wingspan shortest relative to mass among known rhamphorhynchoids. Leg length moderate.

AGE Late Triassic, late Carnian or early Norian.

DISTRIBUTION AND FORMATIONS Italy; lower Dolomia di Forni.

HABITAT Coastal.

NOTES May be most basal known pterosaur. Two very partial specimens, one from lower in formation, may or may not belong to this species. Shared its habitat with *Eudimorphodon ranzii, Seazzadactylus*.

Austriadactylus cristatus (Dalla Vecchia et al. 2002)
1.2 m (4 ft) WS, 0.8 kg (1.8 lb)

FOSSIL REMAINS Two nearly complete skulls and partial skeletons, adult and juvenile.

ANATOMICAL CHARACTERISTICS Large shallow crest from tip of snout decreasing in height to over orbits, teeth numerous, very variable in size including large fangs, many small teeth multicusped.

AGE Late Triassic, middle and/or late Norian.

DISTRIBUTION AND FORMATIONS Austria, Italy; Seefeld, middle Dolomia di Forni.

HABITAT Coastal.

HABITS Fed predominantly on invertebrates, possibly some small vertebrates.

NOTES Shared its habitat with *Eudimorphodon dallavecchiai*.

Preondactylus buffarinii

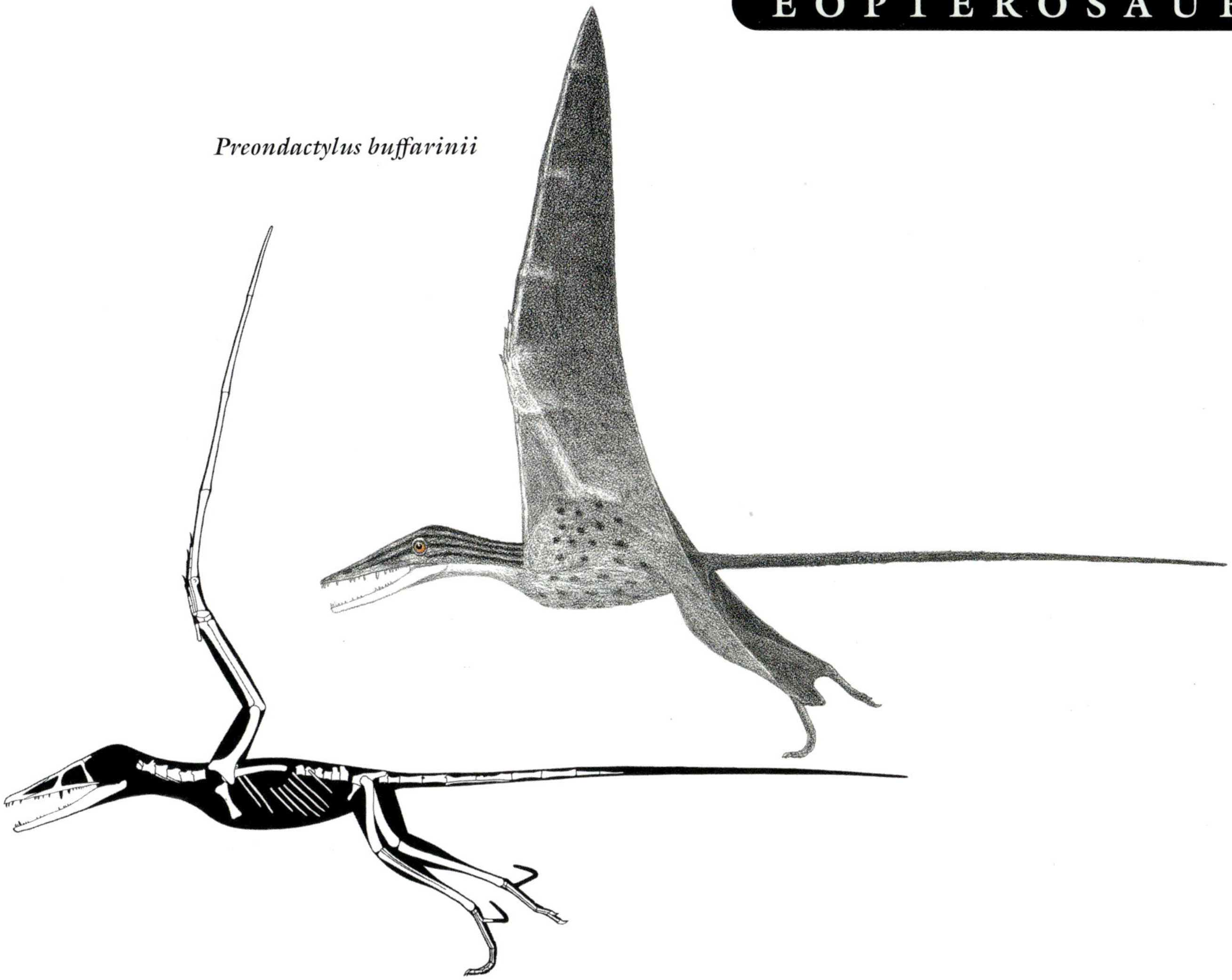

Preondactylus buffarinii

EUDIMORPHODONTOIDS

SMALL EOPTEROSAURS OF THE LATE TRIASSIC OF EUROPE AND GREENLAND

ANATOMICAL CHARACTERISTICS Variable. Head moderately large, teeth numerous, very variable in size including large fangs, small teeth usually multicusped and strongly occluding in adults. Tail stiffened by long, ossified tendons. Wingspan long relative to mass, fingers about as large as toes.

GROWTH AND HABITS Absence of fossils of very small juveniles suggests they did not fly immediately after hatching. Flight performance from relatively limited to well developed.

NOTES Only pterosaurs with complex, occluding teeth.

EUDIMORPHODONTIDS

SMALL EUDIMORPHODONTOIDS OF THE LATE TRIASSIC OF EUROPE AND GREENLAND

ANATOMICAL CHARACTERISTICS Uniform. Head fairly low, subtriangular, modest beak before modest-sized nasal opening, lower jaw shallow, teeth along nearly entire length of jaws, occluding multicusped teeth. Neck length moderate. Legs fairly well developed.

Eudimorphodon shaded skull

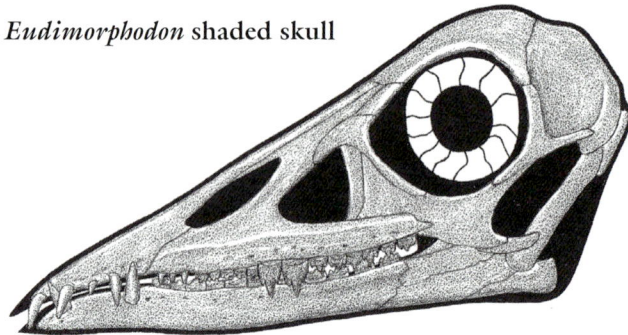

Eudimorphodon ranzii
0.74 m (2.4 ft) WS, 0.065 kg (0.15 lb)

FOSSIL REMAINS Complete mature skull with majority of skeleton, partial immature skull with majority of skeleton.

ANATOMICAL CHARACTERISTICS No gap between frontmost and rest of upper teeth, second lower tooth an especially large fang. Scapula blade strap shaped.

AGE Late Triassic, late Carnian or early Norian.

DISTRIBUTION AND FORMATIONS Italy; lower Dolomia di Forni.

HABITAT Coastal.

HABITS Flight abilities good.

NOTES Pterosaur with the most sophisticated dentition, suitable for chewing, fish remains in rib cage confirm fishing habits. *Carniadactylus rosenfeldi* is probably a juvenile of this species. Shared its habitat with *Preondactylus*, *Seazzadactylus*. May be direct ancestor of *E. wildi*.

Eudimorphodon ranzii

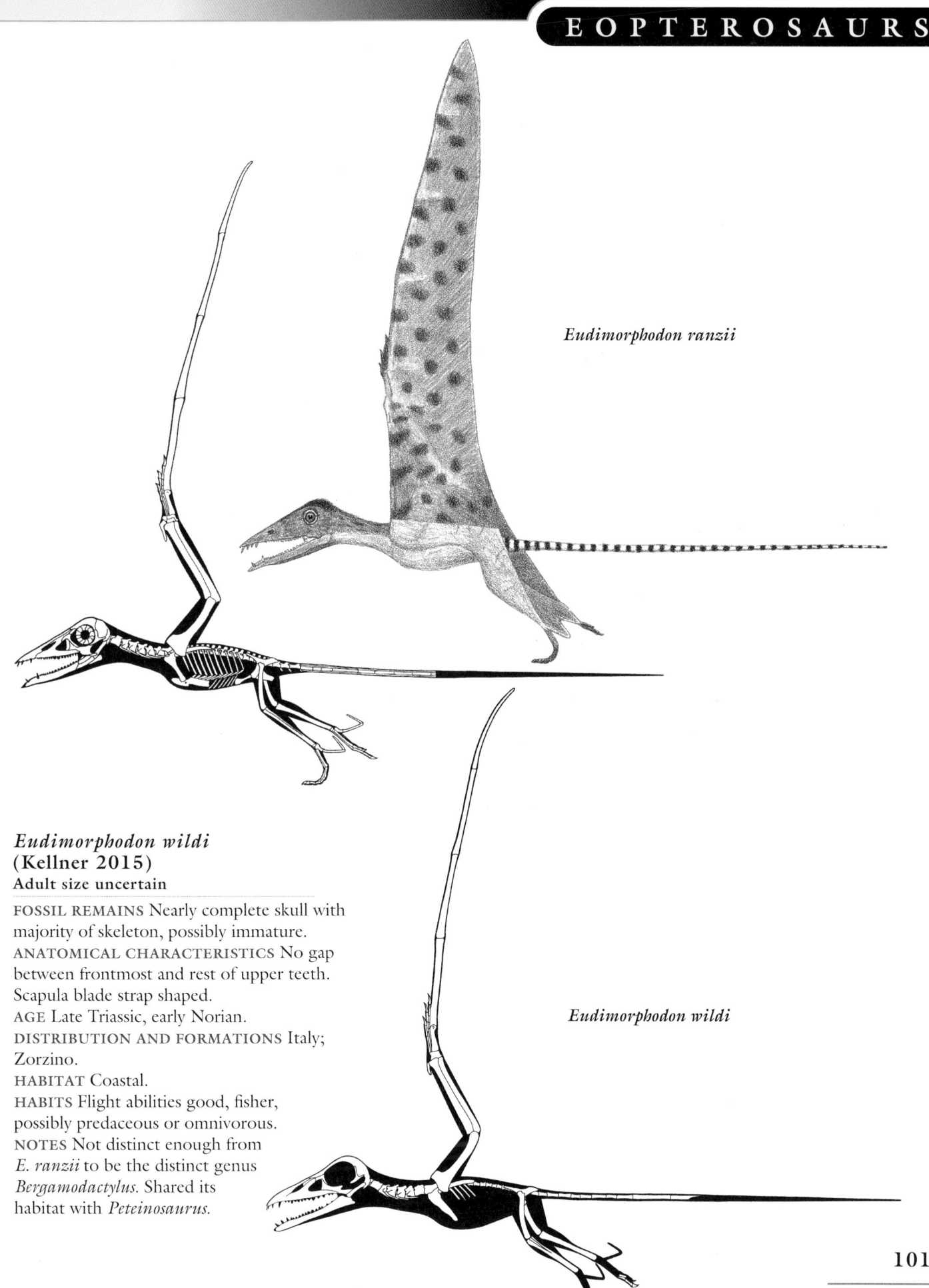

Eudimorphodon ranzii

Eudimorphodon wildi
(Kellner 2015)
Adult size uncertain

FOSSIL REMAINS Nearly complete skull with majority of skeleton, possibly immature.

ANATOMICAL CHARACTERISTICS No gap between frontmost and rest of upper teeth. Scapula blade strap shaped.

AGE Late Triassic, early Norian.

DISTRIBUTION AND FORMATIONS Italy; Zorzino.

HABITAT Coastal.

HABITS Flight abilities good, fisher, possibly predaceous or omnivorous.

NOTES Not distinct enough from *E. ranzii* to be the distinct genus *Bergamodactylus.* Shared its habitat with *Peteinosaurus.*

Eudimorphodon wildi

Eudimorphodon (?Austriadraco) dallavecchiai (Kellner 2015)
0.6 m (1.9 ft) WS, 0.035 kg (0.8 lb)

FOSSIL REMAINS Partial skull and skeleton.
ANATOMICAL CHARACTERISTICS Scapula strap shaped.
AGE Late Triassic, late Norian.
DISTRIBUTION AND FORMATIONS Austria; Seefeld.
HABITAT Coastal.
NOTES May or may not be a distinct genus. Shared its habitat with *Austriadactylus*.

Seazzadactylus venieri (Dalla Vecchia 2019)
0.75 m (2.5 ft) WS, 0.07 kg (0.15 lb)

FOSSIL REMAINS Majority of skull and skeleton.
ANATOMICAL CHARACTERISTICS Long gap between frontmost and rest of upper teeth. Scapula blade fan shaped.
AGE Late Triassic, late Carnian or early Norian.
DISTRIBUTION AND FORMATIONS Italy; lower Dolomia di Forni.
HABITAT Coastal.
NOTES Shared its habitat with *Eudimorphodon ranzii*.

Arcticodactylus cromptonellus (Jenkins et al. 2001)
Adult size uncertain

FOSSIL REMAINS Partial probably juvenile skull and skeleton.
ANATOMICAL CHARACTERISTICS Scapula blade strap shaped. Wings rather short.
AGE Late Triassic, Norian or early Rhaetian.
DISTRIBUTION AND FORMATIONS Greenland; Fleming Fjord.
HABITAT Shallow, sometimes transient lakes.

RAETICODACTYLIDS

SMALL EUDIMORPHODONTOIDS OF THE LATE TRIASSIC OF EUROPE

ANATOMICAL CHARACTERISTICS Head short, deep, subrectangular, nasal opening not large, lower jaw robust, deepest at front, teeth along first two-thirds of jaws. Humerus weak and as slender as femur, wings including inner and outer sections and legs very long.
HABITS Nonpowered flight abilities good, possibly including soaring. May have taken off bipedally rather than with arms.

NOTES Show that basal pterosaurs had become diverse and in some cases specialized as early as the Triassic.

Caviramus schesaplanensis (and/or Raeticodactylus filisurensis) (Fröbisch and Fröbisch 2006)
1.1 m (3.7 ft) WS, 0.18 kg (0.4 lb)

FOSSIL REMAINS Nearly complete adult skull and majority of skeleton, partial juvenile skeleton.
ANATOMICAL CHARACTERISTICS Moderately tall bony crest at least over nose, wingspan longest relative to mass among known rhamphorhynchoids (same as *Campylognathoides*).
AGE Late Triassic, late Norian or early Rhaetian.
DISTRIBUTION AND FORMATIONS Switzerland; lower Kössen.
HABITAT Coastal.
NOTES Very incomplete type specimen of *C. schesaplanensis* is probably a juvenile, and much better-preserved larger specimen is probably adult of same species rather than distinct species or genus *R. filisurensis*.

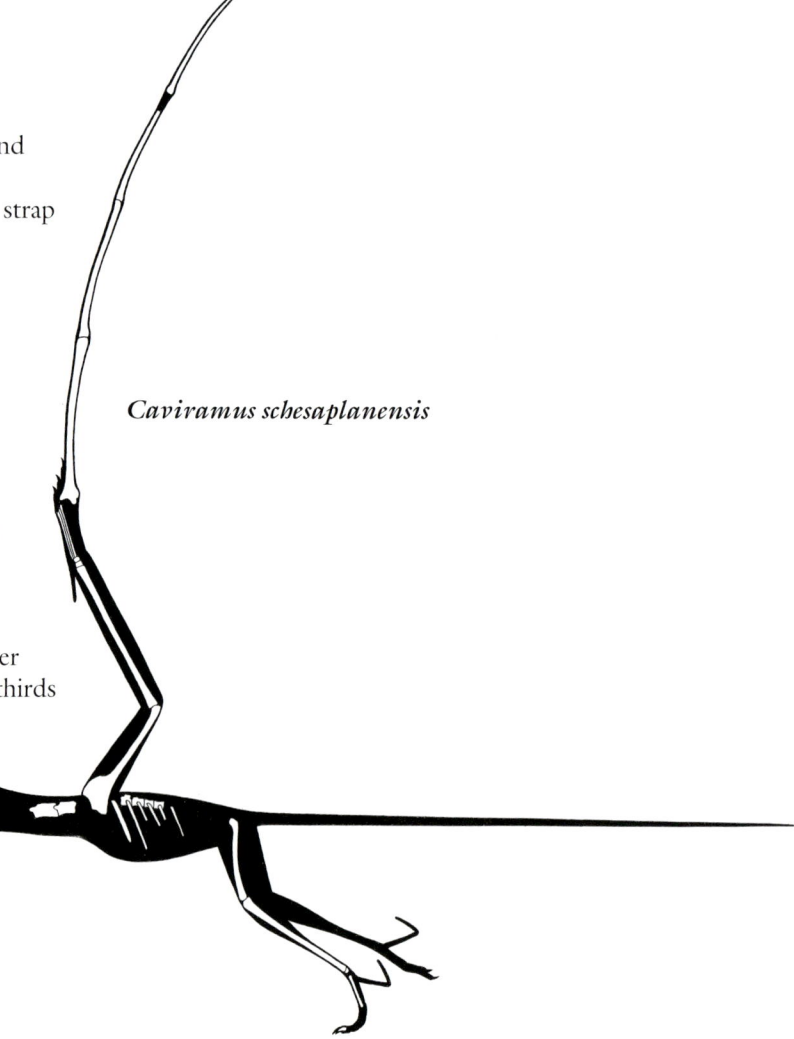

Caviramus schesaplanensis

MACRO-NYCHOPTERANS

SMALL TO GIGANTIC RHAMPHORHYNCHOIDS AND PTERODACTYLOIDS OF THE EARLY JURASSIC TO THE END OF THE MESOZOIC ON ALL CONTINENTS AND OCEANS

ANATOMICAL CHARACTERISTICS Dentaries at least three-quarters of lower jaw length, teeth when present single cusped. Rhamphorhynchoid tail stiffened by long, ossified tendons. When present, fingers larger than toes.

DIMORPHODONTIDS

SMALL AND/OR MEDIUM-SIZED MACRO-NYCHOPTERAN RHAMPHORHYNCHOIDS OF THE LATE TRIASSIC TO EARLY JURASSIC OF EUROPE AND NORTH AMERICA

ANATOMICAL CHARACTERISTICS Uniform. Head large, deep, subtriangular, lightly built, nasal opening very large, upper temporal opening faces more vertically than in other pterosaurs, lower jaw shallow, teeth along first two-thirds or more of jaw, numerous, very variable in size including large fangs, subvertical, throat pouch long. Wingspan rather short compared to mass, not powerfully developed, pteroid fairly long. Legs moderately long.
HABITS Flight performance relatively limited. Small-game predators.
NOTES Absence from at least some other continents probably reflects lack of sufficient sampling.

Dimorphodon shaded skull

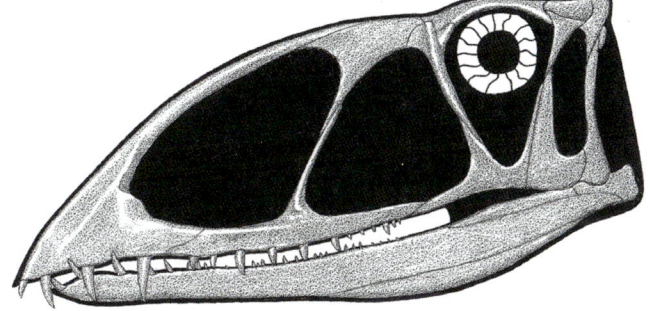

Peteinosaurus zambelli
Adult size uncertain

FOSSIL REMAINS Two or three partial skeletons, all immature.
ANATOMICAL CHARACTERISTICS As for group.
AGE Late Triassic, late Norian.
DISTRIBUTION AND FORMATIONS Italy; Zorzino.
HABITAT Coastal.
NOTES Shared its habitat with *Eudimorphodon wildi*.

Dimorphodon macronyx

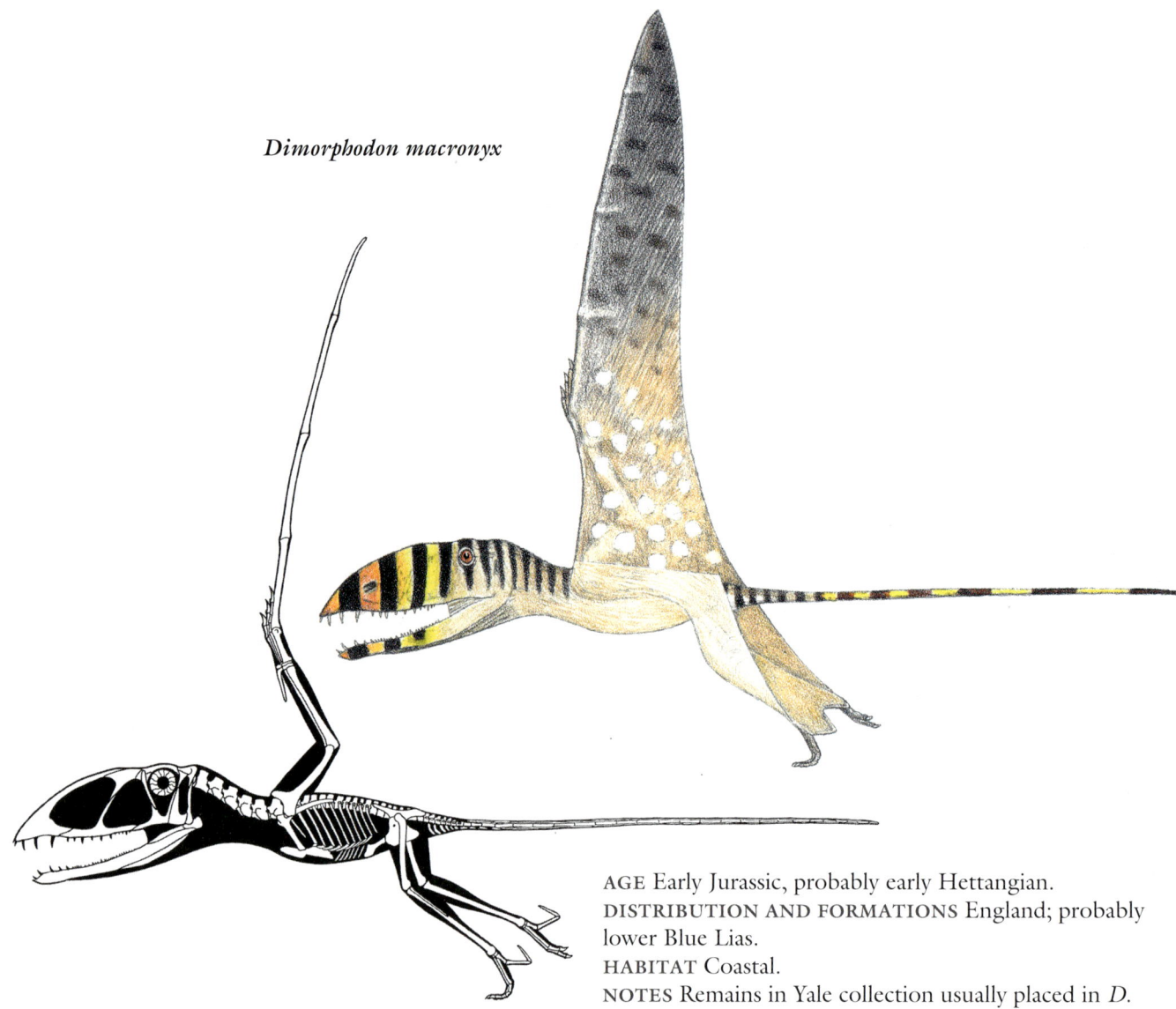

Dimorphodon macronyx

Caelestiventus hanseni (Britt et al. 2018)
1.5 m (5 ft) WS, 1 kg (2 lb)

FOSSIL REMAINS Partial skull and minority of skeleton.
ANATOMICAL CHARACTERISTICS As for group.
AGE Late Triassic, Norian or Rhaetian.
DISTRIBUTION AND FORMATIONS Utah; Nugget Sandstone.
HABITAT Desert lake.
NOTES One of few pterosaurs known from desert deposits, and earliest such, although preservation in lake deposits indicates freshwater habits. Largest known Triassic pterosaur. Bones preserved uncrushed.

Dimorphodon? unnamed species
1.2 m (4 ft) WS, 0.9 m (3 ft) TL, 0.5 kg (1 lb)

FOSSIL REMAINS Partial skull and skeletal remains.
ANATOMICAL CHARACTERISTICS Insufficient information.

AGE Early Jurassic, probably early Hettangian.
DISTRIBUTION AND FORMATIONS England; probably lower Blue Lias.
HABITAT Coastal.
NOTES Remains in Yale collection usually placed in *D. macronyx* appear to be from lower in the Blue Lias, in which case are very probably a different species and perhaps genus.

Dimorphodon macronyx
1.45 m (4.8 ft) WS, 1.1 m (3.5 ft) TL, 1 kg (2 lb)

FOSSIL REMAINS A few skulls and skeletons of varying completeness.
ANATOMICAL CHARACTERISTICS Head very large, arched. Wings rather short.
AGE Early Jurassic, early Sinemurian.
DISTRIBUTION AND FORMATIONS England; upper Blue Lias.
HABITAT Coastal.
NOTES One of the iconic pterosaurs. The probable stratigraphic level and age of the British Museum remains based on sediment type have not previously appeared in a widely distributed publication. This species has not been redescribed in recent decades.

CAMPYLOGNATHOIDIDS

MEDIUM-SIZED MACRONYCHOPTERAN
RHAMPHORHYNCHOIDS OF THE EARLY
JURASSIC OF EUROPE

ANATOMICAL CHARACTERISTICS Head fairly long,
subtriangular, snout pinched in side view, modest beak
before modest-sized nasal opening, larger beak on
lower jaw, teeth along most of jaw length, fairly robust,
subvertical, very variable in size, throat pouch long. Neck
short. Tail moderately long. Wings very long because
wing fingers extremely long, inner wings quite short
but powerfully developed, free fingers small. Leg length
moderate, outer toes small.
HABITS Flight performance very well developed,
probably included soaring. Fishers and/or small game
predators.
NOTES Absence from at least some other continents
probably reflects lack of sufficient sampling.

Campylognathoides zitteli (Padian 2008)
1.6 m (5.5 ft) WS, 0.95 m (3.1 ft) TL, 0.5 kg (1 lb)

FOSSIL REMAINS About ten skulls and skeletons of
varying completeness, adult and large juvenile.
ANATOMICAL CHARACTERISTICS As for group,
wingspan longest relative to mass among known
rhamphorhynchoids (same as *Caviramus*).
AGE Early Jurassic, early Toarcian.
DISTRIBUTION AND FORMATIONS Germany;
Württemberg Lias.
HABITAT Coastal.
GROWTH AND HABITS Absence of fossils of
very small juveniles suggests they did not fly
immediately after hatching.
NOTES *C. liasicus* is probably the juvenile
of this species. Shared its habitat with more
common *Dorygnathus*.

Campylognathoides zitteli
(juvenile, middle)

RHAMPHORHYNCHIDS

SMALL TO MEDIUM-SIZED MACRO-
NYCHOPTERAN RHAMPHORHYNCHOIDS
OF THE EARLY TO LATE JURASSIC OF THE
NORTHERN AND AT LEAST PART OF THE
SOUTHERN HEMISPHERES

ANATOMICAL CHARACTERISTICS Head fairly long,
subtriangular, modest beak before nasal opening, teeth
fairly robust, variable in length within an individual,
throat pouch fairly long. Wings fairly well developed, span
variable relative to mass.
HABITS Flight performance fairly to well developed,
probably included soaring in some cases. Fishers and/or
small-game predators.

Dorygnathus banthensis (Padian 2008)
1.6 m (5 ft) WS, 1 m (3.3 ft) TL, 1.5 kg (3 lb)

FOSSIL REMAINS Two and a half dozen specimens
of varying completeness, adult and large
juvenile.
ANATOMICAL CHARACTERISTICS
Snout shallow, nasal opening
fairly large, front end of lower
jaw upcurved, teeth along almost
entire length of jaw, fangs large.
Neck length moderate, robust.
Tail moderately long. Inner wings
long but powerfully developed,
wingspan low relative to mass, free
fingers fairly large. Legs moderately long, outer toes well
developed.
AGE Early Jurassic, early Toarcian.
DISTRIBUTION AND FORMATIONS Germany;
Württemberg Lias.
HABITAT Coastal.
HABITS Tooth microwear patterns indicate small game
predation.
GROWTH AND HABITS Absence of fossils of very small
juveniles suggests they did not fly immediately after
hatching.
NOTES Shared its habitat with *Campylognathoides*.

Parapsicephalus purdoni
(O'Sullivan and Martill 2017)
Adult size uncertain

FOSSIL REMAINS Partial skull, possibly some wing
elements.
ANATOMICAL CHARACTERISTICS Head fairly broad,
fairly strongly built, teeth along first half of jaw.
AGE Early Jurassic, Toarcian.

Dorygnathus banthensis **muscle study**

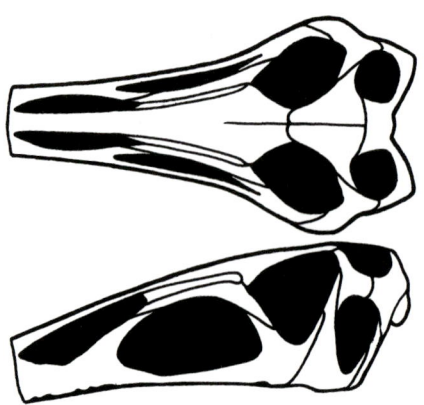

DISTRIBUTION AND FORMATIONS England; Whitby
Mudstone.
HABITAT Coastal marine.
NOTES May be same genus as *Dorygnathus*. Wing
elements may represent individuals larger than the skull.

Parapsicephalus purdoni

Dorygnathus banthensis

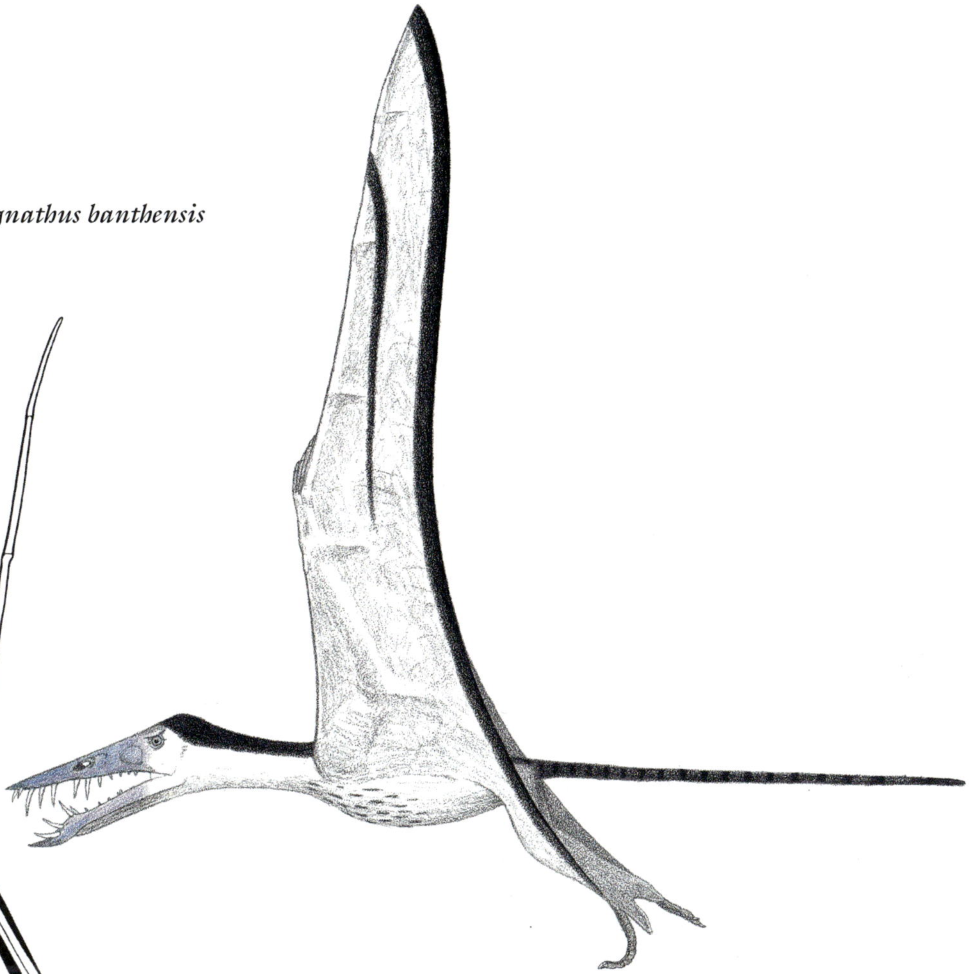

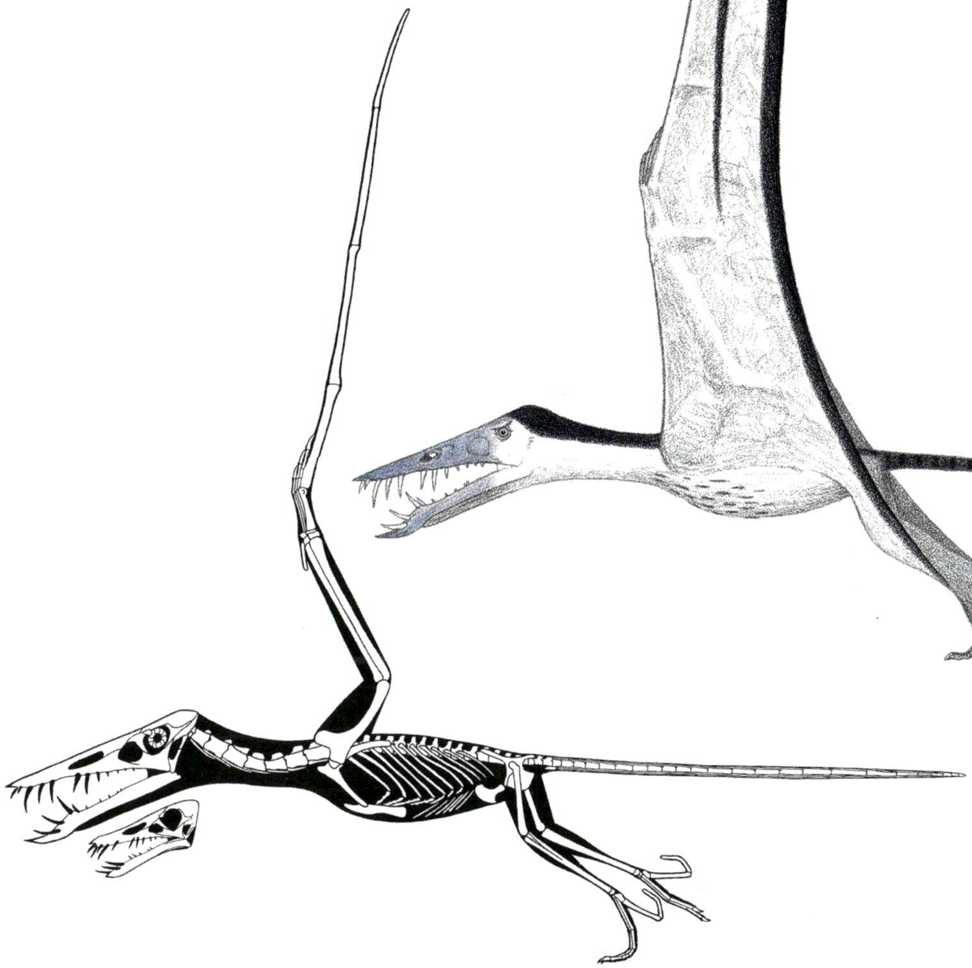

Angustinaripterus longicephalus
1.5 m (5 ft) WS, 1 kg (2 lb)

FOSSIL REMAINS Nearly complete skull.
ANATOMICAL CHARACTERISTICS Head shallow, snout
long, nasal opening long but shallow, teeth along first
two-thirds of jaw, most project forward.
AGE Middle Jurassic, Bathonian.
DISTRIBUTION AND FORMATIONS Central China;
Shaximiao.
HABITAT Heavily forested, lakes.

Angustinaripterus longicephalus

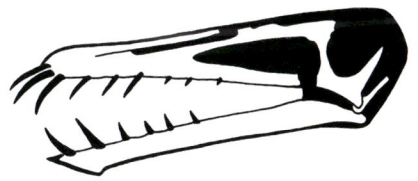

Dorygnathus banthensis and *Campylognathoides zitteli*

Harpactognathus gentryii
(Carpenter et al. 2003)
2.5 m (8 ft) WS, 4 kg (9 lb)

FOSSIL REMAINS Minority of skull.
ANATOMICAL CHARACTERISTICS Snout robust, bony head crest runs to tip of beak, front teeth large, stout, and procumbent.
AGE Late Jurassic, early Tithonian.
DISTRIBUTION AND FORMATIONS Wyoming; middle Morrison.
HABITAT Seasonally dry open woodlands.
NOTES One of the largest rhamphorhynchoids.

Cacibupteryx caribensis
(Gasparini et al. 2004)
2 m (6.5 ft) WS, 2 kg (4.5 lb)

FOSSIL REMAINS Majority of skull.
ANATOMICAL CHARACTERISTICS Head fairly broad, back of skull subvertical, teeth along first two-thirds of jaws.

AGE Late Jurassic, middle or late Oxfordian.
DISTRIBUTION AND FORMATIONS Cuba; Jagua.
HABITAT Coastal marine.
NOTES One of the largest rhamphorhynchoids. Shared its habitat with *Nesodactylus*.

Cacibupteryx caribensis

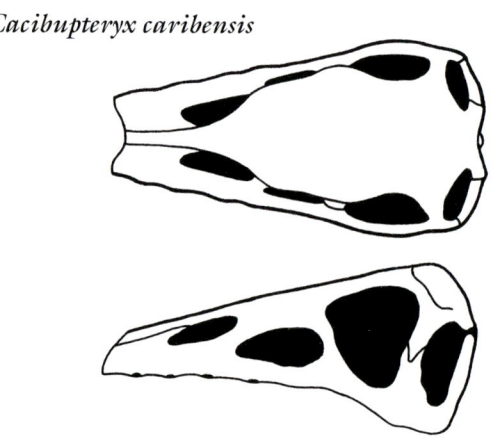

Sordes pilosus

Nesodactylus hesperius
1 m (3.3 ft) WS, 0.2 kg (0.4 lb)

FOSSIL REMAINS Majority of skeleton.
ANATOMICAL CHARACTERISTICS Neck length
moderate. Inner wings long, well developed, free fingers
fairly large. Leg length moderate.
AGE Late Jurassic, middle or late Oxfordian.
DISTRIBUTION AND FORMATIONS Cuba; Jagua.
HABITAT Coastal marine.

Sordes pilosus
**0.7 m (2.2 ft) WS, 0.475 m (1.6 ft) TL, 0.09 kg
(0.2 lb)**

FOSSIL REMAINS A number of skulls and skeletons of
varying completeness, adult and juvenile, extensive soft
tissues including flight membranes and pycnofibers.
ANATOMICAL CHARACTERISTICS Snout shallow,
nasal opening small, teeth few in number and along first
half of jaws, modest sized, orientation variable, throat
pouch short. Neck length moderate, rather slender.

Tail moderately long. Inner wings long, but overall
wingspan rather short relative to mass, free fingers fairly
large. Leg length moderate, outer toes well developed.
Brachiopatagium appears to attach to ankles, uropatagium
sheet between legs not attached to tail, trailing edge has
shallow concave V shape, last quarter of tail adorned with
series of shallow, subrectangular serrated vanes.
AGE Late Jurassic, Oxfordian or Kimmeridgian.
DISTRIBUTION AND FORMATIONS Kazakhstan;
Karabastau Svita.
HABITAT Found in lake deposits.
HABITS Large free fingers and moderate leg size indicate
arboreality.
NOTES First known pterosaur fossils with well-preserved
pycnofibers. Attachment characteristics of inner
brachiopatagium and uropatagium may be widespread
among rhamphorhynchoids. Shared its habitat with
Batrachognathus.

Bellubrunnus rothgaengeri (Hone et al. 2012)
Adult size uncertain

FOSSIL REMAINS Complete juvenile skull and skeleton.
ANATOMICAL CHARACTERISTICS Wing-tip bone curved forward.
AGE Late Jurassic, late Kimmeridgian.
DISTRIBUTION AND FORMATIONS Southern Germany; unnamed.
HABITAT Lagoonal deposits near probably arid, brush-covered islands.
NOTES Low number of teeth and lack of ossified tail tendons are probably juvenile features. Both fingertips share identical forward curve, only known pterosaur with this forward sweep, whether purpose aerodynamic or otherwise not certain.

Bellubrunnus rothgaengeri

Fenghuangopterus lii (Lü et al. 2010)
0.9 m (3 ft) WS, 0.2 kg (0.4 lb)

FOSSIL REMAINS Majority of skull and skeleton.
ANATOMICAL CHARACTERISTICS Head moderate in size, not deep, teeth along first three-quarters of jaws, most teeth large, orientation variable, throat pouch fairly long. Neck length moderate, fairly robust. Inner wings long but powerfully developed, wingspan rather long relative to mass, free fingers fairly small. Legs long and strong, outer toes medium sized.
AGE Late Jurassic, early Oxfordian.
DISTRIBUTION AND FORMATIONS Northeastern China; middle Tiaojishan.
HABITAT Well-watered forests and lakes.
HABITS Well-developed legs and small free fingers suggest considerable time spent on ground.
NOTES *Qinglongopterus guoi* and/or *Jiangchangopterus zhaoianus* may be juveniles of this species. Shared its habitat with *Changchengopterus, Pterorhynchus, Wukongopterus, Douzhanopterus, Dendrorhynchoides, Liaodactylus, Jiangchangnathus.*

Jiangchangnathus robustus

Jiangchangnathus robustus (Cheng et al. 2012)
Adult size uncertain

FOSSIL REMAINS Majority of skull and skeleton, immature.
ANATOMICAL CHARACTERISTICS Head rather large, fairly deep, strongly built, nasal opening small, teeth along first two-thirds of jaws, most teeth large, upper teeth project forward, throat pouch long. Neck length moderate, fairly robust. Inner wings long but powerfully developed and overall wing length moderate, free fingers well developed.

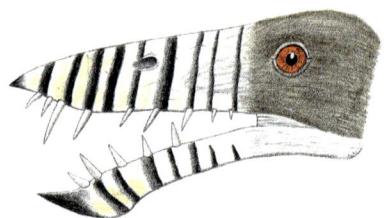

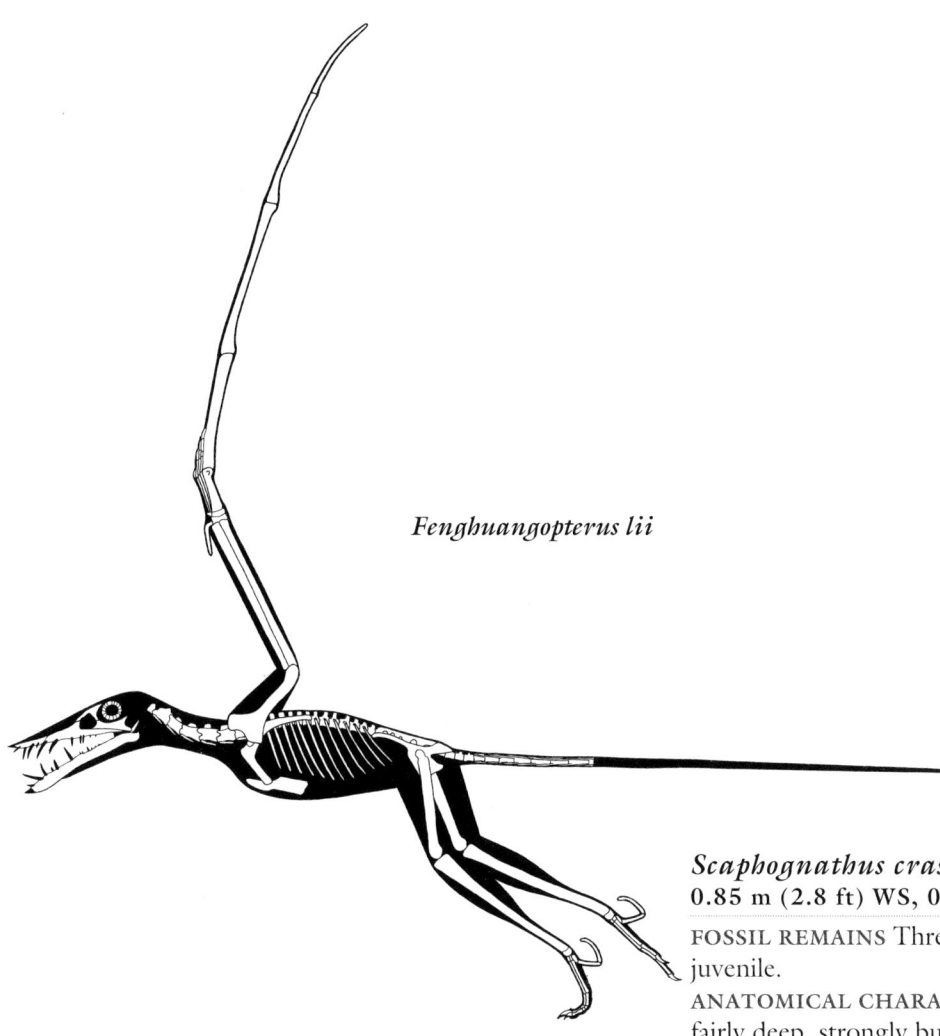

Fenghuangopterus lii

AGE Late Jurassic, early Oxfordian.
DISTRIBUTION AND FORMATIONS Northeastern China; middle Tiaojishan.
HABITAT Well-watered forests and lakes.
HABITS Powerfully built wings and stout skull indicate terrestrial flight in pursuit of land prey, but tooth microwear patterns indicate predominantly a fisher.
NOTES Shared its habitat with *Fenghuangopterus*, *Changchengopterus*, *Pterorhynchus*, *Wukongopterus*, *Douzhanopterus*, *Dendrorhynchoides*, *Liaodactylus*.

Orientognathus chaoyangensis (Lü et al. 2015)
1 m (3 ft) WS, 0.3 kg (6 lb)

FOSSIL REMAINS Partial skull and skeleton.
ANATOMICAL CHARACTERISTICS Robustly constructed.
AGE Late Jurassic.
DISTRIBUTION AND FORMATIONS Northeastern China; Tuchengzi.
HABITAT Well-watered forests and lakes.

Scaphognathus crassirostris
0.85 m (2.8 ft) WS, 0.2 kg (0.4 lb)

FOSSIL REMAINS Three skulls and skeletons, adult and juvenile.
ANATOMICAL CHARACTERISTICS Head rather large, fairly deep, strongly built, nasal opening large, bony orbital bar may have resulted in eagle eye, teeth along first three-quarters of jaws, teeth often large, subvertical, an upper tooth on side of snout migrated down to rest of tooth row with growth, throat pouch fairly long. Neck rather short, robust. Inner wings long and well developed, wingspan moderate relative to mass, free fingers well developed. Leg length moderate.
AGE Late Jurassic, early Tithonian.
DISTRIBUTION AND FORMATIONS Southern Germany; Altmühltal (middle Solnhofen).
HABITAT Lagoonal deposits near probably arid, brush-covered islands.
HABITS Scarcity in lagoonal deposits, powerfully built wings of moderate length, stout skull, and tooth microwear patterns indicate terrestrial flight in pursuit of land invertebrates and small game. Configuration of bony eye rings may indicate diurnal activity.
NOTES Shared its habitat with much more common *Rhamphorhynchus*, as well as *Anurognathus*, *Germanodactylus cristatus*, *Pterodactylus*, *Gnathosaurus*, *Ctenochasma elegans*.

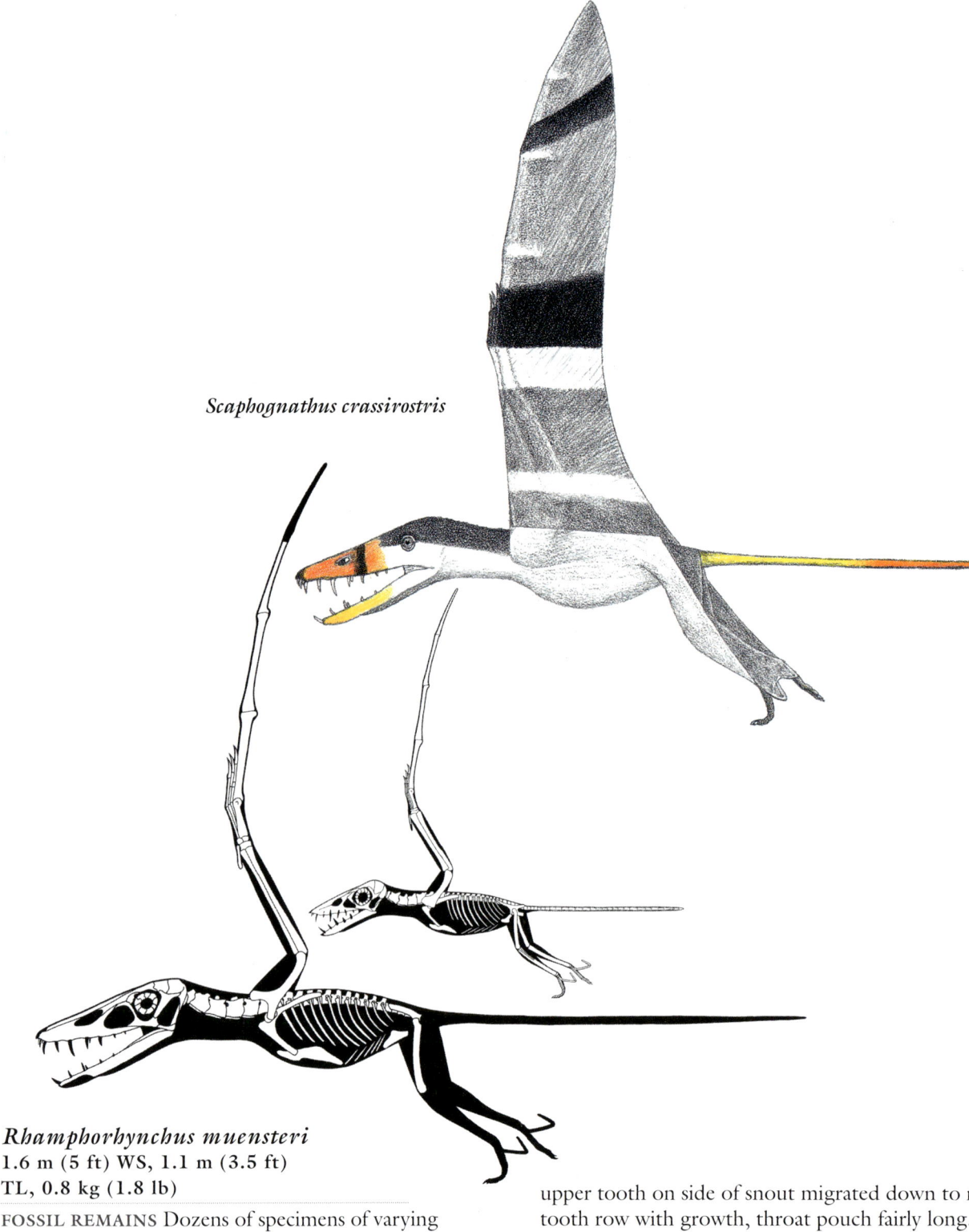

Scaphognathus crassirostris

Rhamphorhynchus muensteri
1.6 m (5 ft) WS, 1.1 m (3.5 ft)
TL, 0.8 kg (1.8 lb)

FOSSIL REMAINS Dozens of specimens of varying completeness, adult and juvenile, extensive soft tissue including wing membranes and beak tip sheaths, a coprolite.
ANATOMICAL CHARACTERISTICS Head rather large, narrow, snout shallow, nasal opening small, front end of lower jaw upcurved, teeth along almost entire length of jaws, most teeth large fangs, project strongly forward, an

upper tooth on side of snout migrated down to rest of tooth row with growth, throat pouch fairly long. Neck length moderate, rather slender. Tail long. Inner wings rather short but fairly powerfully developed, wingspan very long relative to mass, free fingers rather short. Legs short, fifth toes small. Tail tip bears small fan that alters from shallow diamond in small juveniles to deeper diamond in medium-sized juveniles to a triangle in adults.

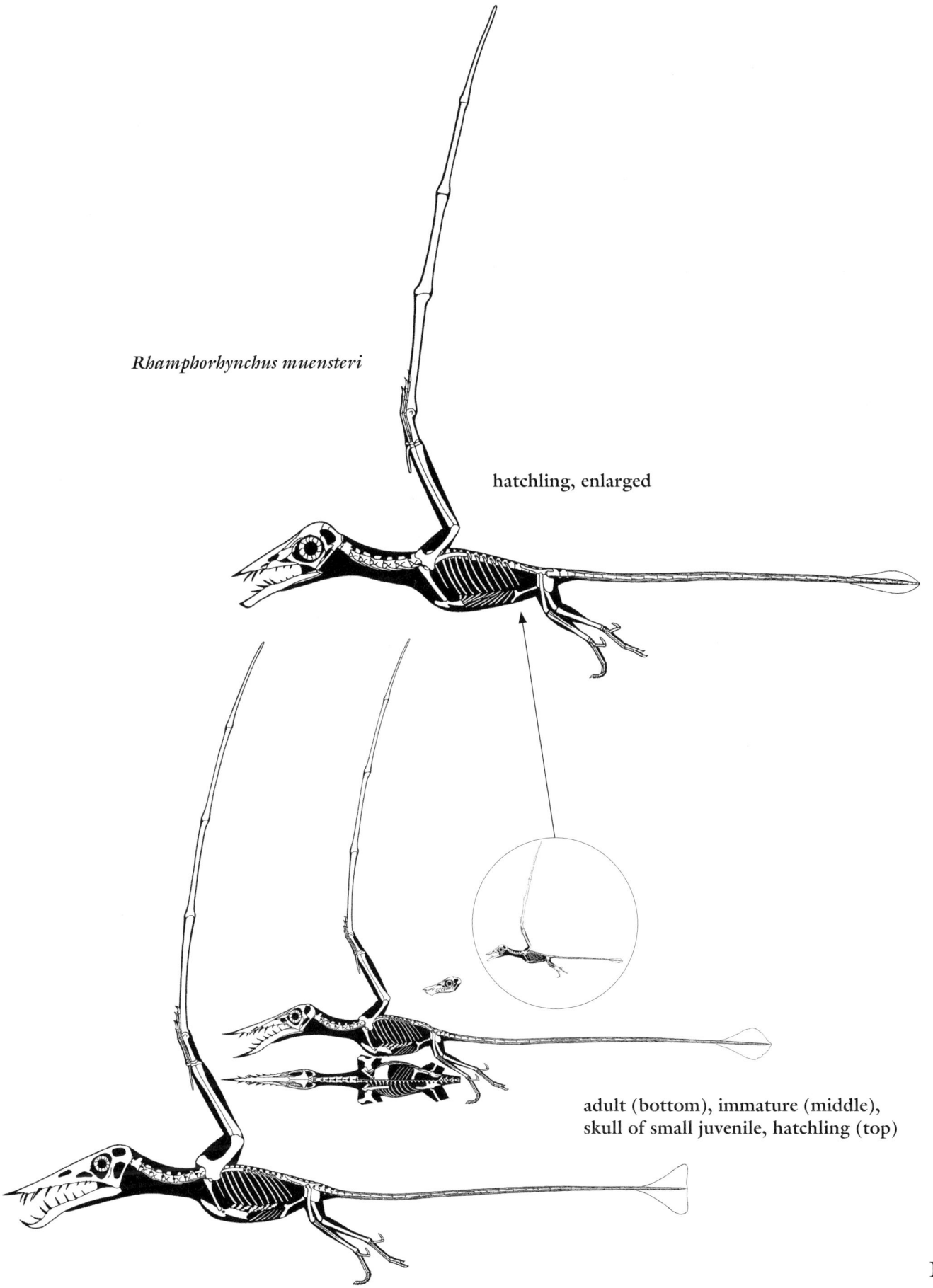

Rhamphorhynchus muensteri

hatchling, enlarged

adult (bottom), immature (middle),
skull of small juvenile, hatchling (top)

113

Rhamphorhynchus muensteri

MONOFENESTRATANS

SMALL TO GIGANTIC DERIVED MACRO-
NYCHOPTERAN RHAMPHORHYNCHOIDS AND
PTERODACTYLOIDS OF THE MIDDLE JURASSIC
TO THE END OF THE MESOZOIC ON ALL
CONTINENTS AND OCEANS

ANATOMICAL CHARACTERISTICS Nasal and preorbital
openings usually or always at least partly confluent. Neck
vertebrae at least somewhat long, neck ribs reduced or
absent.

AGE Late Jurassic, early Tithonian.
DISTRIBUTION AND FORMATIONS Southern Germany;
Altmühltal (middle Solnhofen).
HABITAT Lagoonal deposits near probably arid, brush-
covered islands.
GROWTH AND HABITS Presence of fossil hatchlings far
away from nests indicates they were active independent
fliers upon hatching, juveniles most abundant. Small free
fingers and legs suggest little time spent on ground or
climbing, highly aerial, probably over water. Configuration
of bony eye rings may indicate nocturnal activity.
NOTES One of the iconic pterosaurs and the classic
rhamphorhynchoid. A number of species probably
represent growth stages of *R. muensteri*. This species has
not been redescribed in recent years.

Sericipterus wucaiwanensis
(Andres et al. 2010)
1.7 m (5.5 ft) WS, 1 kg (2 lb)

FOSSIL REMAINS Partial skull and skeleton.
ANATOMICAL CHARACTERISTICS Shallow bony head
crest, teeth long, slender, project strongly forward. Neck
length moderate.
AGE Late Jurassic, early Oxfordian.
DISTRIBUTION AND FORMATIONS Northwestern
China; upper Shishugou.

MONOFENESTRATAN
RHAMPHORHYNCHOID
MISCELLANEA

Changchengopterus pani (Lü 2009)
0.7 m (2 ft) WS

FOSSIL REMAINS Minority of two skulls and majority of
two skeletons, adult and juvenile.
ANATOMICAL CHARACTERISTICS Insufficient
information.
AGE Late Jurassic, early Oxfordian.
DISTRIBUTION AND FORMATIONS Northeastern China;
middle Tiaojishan.
HABITAT Well-watered forests and lakes.
NOTES Whether the two skeletons are the same
species is uncertain, as are their relationships to other
monofenestratan rhamphorhynchoids. Shared its habitat
with *Fenghuangopterus, Jiangchangnathus, Pterorhynchus,
Wukongopterus, Douzhanopterus, Dendrorhynchoides,
Liaodactylus*.

DARWINOPTERANS

SMALL TO MEDIUM-SIZED MONO-FENESTRATAN RHAMPHORHYNCHOIDS OF THE LATE JURASSIC OF EURASIA

ANATOMICAL CHARACTERISTICS Head long and shallow, teeth few in number, limited to about first half or less of jaws, teeth widely spaced, medium sized, fairly stout. Neck length moderate. Inner wing long but overall wing length moderate, not powerfully developed, free fingers fairly well developed. Leg length moderate, fairly equal to inner wing length.
HABITS Flight performance mediocre. Probably predominantly terrestrial. Small-game predators, fishers.
NOTES Absence from at least some other continents probably reflects lack of sufficient sampling.

PTERORHYNCHIANS

SMALL DARWINOPTERANS OF THE LATE JURASSIC OF ASIA

ANATOMICAL CHARACTERISTICS Tail long.
NOTES Absence from at least some other continents probably reflects lack of sufficient sampling.

Pterorhynchus wellnhoferi (Czerkas and Ji 2002)
0.83 m (2.7 ft) WS, 0.7 m (2.3 ft) TL, 0.9 kg (2 lb)
FOSSIL REMAINS Majority of skull and skeleton including nonbony crest, wing membranes, pycnofibers.
ANATOMICAL CHARACTERISTICS Head large, long, shallow bony crest anchored large half-moon-shaped soft tissue crest over three-quarters head length from nasal opening to back of head, teeth widely spaced, moderate in size, subvertical, throat pouch short. Neck moderately long, fairly robust. Tail very long. Wingspan very long relative to mass. Legs rather short. Head crest has subvertical arced color banding, pycnofibers appear to cover much of top of wing membranes, over half of tail adorned with series of shallow, subrectangular serrated vanes.
AGE Late Jurassic, early Oxfordian.
DISTRIBUTION AND FORMATIONS Northeastern China; middle Tiaojishan.
HABITAT Well-watered forests and lakes.
NOTES Longest proportional tail among known pterosaurs. Shared its habitat with *Fenghuangopterus, Jiangchangnathus, Changchengopterus, Wukongopterus, Douzhanopterus, Dendrorhynchoides, Liaodactylus.*

Pterorhynchus wellnhoferi

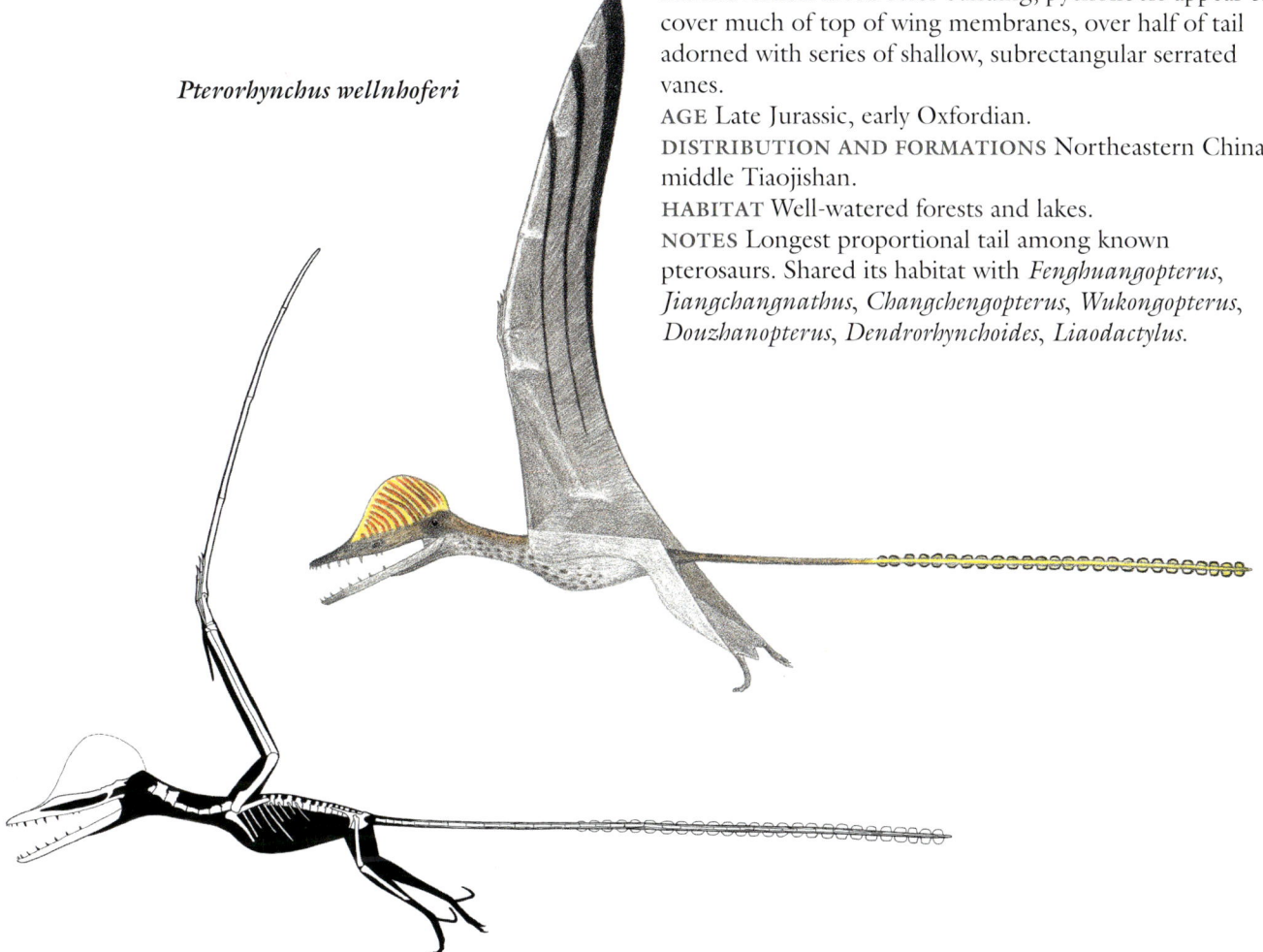

WUKONGOPTERIDS

SMALL TO MEDIUM-SIZED DARWINOPTERANS
OF THE LATE JURASSIC OF EURASIA

ANATOMICAL CHARACTERISTICS Head very long,
shallow, teeth only along first third of jaws, throat pouch
long. Tail shorter than typical in rhamphorhynchoids. Size
of fifth toes moderate.
GROWTH AND HABITS Absence of fossils of very small
juveniles suggests they did not fly immediately after
hatching.

Wukongopterus

NOTES Absence from at least some other continents
probably reflects lack of sufficient sampling. Reduced tails
indicate these are close to short-tailed pterodactyloids,
which then displaced their wukongopterid relatives before
end of Jurassic. Immature *Changchengopterus* may be a
juvenile of one of Tiaojishan taxa.

Wukongopterus lii (Wang et al. 2009)
0.85 m (2.8 ft) WS, 0.6 m (2 ft) TL, 0.2 kg (0.45 lb)

FOSSIL REMAINS A number of skulls and skeletons of
varying completeness, some including extensive soft
tissues including nonbony crests, wing membranes,
pycnofibers, eggs.

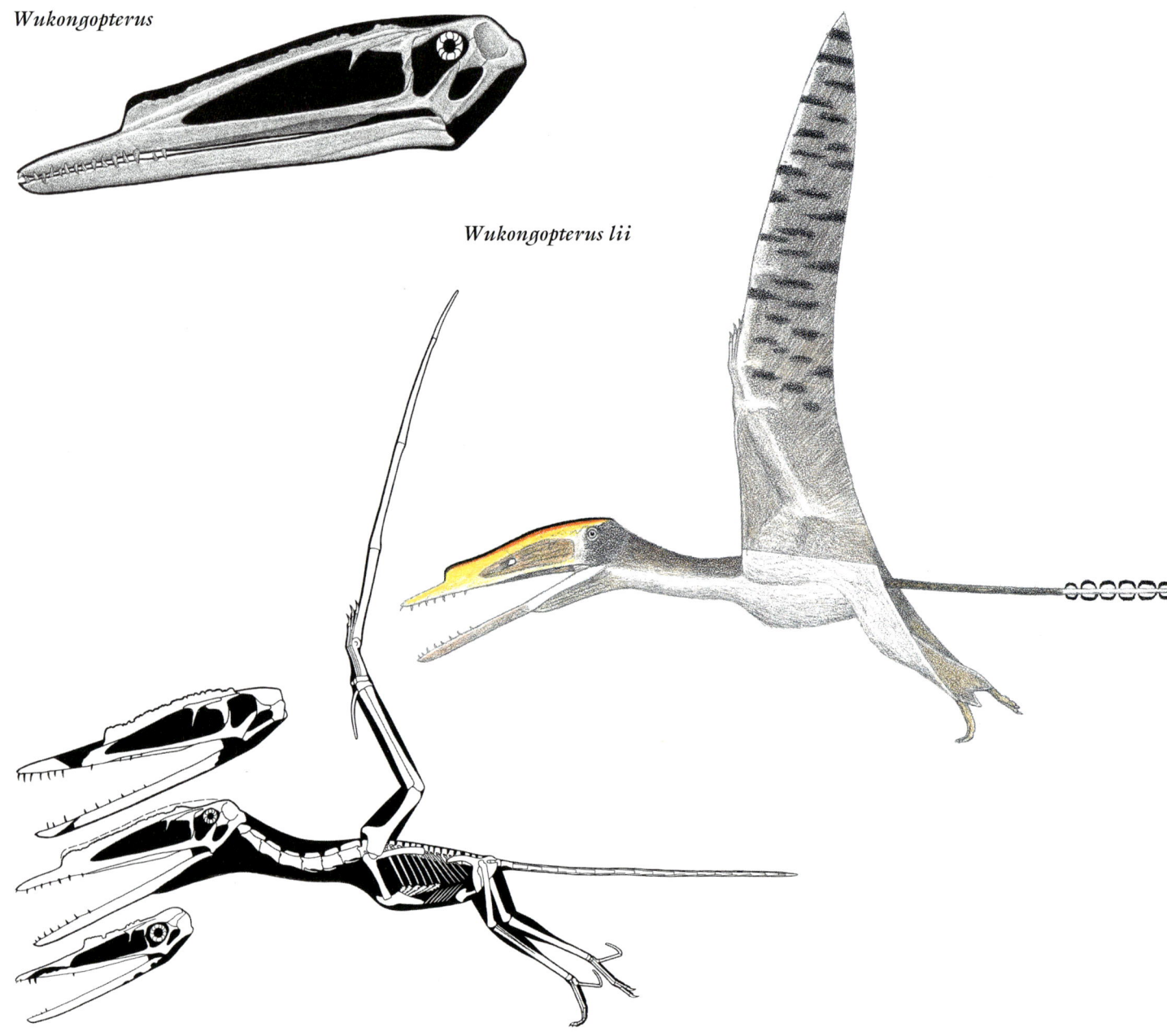

Wukongopterus lii

ANATOMICAL CHARACTERISTICS Head exceptionally large, preorbital opening very long, a long, shallow, irregular bony head crest atop most of head anchored soft tissue crest of uncertain dimensions in at least males and/or adults, teeth robust in at least some adults. Wingspan moderate relative to mass.

AGE Late Jurassic, early Oxfordian.

DISTRIBUTION AND FORMATIONS Northeastern China; middle Tiaojishan.

HABITAT Well-watered forests and lakes.

HABITS Fishers according to tooth microwear patterns.

NOTES Original incomplete immature specimen is not sufficiently distinguishable from more completely known *Darwinopterus*. *D. modularis*, *D. linglongtaensis*, *D. robustodens*, *Kunpengopterus sinensis*, *K. antipollicatus*, *Archaeoistiodactylus linglongtaensis* are based largely on differences in tooth robustness that are minor or difficult to confirm because of damage to the fossils and may represent age and sexual differences. Presence of opposable thumb in *K. antipollicatus* uncertain. Possible regurgitated fish remains suggest fishing.

Skeleton associated with an egg near its pelvis is the only pterosaur specimen that can be firmly determined to be female. Shared its habitat with *Fenghuangopterus*, *Jiangchangnathus*, *Changchengopterus*, *Pterorhynchus*, *Dendrorhynchoides*, *Douzhanopterus*, *Liaodactylus*.

Cuspicephalus scarfi (Martill and Etches 2013)
1.8 m (6 ft) WS, 2 kg (4.5 lb)

FOSSIL REMAINS Majority of skull.

ANATOMICAL CHARACTERISTICS Snout very shallow, with long, shallow, bony ridge atop middle of head anchoring shallow soft tissue crest.

AGE Late Jurassic, early Kimmeridgian.

DISTRIBUTION AND FORMATIONS Southern England; Kimmeridge Clay.

HABITAT Coastal marine.

Cuspicephalus scarfi

Wukongopterus lii

DOUZHAN-OPTERIANS

SMALL MONOFENESTRATAN RHAMPHORHYNCHOIDS OF THE MIDDLE TO LATE JURASSIC OF EURASIA

ANATOMICAL CHARACTERISTICS Neck fairly long. Inner wings long but overall wing length moderate, not powerfully developed, inner hand length intermediate to other rhamphorhynchoids versus pterodactyloids, pteroids long. Leg length moderate, outer toes small.

Douzhanopterus zhengi (Wang et al. 2017)
0.75 m (2.5 ft), 0.3 kg (0.6 lb)

FOSSIL REMAINS Complete skeleton lacking skull.
ANATOMICAL CHARACTERISTICS Tail length intermediate to wukongopterids versus anurognathids and pterodactyloids.
AGE Late Jurassic, early Oxfordian.
DISTRIBUTION AND FORMATIONS Northeastern China; middle Tiaojishan.
HABITAT Well-watered forests and lakes. Shared its habitat with *Fenghuangopterus*, *Changchengopterus*, *Pterorhynchus*, *Wukongopterus*, *Dendrorhynchoides*, *Jiangchangnathus*, *Liaodactylus*.

Unnamed genus and species
Adult size uncertain

FOSSIL REMAINS Complete skull and skeleton, juvenile.
ANATOMICAL CHARACTERISTICS Tail short.
AGE Late Jurassic, early Tithonian.
DISTRIBUTION AND FORMATIONS Southern Germany; Painten (middle Solnhofen).
HABITAT Lagoonal deposits near probably arid, brush-covered islands.
NOTES Relationships uncertain.

PTERODACTYLI-FORMES

DERIVED MONOFENESTRATAN RHAMPHORHYNCHOIDS AND ALL PTERODACTYLOIDS OF THE MIDDLE JURASSIC TO THE END OF THE MESOZOIC ON ALL CONTINENTS AND OCEANS

ANATOMICAL CHARACTERISTICS Tail shorter than in wukongopterids.

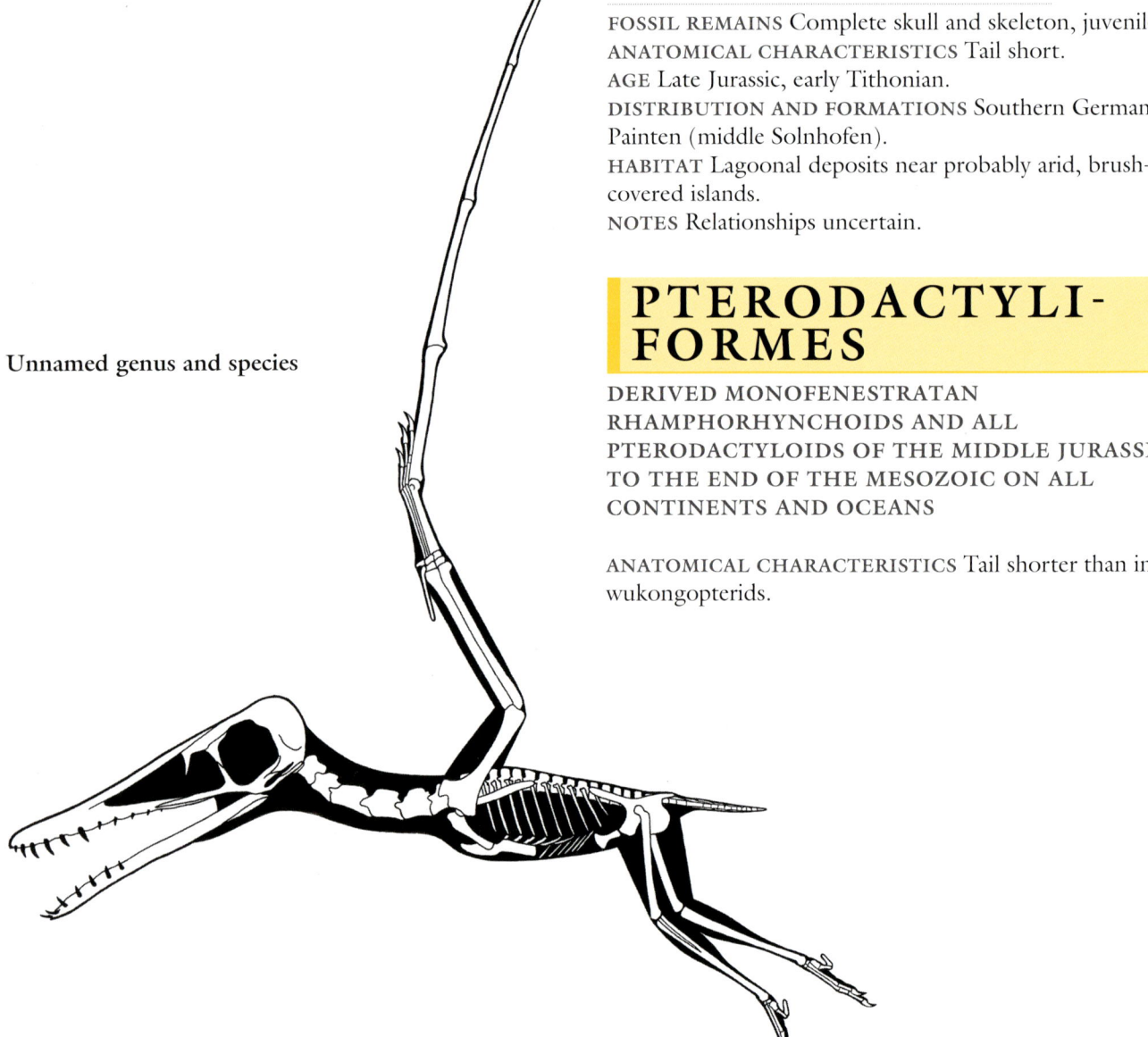

Unnamed genus and species

ANUROGNATHIDS

SMALL PTERODACTYLIFORMES OF
THE MIDDLE JURASSIC TO THE EARLY
CRETACEOUS OF THE NORTHERN
HEMISPHERE

ANATOMICAL CHARACTERISTICS Frog-like head short, fairly deep, very broad, broader than long, beakless, nasal and preorbital openings probably confluent, eyes very large and face partly forward and upward, back of head vertical rather than sloped down and forward, jaws can open extremely wide, teeth widely spaced, small pin tips, subvertical. Neck short, broad. Body broad. Wings strongly built, wingspan moderate relative to mass, pteroids short, free fingers long, fourth wing fingertip bone reduced or absent. Legs and outer toes well developed. Pycnofibers start at tip of snout.

GROWTH AND HABITS Absence of fossils of very small juveniles suggests they did not fly immediately after hatching. Powered flight performance highest among pterosaurs, including burst takeoff, rapid climb, aerial agility, ability to sustain vigorous powered flight. Some paleoresearchers suggest these outmaneuvered aerial insects with slow, hard-turning flight like bats. But narrow-chorded wings lacking long pteroids to increase leading edge camber during turns indicate anurognathids were high-speed fliers that, like narrow-winged swallows, swifts, and nightjars, snatched insects by approaching them so fast the insects did not have time to take evasive action. Like some birds, some anurognathids may have cruised low over ground or water to facilitate spotting higher-flying insects against the sky and attacking them from below, where insects would be harder pressed to see the approaching, possibly camouflaged pterosaur against the terrestrial background. Fuzzy trailing edges of wings in at least some examples may have made attacks acoustically stealthier while reducing wing-tip vortex-induced drag. Possible fossil evidence of attacks on both

diurnal and nocturnal insects leaves preferred hunting time uncertain. May have been capable on the ground, and may have been arboreal branch huggers when resting.

NOTES Anurognathids retain anatomical characteristics not found in other monofenestratans, which means anurognathids are instead more basal rhamphorhynchoids that paralleled pterodactyloids in the loss of a long tail, or that wukongopterid rhamphorhynchoids developed some monofenestratan characteristics in parallel with pterodactyloids—uncertain separation or joining of nasal and preorbital openings favors first and second possibilities, respectively. If anurognathids are not monofenestratans, they may be basal rhamphorhynchoids, or more probably derived rhamphorhynchoids close to monofenestratans. Broad skulls are always preserved severely flattened dorsoventrally, and only that of *Anurognathus* has been restored in detail, skulls are based in part on that. Lack of preserved facial bristles makes their presence unlikely. Fragmentary remains suggest presence in Middle Jurassic of Asia, fragmentary *Mesadactylus* may be a North American Late Jurassic anurognathid. Absence from Southern Hemisphere may reflect lack of sufficient sampling. Had shortest and broadest heads and necks among pterosaurs. Last of rhamphorhynchoids, and last known pterosaurs with a short pteroid, survival of group into the Late Cretaceous uncertain.

Anurognathus

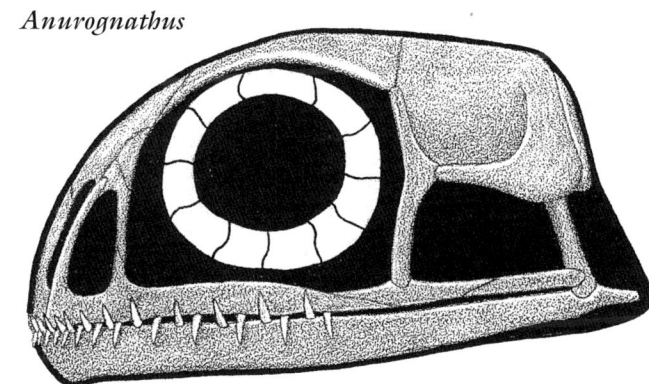

Anurognathus muscle study

*Dendrorhynchoides
curvidentatus*

Dendrorhynchoides curvidentatus (Hone 2020)
0.72 m (2.4 ft) WS, 0.18 m (0.6 ft) TL, 0.11 kg (0.25 lb)

FOSSIL REMAINS Four skulls and skeletons, adult and juvenile, soft tissues including membranes, foot webbing, pycnofibers.
ANATOMICAL CHARACTERISTICS Inner hand length moderate. Leg length moderate. Pycnofibers appear to have covered entire head beginning with snout, pycnofiber tuft only small percentage of outer membrane at wing tip, pigment capsules indicate brown with reddish tinge.
AGE Late Jurassic, early Oxfordian.
DISTRIBUTION AND FORMATIONS Northeastern China; middle Tiaojishan.
HABITAT Well-watered forests and lakes.
HABITS Wing-tip tufts may have made flapping quieter in order to better hide approach to flying insects.
NOTES Original specimen was first thought, apparently errantly, to be from Early Cretaceous deposits and to have a long tail. Small specimens placed in *D. (Luopterus) mutoudengensis, Jeholopterus ninchengensis*, and *Sinomacrops bondei* are probably growth stages of this species. Oher anurognathids may have had furry snout tips and tufts at wing tips. Shared its habitat with *Fenghuangopterus, Jiangchangnathus, Changchengopterus, Pterorhynchus, Wukongopterus, Douzhanopterus, Liaodactylus.*

Batrachognathus volans (Hone 2020)
0.5 m (2.5 ft) WS, 0.04 kg (0.09 lb)

FOSSIL REMAINS Majority of skull and skeleton.
ANATOMICAL CHARACTERISTICS Standard for group.
AGE Late Jurassic, Oxfordian or Kimmeridgian.
DISTRIBUTION AND FORMATIONS Kazakhstan; Karabastau Svita.
HABITAT Found in lake deposits.
NOTES Shared its habitat with *Sordes.*

Anurognathus ammoni (Bennett 2007b)
0.45 m (1.5 ft) WS, 0.12 m (0.4 ft) TL,
0.035 kg (0.08 lb)

FOSSIL REMAINS Two skulls and skeletons, adult and juvenile, soft tissues including wing membranes.
ANATOMICAL CHARACTERISTICS Inner hands very short, wing fingers reduced to three elements. Legs large.
AGE Late Jurassic, early Tithonian.
DISTRIBUTION AND FORMATIONS Southern Germany; Altmühltal (middle Solnhofen).
HABITAT Lagoonal deposits near probably arid, brush-covered islands.
HABITS Large legs suggest were more terrestrial than other anurognathids.
NOTES If largest specimen is mature, then is smallest known pterosaur, with mass equivalent to cardinal or vampire bat. Reduction of wing finger to three elements is an extreme streamlining feature otherwise found only in nyctosaurs. Shared its habitat with *Scaphognathus*, *Rhamphorhynchus*, *Germanodactylus cristatus*, *Pterodactylus*, *Gnathosaurus*, *Ctenochasma elegans*.

Vesperopterylus lamadongensis (Lü et al. 2017)
1 m (3.3 ft.) WS, 0.3 kg (0.6 lb)

FOSSIL REMAINS Nearly complete skull and skeleton.
ANATOMICAL CHARACTERISTICS Inner toe may be reversed.
AGE Early Cretaceous, early or middle Aptian.
DISTRIBUTION AND FORMATIONS Northeastern China; Jiufotang.
HABITAT Well-watered highland forests and lakes, winters chilly with some snow.
HABITS May have been highly arboreal.
NOTES Geologically last known rhamphorhynchoid and only species yet known from Cretaceous. Reversal of inner toe uncertain. Shared its habitat with *Forfexopterus*, *Hongshanopterus*, *Nurhachius*, *Istiodactylus sinensis*, *Sinopterus*, *Chaoyangopterus*, *Shenzhoupterus*, *Guidraco*, *Liaoningopterus*, *Ikrandraco*.

Anurognathus ammoni

Anurognathus ammoni

PTERODACTYLOIDS

SMALL TO GIGANTIC PTERODACTYLIFORM PTEROSAURS OF THE MIDDLE/LATE JURASSIC TO THE END OF THE MESOZOIC ON ALL CONTINENTS AND OCEANS

ANATOMICAL CHARACTERISTICS Fairly variable. Nasal and preorbital openings always confluent. Tail lacks long, ossified tendons. Inner hands at least fairly long, pteroids consistently long. Fifth toes very reduced. Less bipedal than rhamphorhynchoids.
HABITAT From strongly continental to oceanic.
HABITS Highly variable. Burst-powered fliers to extreme soarers. Predators, fishers, filter feeders, frugivores.

KRYPTODRAKONS

SMALL BASAL PTERODACTYLOIDS OF THE MIDDLE/LATE JURASSIC OF ASIA
ANATOMICAL CHARACTERISTICS Skeleton nonpneumatic.
HABITS Insufficient information.
NOTES Earliest and most basal known pterodactyloids. Absence from at least some other continents probably reflects lack of sufficient sampling.

Kryptodrakon progenitor (Andres et al. 2014)
1.5 m (4.8 ft) WS, 1.5 kg (3.5 lb)

FOSSIL REMAINS Minority of skeleton.
AGE Latest Middle or earliest Late Jurassic, late Callovian or early Oxfordian.
DISTRIBUTION AND FORMATIONS Northwestern China; lower Shishugou.

LOPHOCRATIANS

SMALL TO GIGANTIC PTERODACTYLOIDS OF THE LATE JURASSIC TO THE END OF THE MESOZOIC ON ALL CONTINENTS AND OCEANS

ANATOMICAL CHARACTERISTICS Skeleton strongly pneumatic. Neck length at least moderate.
HABITAT From strongly continental to coastal.

ARCHAEOPTERO-DACTYLOIDS

SMALL TO GIGANTIC LOPHOCRATIANS OF THE LATE JURASSIC TO THE END OF THE MESOZOIC ON MOST OR ALL CONTINENTS

ANATOMICAL CHARACTERISTICS Head never deep, teeth always present. Neck slender. Wings not long or powerfully developed, inner hands moderately long. Leg length moderate.

HABITAT Generally shorelines, marine and freshwater.
HABITS Flight performance, including takeoffs, modest to mediocre, some burst fliers. Generally consumed aquatic organisms.
NOTES Remains, including fragmentary specimens, indicate wide paleogeographic presence.

PTERODACTYLIDS

SMALL AND MEDIUM-SIZED ARCHAEOP-TERODACTYLOIDS OF THE LATE JURASSIC OF EUROPE

ANATOMICAL CHARACTERISTICS Inner brachiopatagium attached to thigh. Wingspan variable relative to mass.
HABITAT Shorelines and nearshore areas.
HABITS Small terrestrial and aquatic prey, some omnivory. Possibly prone toward daytime activity.
NOTES Absence from at least some other continents probably reflects lack of sufficient sampling, fragmentary remains may indicate presence in Africa and/or South America.

Pterodactylus antiquus
0.9 m (3 ft) WS, 0.475 m (1.5 ft) TL, 0.35 kg (0.8 lb)

FOSSIL REMAINS Two and a half dozen specimens of varying completeness, adult and juvenile, soft tissues including wing and leg membranes, throat pouches, foot webbing.
ANATOMICAL CHARACTERISTICS Head profile shallow and straight, head narrow, preorbital opening size modest, very shallow bony crest atop much of head probably anchored soft tissue crest, teeth nearly to tips of jaws, teeth rather small, project a little forward, preserved throat pouch short. Neck moderately long. Wingspan short relative to mass.
AGE Late Jurassic, early Tithonian.
DISTRIBUTION AND FORMATIONS Southern Germany; Altmühltal (middle Solnhofen).

Pterodactylus muscle study

Pterodactylus antiquus, from bottom:
adult, immature, small juvenile,
hatchling, and adult skull (see
opposite for enlarged small juvenile
and hatchling)

HABITAT Lagoonal deposits near probably arid, brush-covered islands.

HABITS Picked up and grubbed for small burrowing organisms on and in mud and sand flats, other invertebrates, small fish, and small game.

GROWTH AND HABITS Presence of fossil hatchlings far away from nests indicates they were active independent fliers upon hatching, juveniles most abundant to degree that fully mature adults are very rare. Configuration of bony eye rings may indicate diurnal activity.

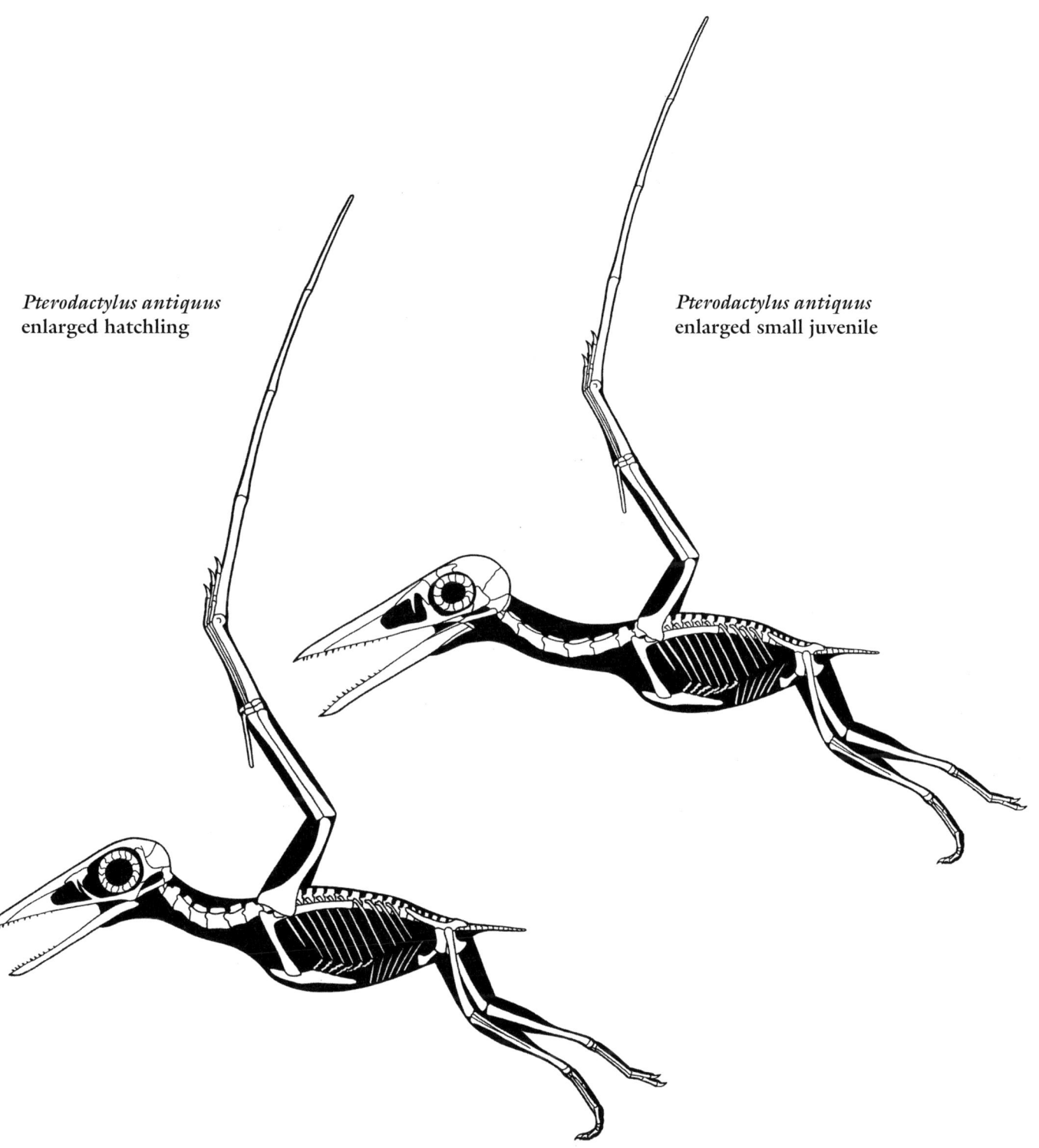

Pterodactylus antiquus
enlarged hatchling

Pterodactylus antiquus
enlarged small juvenile

Pterodactylus antiquus

NOTES One of the iconic pterosaurs. A large number of species probably represent growth stages of one species, only one partial skull represents a fully grown adult. This species has not been redescribed in recent years. Shared its habitat with *Scaphognathus*, *Rhamphorhynchus*, *Anurognathus*, *Gnathosaurus*, *Ctenochasma elegans*, and *Germanodactylus cristatus*.

Germanodactylus cristatus (Bennett 2006)
1 m (3.3 ft) WS, 0.35 m (1.1 ft) TL, 0.4 kg (0.9 lb)

FOSSIL REMAINS Skulls and skeletons, adult and juvenile.

ANATOMICAL CHARACTERISTICS Head profile fairly shallow and straight, preorbital opening large, slender, spike-like beak lacks teeth at front end, shallow bony crest atop last two-thirds of head anchored shallow soft tissue crest, teeth along first half of jaws, teeth rather small, project a little forward, throat pouch fairly long. Neck moderately long. Wingspan short relative to mass.

AGE Late Jurassic, early Tithonian.

DISTRIBUTION AND FORMATIONS Southern Germany; Altmühltal (middle Solnhofen).

HABITAT Lagoonal deposits near probably arid, brush-covered islands.

HABITS Picked up and grubbed for small burrowing organisms on and in mud and sand flats, other invertebrates, small fish, and small game.

GROWTH AND HABITS Presence of fossil hatchlings far away from nests indicates they were active independent fliers upon hatching.

NOTES Some paleozoologists consider germanodactyls to be relatives of or basal dsungaripterids.

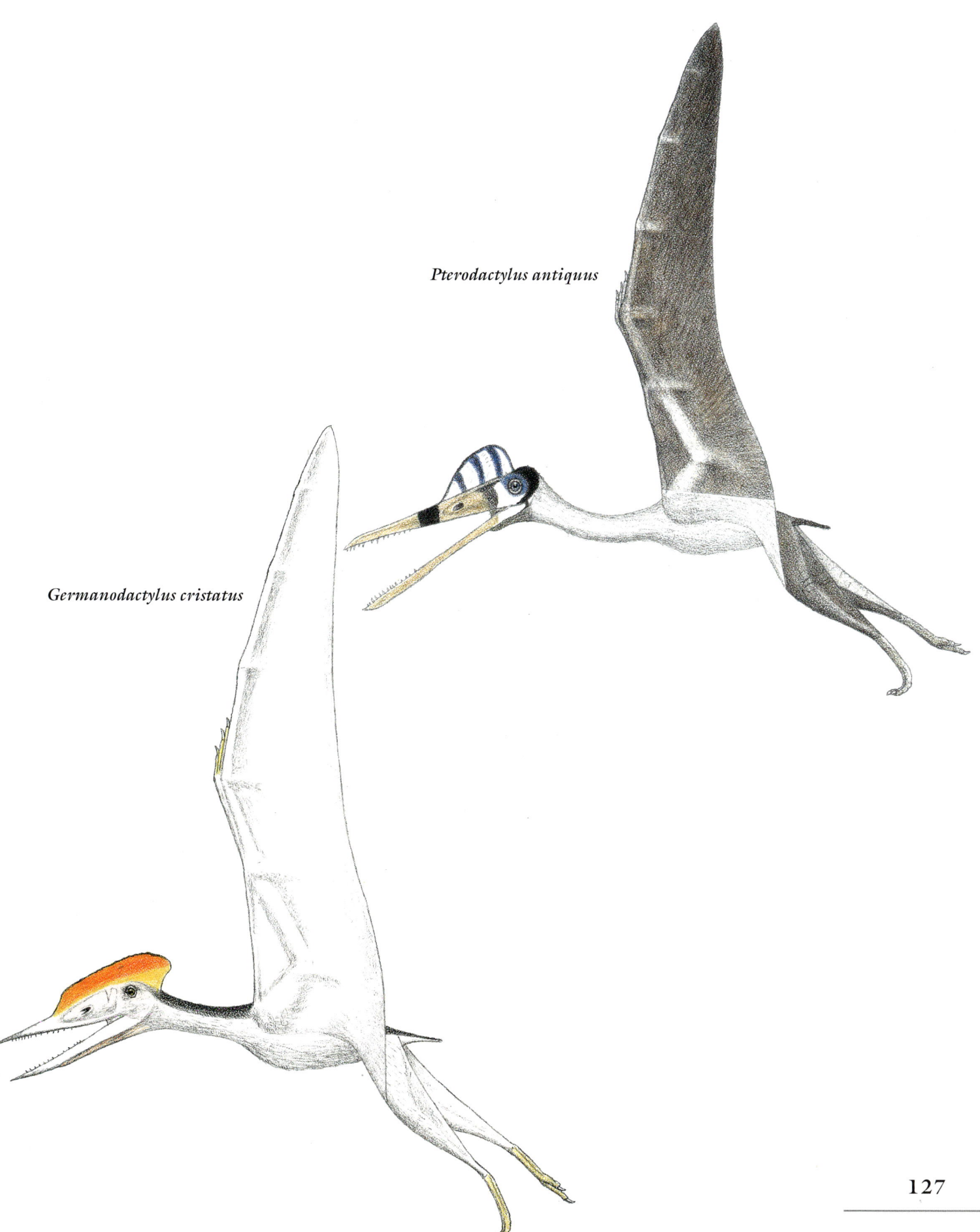

Pterodactylus antiquus

Germanodactylus cristatus

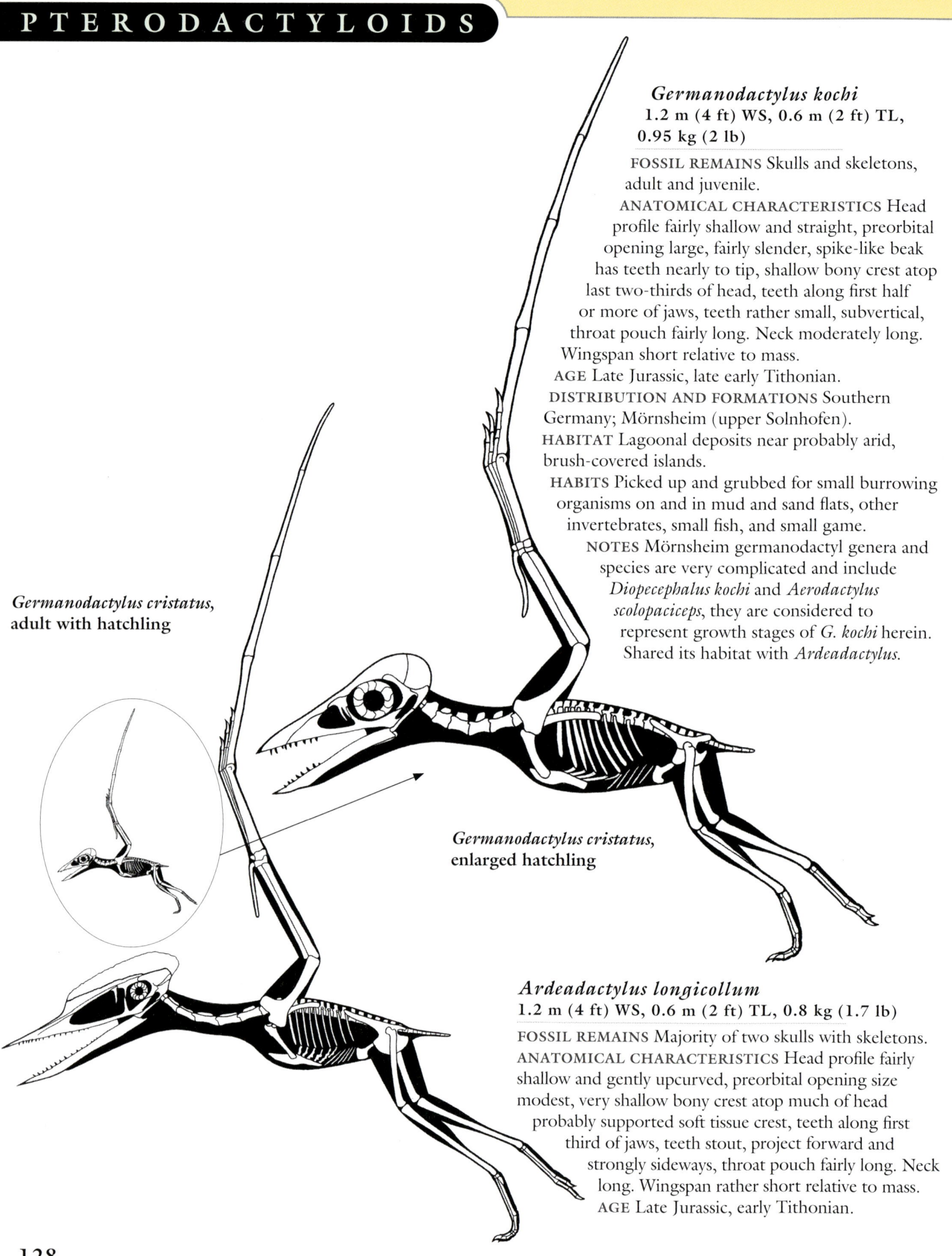

Germanodactylus kochi
1.2 m (4 ft) WS, 0.6 m (2 ft) TL, 0.95 kg (2 lb)

FOSSIL REMAINS Skulls and skeletons, adult and juvenile.

ANATOMICAL CHARACTERISTICS Head profile fairly shallow and straight, preorbital opening large, fairly slender, spike-like beak has teeth nearly to tip, shallow bony crest atop last two-thirds of head, teeth along first half or more of jaws, teeth rather small, subvertical, throat pouch fairly long. Neck moderately long. Wingspan short relative to mass.

AGE Late Jurassic, late early Tithonian.

DISTRIBUTION AND FORMATIONS Southern Germany; Mörnsheim (upper Solnhofen).

HABITAT Lagoonal deposits near probably arid, brush-covered islands.

HABITS Picked up and grubbed for small burrowing organisms on and in mud and sand flats, other invertebrates, small fish, and small game.

NOTES Mörnsheim germanodactyl genera and species are very complicated and include *Diopecephalus kochi* and *Aerodactylus scolopaciceps*, they are considered to represent growth stages of *G. kochi* herein. Shared its habitat with *Ardeadactylus*.

Germanodactylus cristatus,
adult with hatchling

Germanodactylus cristatus,
enlarged hatchling

Ardeadactylus longicollum
1.2 m (4 ft) WS, 0.6 m (2 ft) TL, 0.8 kg (1.7 lb)

FOSSIL REMAINS Majority of two skulls with skeletons.

ANATOMICAL CHARACTERISTICS Head profile fairly shallow and gently upcurved, preorbital opening size modest, very shallow bony crest atop much of head probably supported soft tissue crest, teeth along first third of jaws, teeth stout, project forward and strongly sideways, throat pouch fairly long. Neck long. Wingspan rather short relative to mass.

AGE Late Jurassic, early Tithonian.

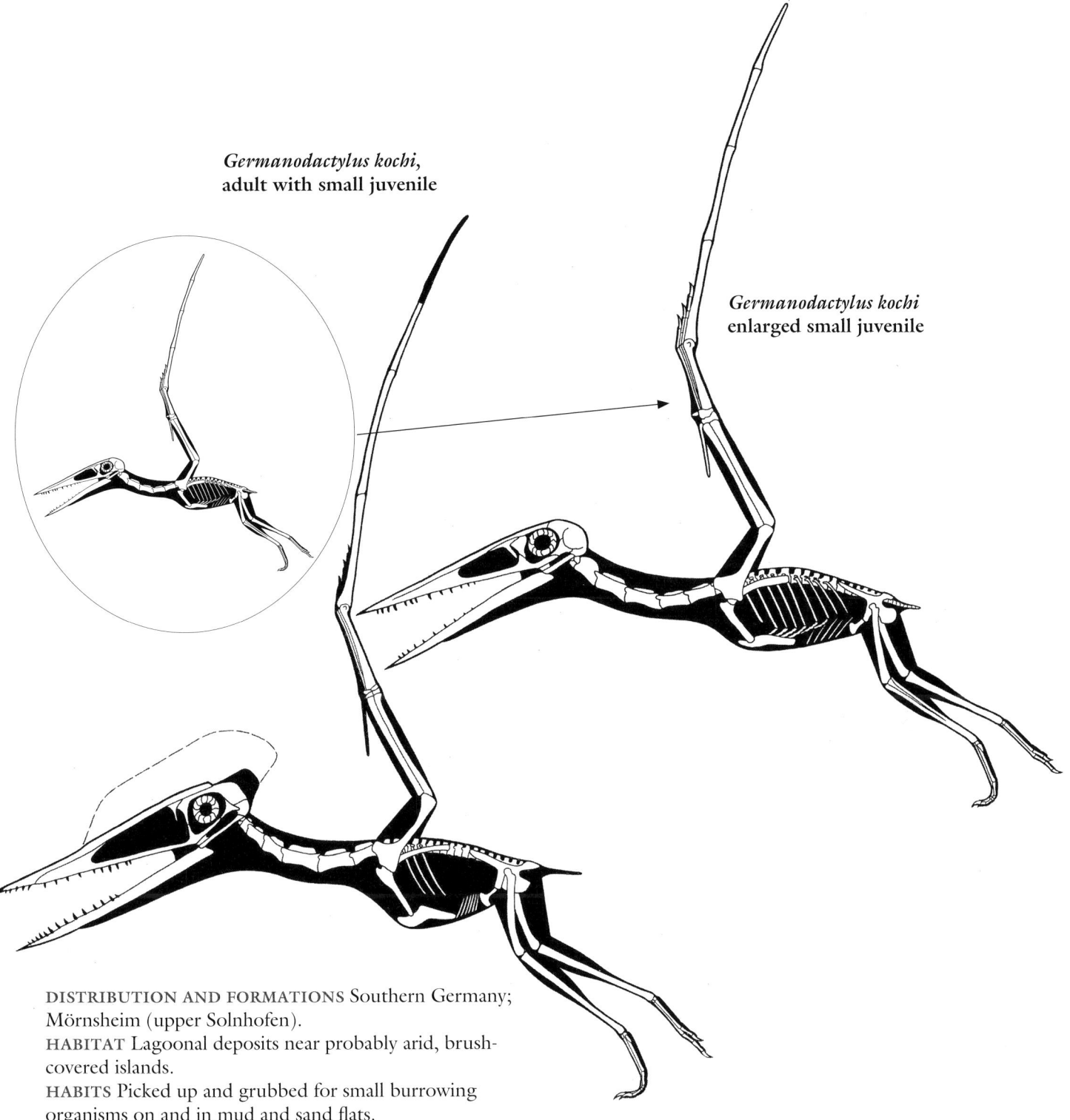

Germanodactylus kochi,
adult with small juvenile

Germanodactylus kochi
enlarged small juvenile

DISTRIBUTION AND FORMATIONS Southern Germany;
Mörnsheim (upper Solnhofen).
HABITAT Lagoonal deposits near probably arid, brush-
covered islands.
HABITS Picked up and grubbed for small burrowing
organisms on and in mud and sand flats.

Cycnorhamphus suevicus (Bennett 2013)
2.3 m (7.7 ft) WS, 0.75 m (2.5 ft) TL, 2 kg (4.5 lb)

FOSSIL REMAINS A few skulls and skeletons, adult and
juvenile.
ANATOMICAL CHARACTERISTICS Head fairly robust,
short bony crest projects from back of head, preorbital

opening size modest, shallow soft tissue crest atop most
of head, upper beak bears hard nonbony downward
projection that fits into upcurved lower beak, which is
increasingly strongly flexed during growth, a few blunt,
forward-projecting teeth are limited to front of jaws,
throat pouch long. Neck moderately long. Pteroid very

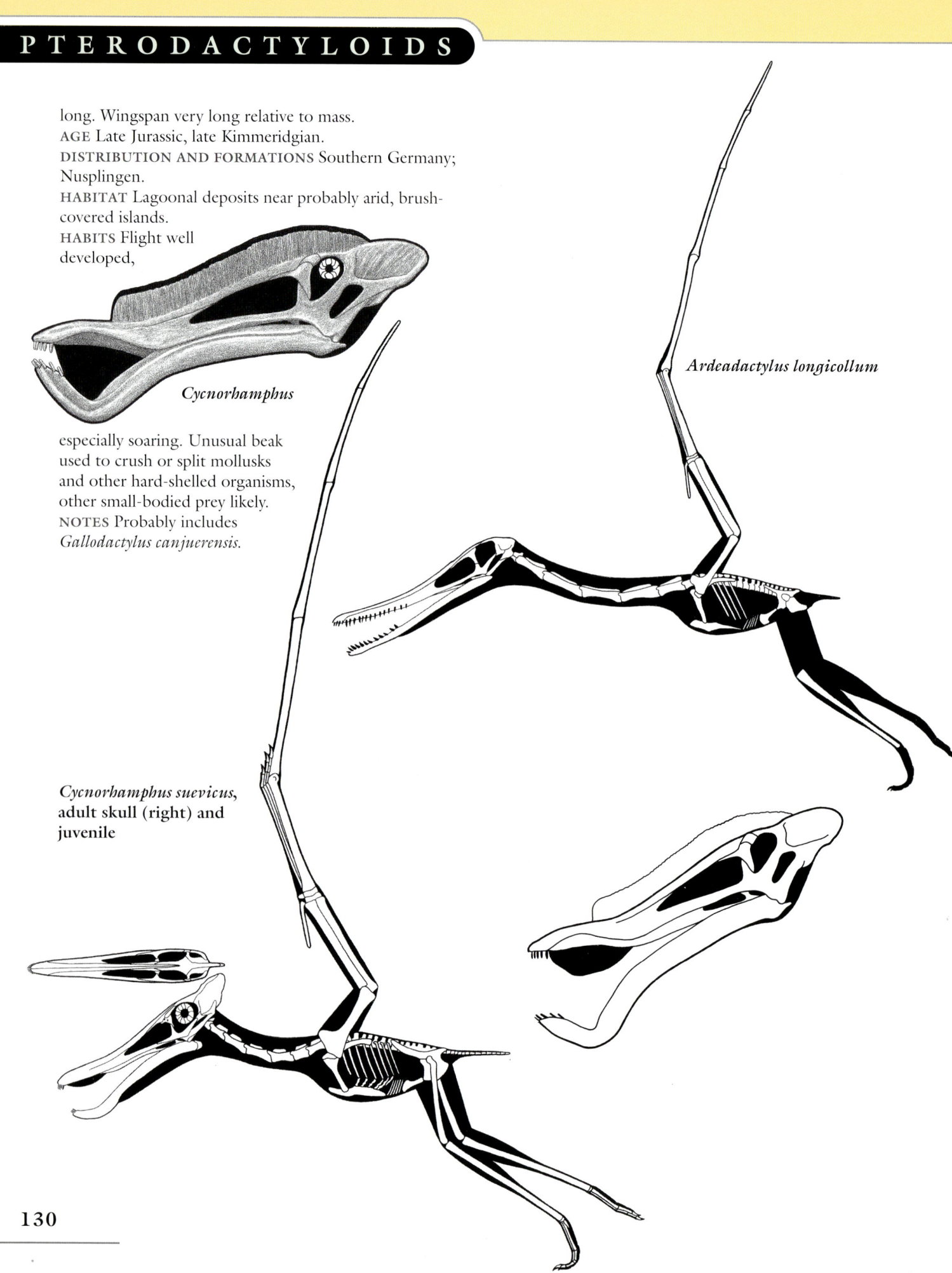

long. Wingspan very long relative to mass.
AGE Late Jurassic, late Kimmeridgian.
DISTRIBUTION AND FORMATIONS Southern Germany;
Nusplingen.
HABITAT Lagoonal deposits near probably arid, brush-covered islands.
HABITS Flight well
developed,

Cycnorhamphus

especially soaring. Unusual beak
used to crush or split mollusks
and other hard-shelled organisms,
other small-bodied prey likely.
NOTES Probably includes
Gallodactylus canjuerensis.

Ardeadactylus longicollum

**Cycnorhamphus suevicus,
adult skull (right) and
juvenile**

CTENOCHASMATIDS

MEDIUM-SIZED TO LARGE ARCHAEOPTERO-
DACTYLOIDS OF THE LATE JURASSIC TO THE
EARLY CRETACEOUS OF THE NORTHERN AND
SOUTHERN HEMISPHERES

ANATOMICAL CHARACTERISTICS Head long and
shallow, preorbital openings small, teeth slender.
Wingspan not long relative to mass.
HABITS Variable, largely patrolling shoreline and wading
in shallow water, with an emphasis on water straining and
filter feeding, may have been prone to nighttime activity.
NOTES Absence from at least some other continents may
reflect lack of sufficient sampling.

Kepodactylus insperatus (Harris and Carpenter 1996)
2.5 m (8 ft) WS, 10 kg (20 lb)

FOSSIL REMAINS Minority of skeleton.
ANATOMICAL CHARACTERISTICS Insufficient
information.
AGE Late Jurassic, early Tithonian.
DISTRIBUTION AND FORMATIONS Wyoming; middle
Morrison.
HABITAT Seasonally dry open woodlands.

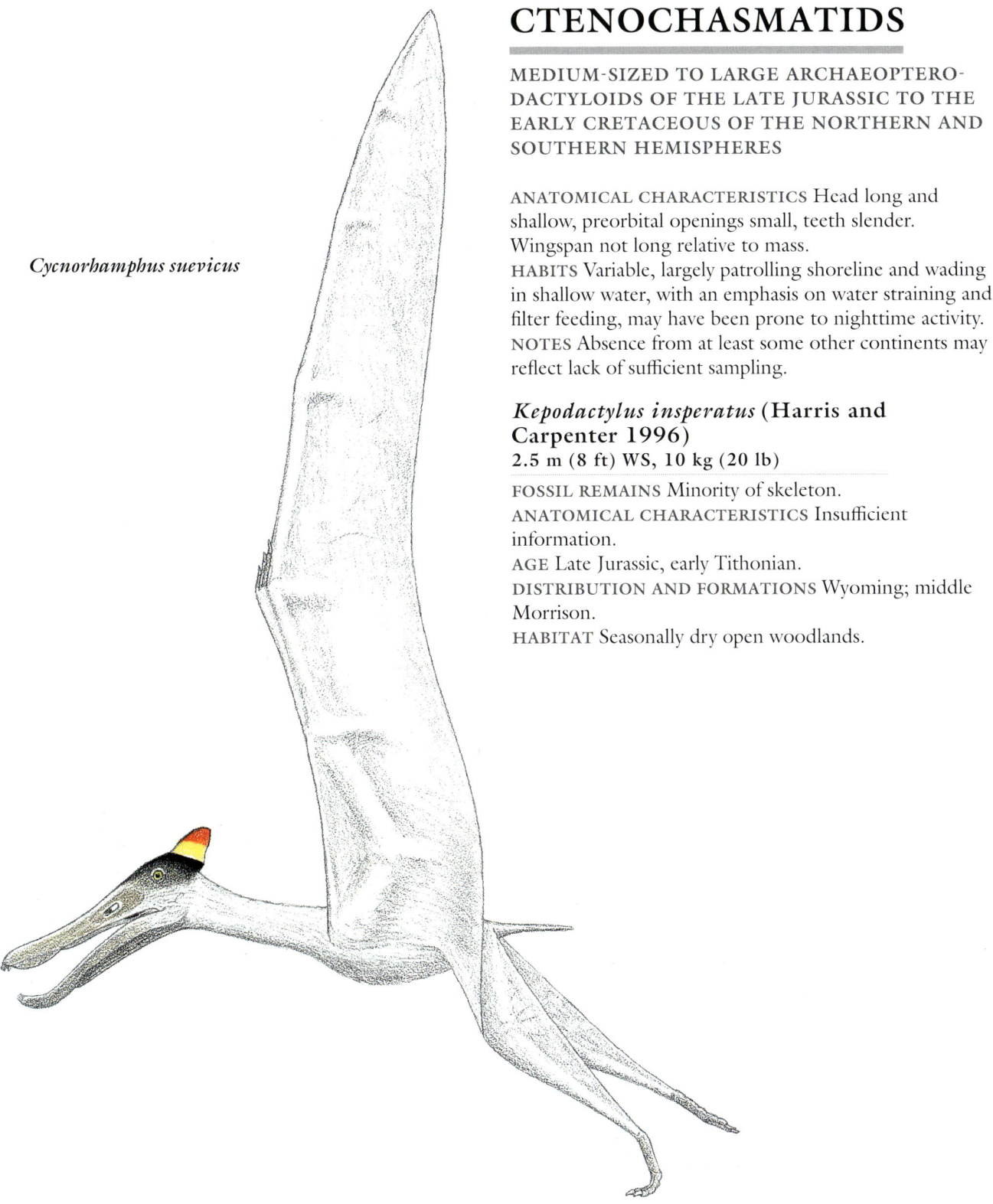

Cycnorhamphus suevicus

Huanhepterus quingyangensis
2.4 m (8 ft) WS, 1.6 m (5 ft) TL, 9 kg (20 lb)

FOSSIL REMAINS Majority of skull and skeleton.
ANATOMICAL CHARACTERISTICS Head very long
and slender, bony crest atop most of first two-thirds of
head, teeth along first half of jaws, teeth large spikes,
subvertical, throat pouch length moderate. Neck very
long. Wingspan very short relative to mass.
AGE Late Jurassic.
DISTRIBUTION AND FORMATIONS Northeastern China;
Huachihuanhe.
HABITS Small-game predator.
NOTES Earliest known long-necked pterosaur.

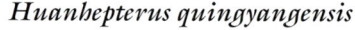

Huanhepterus quingyangensis

Elanodactylus prolatus (Andres and Ji 2008)
2.5 m (8 ft) WS, 10 kg (20 lb)

FOSSIL REMAINS Partial skeleton.
ANATOMICAL CHARACTERISTICS Neck long.
AGE Early Cretaceous, early Aptian.
DISTRIBUTION AND FORMATIONS Northeastern China;
lower Yixian.
HABITAT Well-watered highland forests and lakes, winters
chilly with some snow.
NOTES Shared its habitat with *Eosipterus, Beipiaopterus,
Haopterus, Boreopterus, Gegepterus, Pterofiltrus,
Moganopterus.*

Moganopterus zhuiana or Feilongus youngi
(Lü et al. 2012)
3 m (10 ft) WS, 15 kg (35 lb)

FOSSIL REMAINS Skulls, adult and juvenile.
ANATOMICAL CHARACTERISTICS Head hyperlong
and extremely slender, shallow bony crest may be atop
tip of jaw, moderate-length bony crest projects straight
backward from head, eyes not large, teeth along first third
of jaws, slender, rather small, project a little forward.
Neck long.

Huanhepterus quingyangensis (bottom) and
Boreopterus cuiae

AGE Early Cretaceous, earliest Aptian.
DISTRIBUTION AND FORMATIONS China; lower Yixian.
HABITAT Well-watered highland forests and lakes, winters chilly with some snow.
HABITS Uncertain.
NOTES Juvenile *Feilongus youngi* may be same species, in which case that name has priority. Length of rear crest uncertain. Possessed the most elongated and slender head known among vertebrates, longest known head of a toothed pterosaur. Estimates of dramatically greater wingspans are errant.

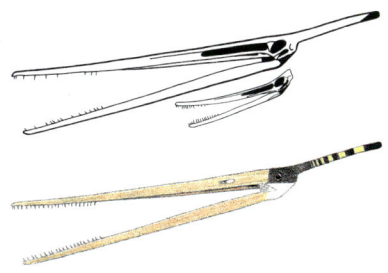

Magonpterus zhuiana or **Feilongus youngi**
adult and juvenile

Forfexopterus jeholensis

Forfexopterus jeholensis (Jiang et al. 2016)
2.2 m (7.5 ft) WS, 1.1 m (3.5 ft) TL, 5.5 kg (12 lb)

FOSSIL REMAINS Majority of skull and skeleton, minority of skeleton.
ANATOMICAL CHARACTERISTICS Head very long and slender, a shallow, subrectangular bony crest atop back of head, teeth along first third of jaws, teeth variable in length, fairly slender, project a little forward. Neck very long. Wingspan short relative to mass.
AGE Early Cretaceous, early or middle Aptian.
DISTRIBUTION AND FORMATIONS Northeastern China; Jiufotang.
HABITAT Well-watered highland forests and lakes, winters chilly with some snow.
HABITS Fisher.
NOTES Shared its habitat with *Vesperopterylus*, *Hongshanopterus*, *Nurhachius*, *Istiodactylus sinensis*, *Sinopterus*, *Chaoyangopterus*, *Shenzhoupterus*, *Guidraco*, *Liaoningopterus*, *Ikrandraco*.

Pterofiltrus qiui (Jiang and Wang 2011)
1.5 m (5 ft) WS, 1.5 kg (3 lb)

FOSSIL REMAINS Majority of skull.
ANATOMICAL CHARACTERISTICS Beak shallow, back of head moderately deep, long, slender, vertical teeth limited to front half of jaws.
AGE Early Cretaceous, early Aptian.
DISTRIBUTION AND FORMATIONS Northeastern China; lower Yixian.
HABITAT Well-watered highland forests and lakes, winters chilly with some snow.
HABITS Wading and/or floating water-straining feeder.
NOTES Skull too disarticulated to restore. Shared its habitat with *Elanodactylus*, *Eosipterus*, *Beipiaopterus*, *Haopterus*, *Boreopterus*, *Moganopterus*, *Gegepterus*.

**Gladocephaloideus jingangshanensis,
adult (skull) and juvenile**

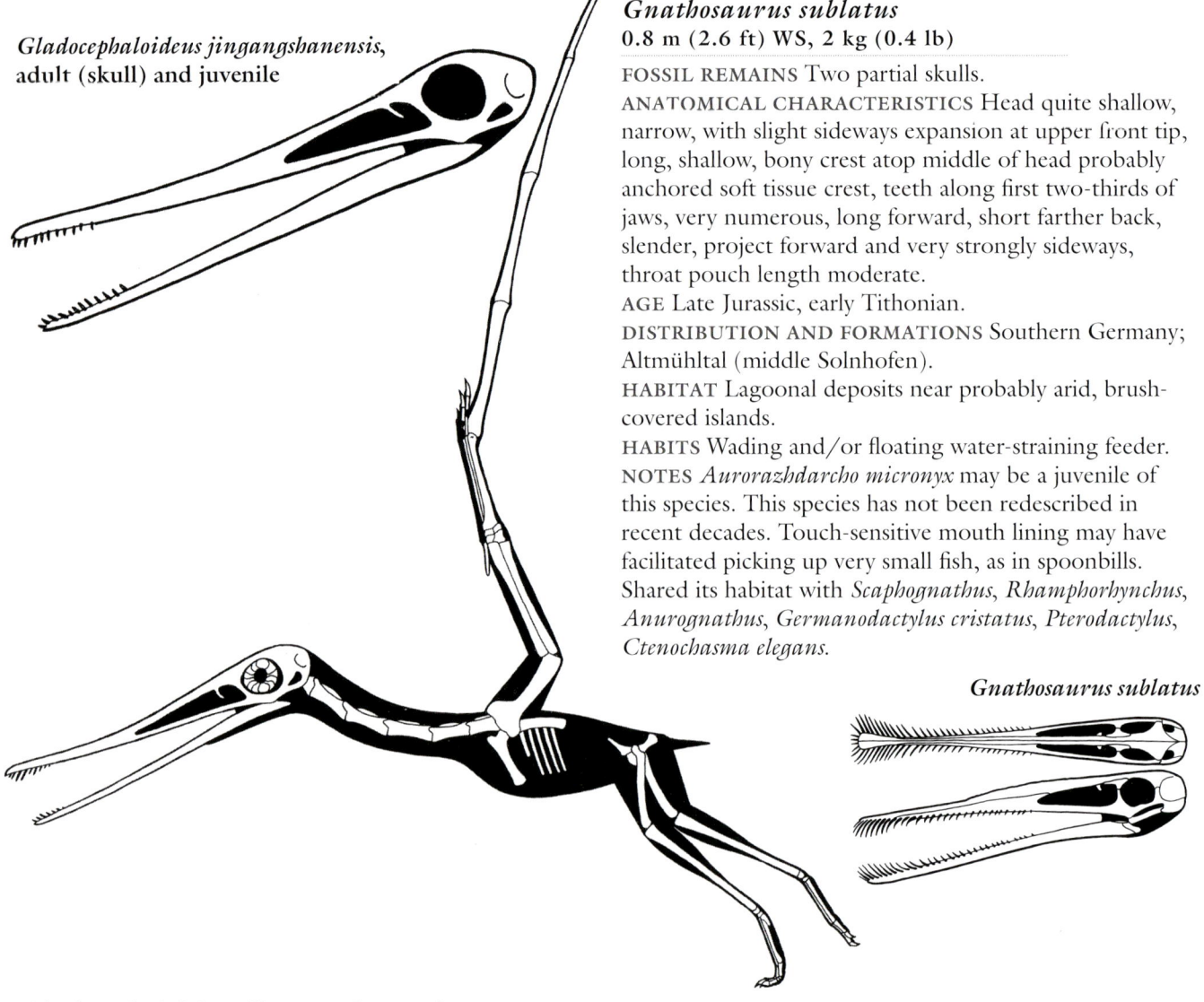

Gnathosaurus sublatus
0.8 m (2.6 ft) WS, 2 kg (0.4 lb)

FOSSIL REMAINS Two partial skulls.

ANATOMICAL CHARACTERISTICS Head quite shallow,
narrow, with slight sideways expansion at upper front tip,
long, shallow, bony crest atop middle of head probably
anchored soft tissue crest, teeth along first two-thirds of
jaws, very numerous, long forward, short farther back,
slender, project forward and very strongly sideways,
throat pouch length moderate.

AGE Late Jurassic, early Tithonian.

DISTRIBUTION AND FORMATIONS Southern Germany;
Altmühltal (middle Solnhofen).

HABITAT Lagoonal deposits near probably arid, brush-
covered islands.

HABITS Wading and/or floating water-straining feeder.

NOTES *Aurorazhdarcho micronyx* may be a juvenile of
this species. This species has not been redescribed in
recent decades. Touch-sensitive mouth lining may have
facilitated picking up very small fish, as in spoonbills.
Shared its habitat with *Scaphognathus, Rhamphorhynchus,
Anurognathus, Germanodactylus cristatus, Pterodactylus,
Ctenochasma elegans.*

Gnathosaurus sublatus

Gladocephaloideus jingangshanensis
(Lü et al. 2016)
0.85 m (2.8 ft) WS, 0.3 kg (7 lb)

FOSSIL REMAINS Two complete skulls, one with majority
of skeleton.

ANATOMICAL CHARACTERISTICS Head shallow,
preorbital opening larger, low number of procumbent,
modest-sized, nonsharp teeth limited to front tip of jaws.
Neck moderately long. Wingspan short relative to mass,
pteroids short.

AGE Early Cretaceous, early Aptian.

DISTRIBUTION AND FORMATIONS Northeastern China;
upper Yixian.

HABITAT Well-watered highland forests and lakes, winters
chilly with some snow.

HABITS Used short tooth array to grub for small
burrowing organisms in mud and sand flats. Shared its
habitat with *Yixianopterus, Luchibang, Eopteranodon.*

Ctenochasma? elegans (Bennett 2007a)
1.8 m (6 ft) WS, 0.8 m (2.5 ft) TL, 2.5 kg (5 lb)

FOSSIL REMAINS Over a dozen skulls, many with
skeletons, juvenile and adult.

ANATOMICAL CHARACTERISTICS Head very shallow,
narrow, very shallow bony crest atop back of head probably
anchored soft tissue crest, teeth along first three-quarters of
jaws, number about 400 in adults, limited to first quarter
of jaws and much less numerous in juveniles, long forward,
short farther back, teeth slender bristles, project forward
and very strongly sideways, throat pouch rather short.
Neck moderately long. Wingspan short relative to mass.

AGE Late Jurassic, early Tithonian.

DISTRIBUTION AND FORMATIONS Southern Germany;
Altmühltal (middle Solnhofen).

HABITAT Lagoonal deposits near probably arid, brush-
covered islands.

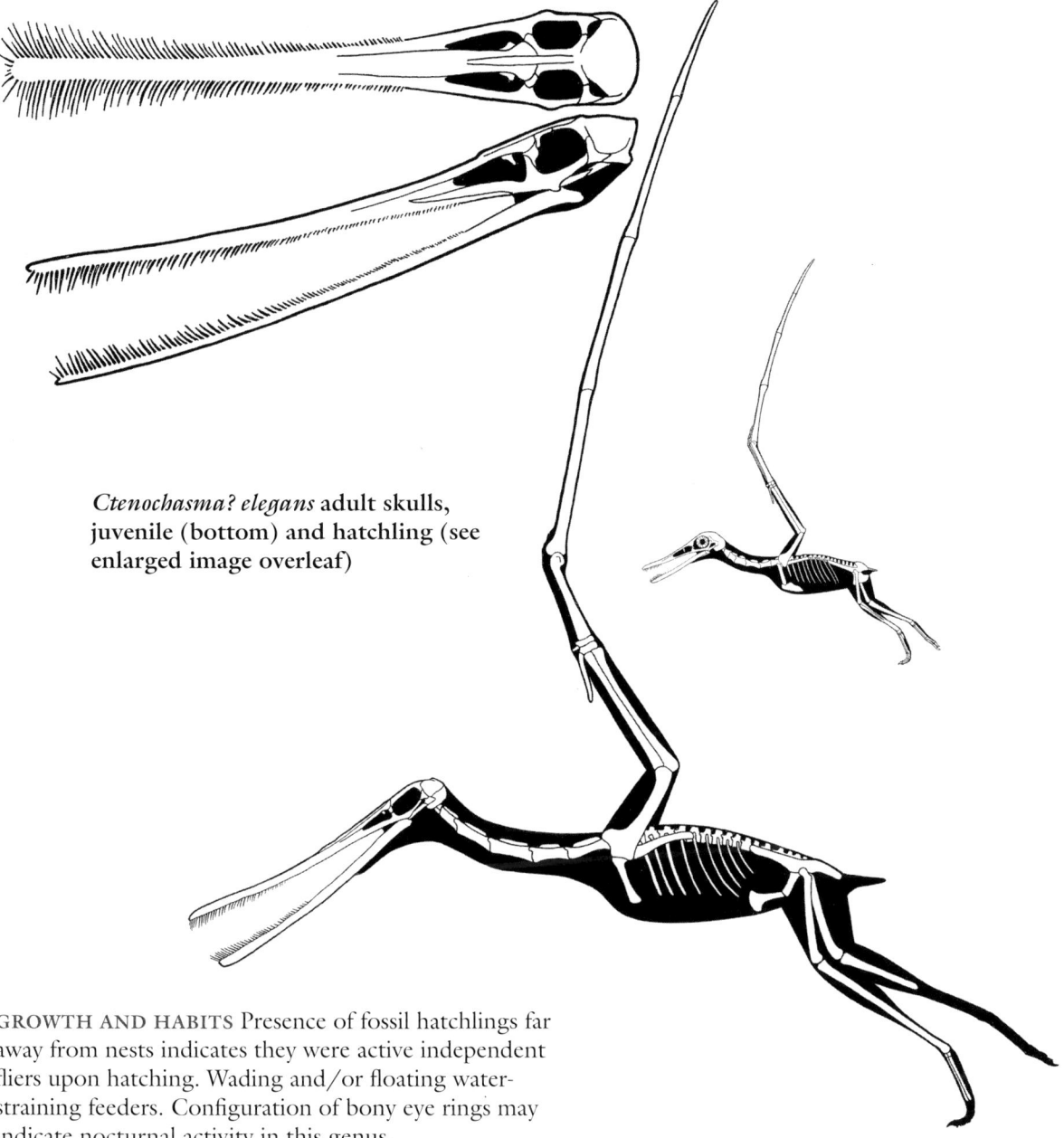

Ctenochasma? elegans adult skulls, juvenile (bottom) and hatchling (see enlarged image overleaf)

GROWTH AND HABITS Presence of fossil hatchlings far away from nests indicates they were active independent fliers upon hatching. Wading and/or floating water-straining feeders. Configuration of bony eye rings may indicate nocturnal activity in this genus.
NOTES Whether this is the same genus as a very fragmentary and later *C. roemeri* is not certain.

Ctenochasma roemeri
2.5 m (8 ft) WS, 7 kg (15 lb)

FOSSIL REMAINS Tip of lower jaw.
ANATOMICAL CHARACTERISTICS Teeth are very numerous slender bristles, long forward, project forward and strongly sideways.
AGE Late Jurassic, early Tithonian.
DISTRIBUTION AND FORMATIONS Northern Germany; Purbeck Kalk.
HABITAT Lagoonal deposits near probably arid, brush-covered islands.

HABITS Wading and/or floating water-straining feeder.
NOTES Whether this is a valid taxon is dubious.

Ctenochasma? taqueti
1.2 m (4 ft) WS, 0.8 kg (1.8 lb)

FOSSIL REMAINS Majority of skull.
ANATOMICAL CHARACTERISTICS Head very shallow, very shallow bony crest atop middle of head probably anchored soft tissue crest, teeth slender bristles, subvertical, throat pouch length moderate.
AGE Late Jurassic, early Tithonian.
DISTRIBUTION AND FORMATIONS France; Calcaires tachetés.

HABITAT Lagoonal deposits near probably arid, brush-covered islands.
HABITS Wading and/or floating water-straining feeder.
NOTES Whether this is the same genus as a very fragmentary *C. roemeri* is not certain.

Liaodactylus primus (Zhou et al. 2017)
0.75 m (2.5 ft) WS, 0.2 kg (0.4 lb)

FOSSIL REMAINS Majority of skull and minority of skeleton.
ANATOMICAL CHARACTERISTICS Head very shallow, teeth numerous, fairly slender, forward projecting.
AGE Late Jurassic, early Oxfordian.
DISTRIBUTION AND FORMATIONS Northeastern China; middle Tiaojishan.
HABITAT Well-watered forests and lakes, winters chilly with some snow.
NOTES Shared its habitat with *Fenghuangopterus, Changchengopterus, Pterorhynchus, Wukongopterus, Douzhanopterus, Dendrorhynchoides, Jiangchangnathus.*

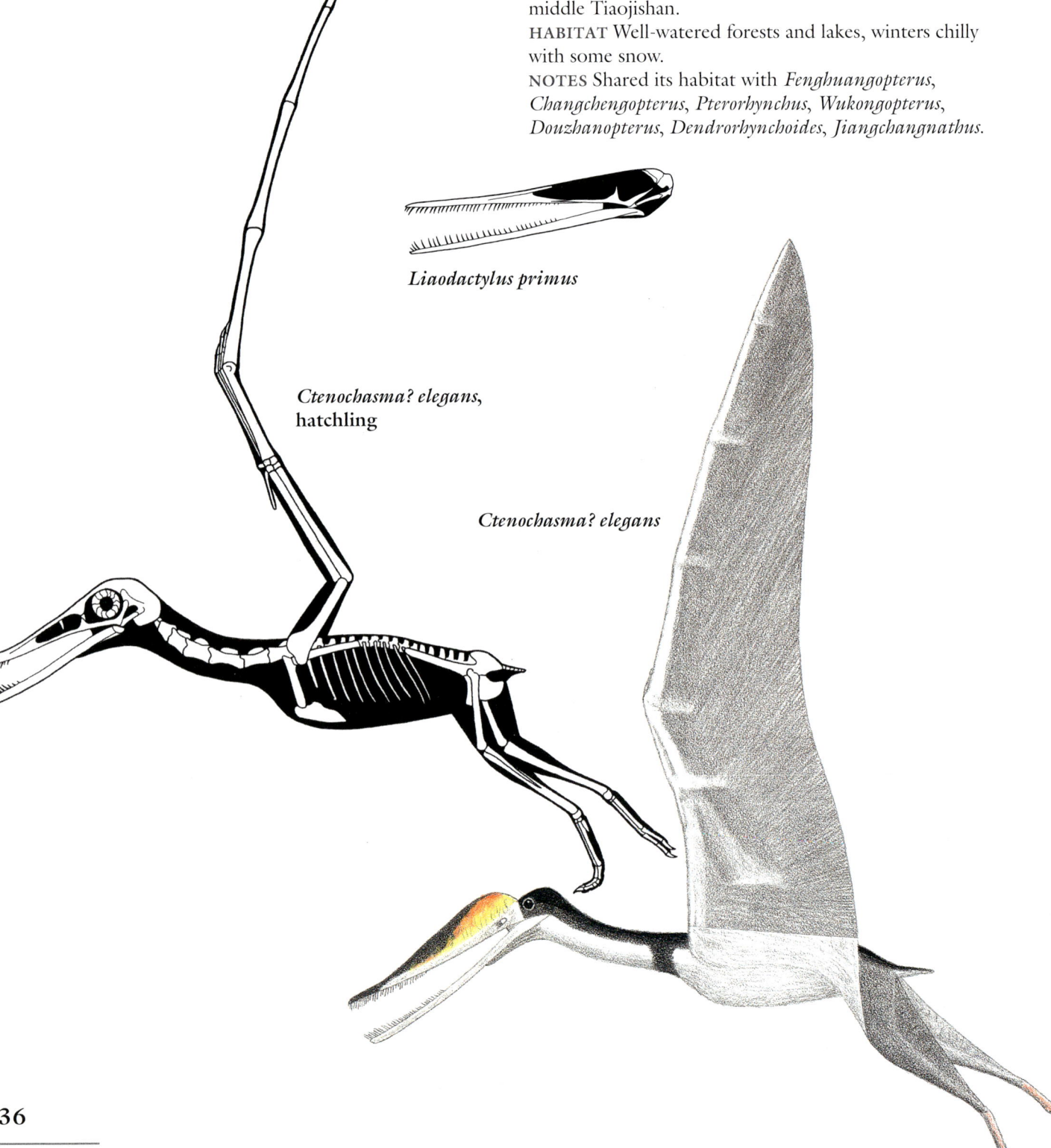

Liaodactylus primus

Ctenochasma? elegans,
hatchling

Ctenochasma? elegans

Unnamed genus and species
1.2 m (4 ft) WS, 0.4 m (1.3 ft) TL, 0.55 kg (1.2 lb)

FOSSIL REMAINS Complete skull and skeleton.
ANATOMICAL CHARACTERISTICS Head very shallow, snout strongly upcurved, very shallow bony crest atop back of head probably anchored soft tissue crest, teeth along first three-quarters of upper jaw, less on lower, teeth longer on upper jaw than on lower, teeth very fine, densely packed bristles. Neck moderately long. Wingspan moderate relative to mass.
AGE Late Jurassic, late Kimmeridgian.
DISTRIBUTION AND FORMATIONS Southern Germany; Wattendorf.
HABITAT Lagoonal deposits near probably arid, brush-covered islands.
HABITS Wading and/or floating filter feeder.

Eosipterus yangi (Ji and Ji 1997)
Adult size uncertain

FOSSIL REMAINS Majority of skeleton, immature.
ANATOMICAL CHARACTERISTICS Neck length moderate.
AGE Early Cretaceous, early Aptian.

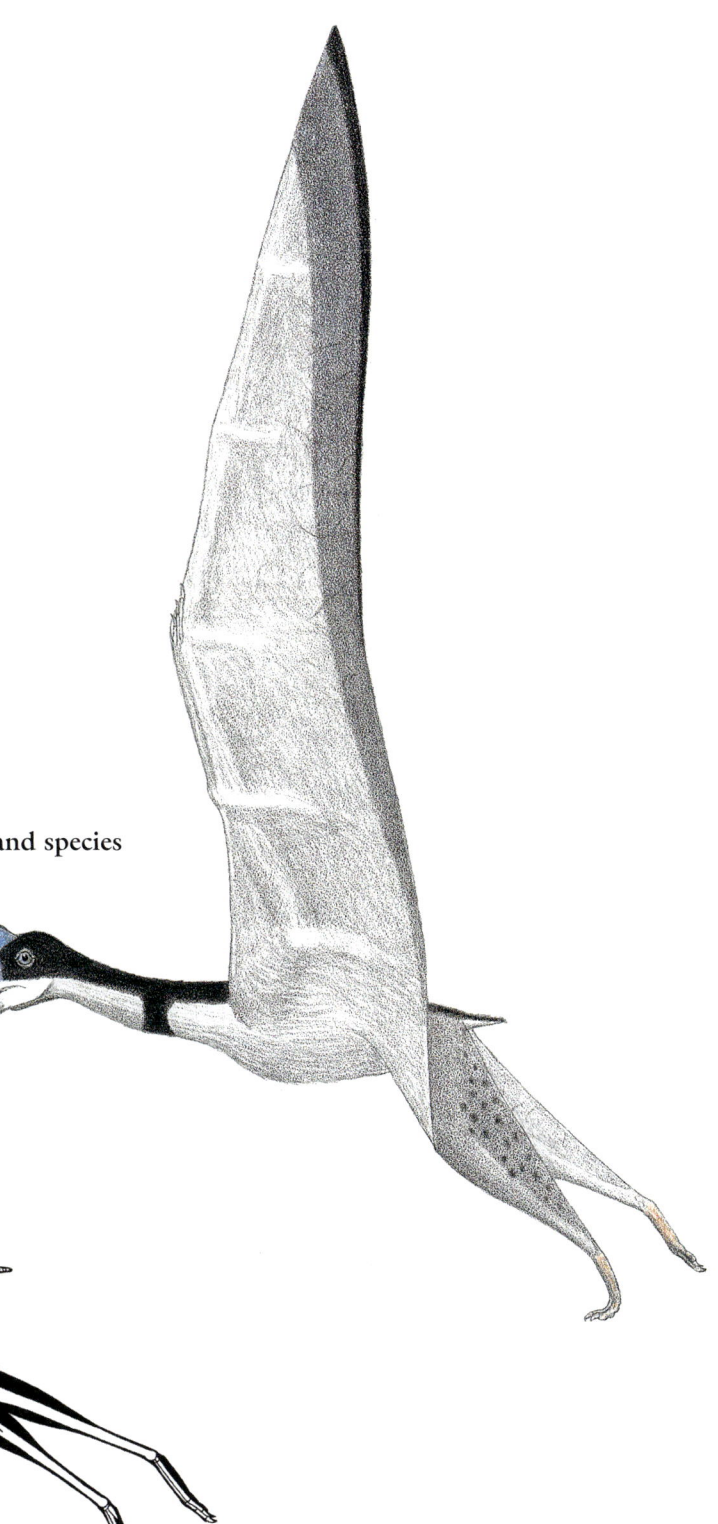

Unnamed genus and species

DISTRIBUTION AND FORMATIONS Northeastern China; lower Yixian.

HABITAT Well-watered highland forests and lakes, winters chilly with some snow.

NOTES Shared its habitat with *Elanodactylus*, *Haopterus*, *Boreopterus*, *Moganopterus*, *Pterofiltrus*, *Gegepterus*, *Beipiaopterus*.

Beipiaopterus chenianus (Lü 2003)
Adult size uncertain

FOSSIL REMAINS Majority of skeleton with soft tissues including wing membranes, foot webbing, possibly immature.

ANATOMICAL CHARACTERISTICS Neck moderately long.

AGE Early Cretaceous, early Aptian.

DISTRIBUTION AND FORMATIONS Northeastern China; lower Yixian.

HABITAT Well-watered highland forests and lakes, winters chilly with some snow.

Gegepterus changi (Wang et al. 2007)
Adult size uncertain

FOSSIL REMAINS Two partial skulls and skeletons, immature.

ANATOMICAL CHARACTERISTICS Head very shallow, long snout gently upcurved, modest sized, fairly slender, not densely packed, procumbent teeth limited to first three-fifths of jaws.

AGE Early Cretaceous, earliest Aptian.

DISTRIBUTION AND FORMATIONS Northeastern China; lower Yixian.

HABITAT Well-watered highland forests and lakes, winters chilly with some snow.

Gegepterus changi

Pterodaustro guinazui
2.5 m (8 ft) WS, 1.3 m (4.3 ft) TL, 12 kg (25 lb)

FOSSIL REMAINS Hundreds of specimens, adults and juveniles, including embryos in eggs.

ANATOMICAL CHARACTERISTICS Head extremely shallow, back of head and orbit small, slender snout strongly upcurved, possible very low bony crest anchored soft tissue crest above back of head, teeth along nearly entire jaws, upper teeth extremely reduced, lower teeth about 1,000 extremely fine, densely packed, flexible, subvertical bristles set in a continuous groove projecting above top of upper jaw when mouth closed, throat pouch small. Wingspan shortest relative to mass among known pterosaurs.

AGE Early Cretaceous, Albian.

DISTRIBUTION AND FORMATIONS Central Argentina; Lagarcito.

HABITAT Shallow lakes.

GROWTH AND HABITS Takeoff and flight abilities possibly mediocre. Wading and/or floating filter feeder, gizzard stones used to process hard-shelled crustaceans. Configuration of bony eye rings may indicate nocturnal activity in this genus.

NOTES Most extremely evolved known filter-feeding pterosaur, possessed the slenderest known teeth of any animal. May have had a rapid-cycling water-pumping system to strain and ingest tiny organisms. Has not been redescribed in recent years, numerous juveniles have yet to be described in detail.

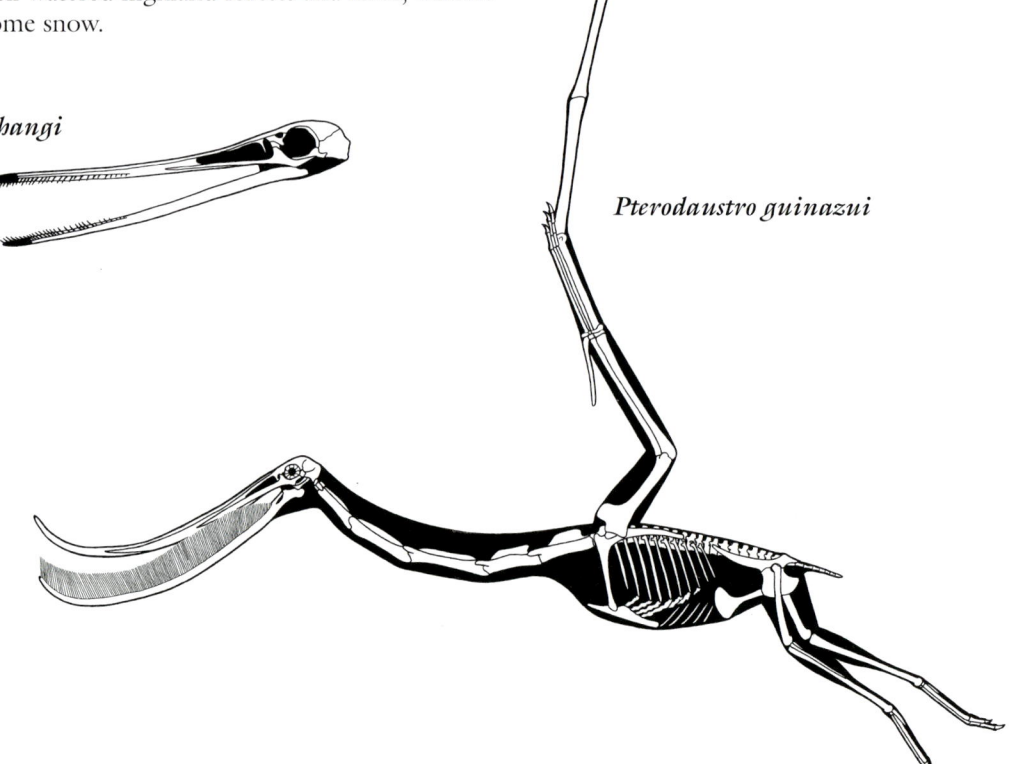

Pterodaustro guinazui

Pterodaustro guinazui

Pterodaustro guinazui

DSUNGARIPTERIDS

LARGE ARCHAEOPTERODACTYLOIDS OF THE EARLY CRETACEOUS OF ASIA

ANATOMICAL CHARACTERISTICS Head large, fairly deep at back, very robustly built, preorbital opening fairly large, shallow bony crest along most of length of head, most prominent at front, modest spike projects backward, orbit small and high set because lower orbit closed off, spike-like toothless beak, teeth stout, subtriangular, increasingly rounded, back teeth largest, jaw muscles exceptionally powerful, throat pouch length moderate. Skeletal elements more robust walled than usual in pterodactyloids. Neck length moderate. Notarium present. Wingspan long relative to mass, well developed, pteroids long. Legs very long, feet large.

HABITS Very well-developed fliers, apparently for both flapping and soaring. Well adapted for ground locomotion. Strongly built spike-like beak, powerful skull, and stout teeth indicate digging for and breaking up of tough food items, probably hard-shelled invertebrates found along shorelines and in shallow waters. Also fished with spike.

NOTES Had longest legs known among pterosaurs. Fragmentary remains may indicate presence in South America and/or Africa. Relationships uncertain, some consider these close relations of germanodactyls, others azhdarchoids, others believe are a distinct group. If the latter, despite sophisticated specializations apparently a not very successful group.

Dsungaripterus

Dsungaripterus weii (Chen et al. 2020)
4.2 m (14 ft) WS, 1.3 m (4.5 ft) TL, 19 kg (40 lb)

FOSSIL REMAINS A few skulls and skeletons of varying completeness.
ANATOMICAL CHARACTERISTICS Beak increasingly upturned and teeth stouter with growth.
AGE Early Cretaceous.

Dsungaripterus weii, adult (bottom and left skull), large juvenile skull, juvenile or *Noripterus complicidens* (smaller skeleton)

Dsungaripterus weii

EUPTERO-DACTYLOIDS

MEDIUM-SIZED TO GIGANTIC ARCHAEOP-TERODACTYLOIDS OF THE EARLY CRETACEOUS TO THE END OF THE MESOZOIC ON ALL CONTINENTS AND OCEANS

ANATOMICAL CHARACTERISTICS Inner wings markedly longer than legs.
HABITAT From continental to oceanic.
HABITS Highly variable.

EUPTERODACTYLOID MISCELLANEA

Ikrandraco avatar (Wang et al. 2014)
2.5 m (8 ft) WS, 5 kg (10 lb)

FOSSIL REMAINS Two skulls and minority of skeletons.
ANATOMICAL CHARACTERISTICS Head shallow, beak long, lunate keel crest beneath front of lower jaw, teeth fairly widely spaced, short, subconical, subvertical, throat pouch about half jaw length.
AGE Early Cretaceous, early or middle Aptian.
DISTRIBUTION AND FORMATIONS Northeastern China; Jiufotang.
HABITAT Well-watered highland forests and lakes, winters chilly with some snow.
HABITS Possible surface skimmer.

Dsungaripterus weii

DISTRIBUTION AND FORMATIONS Northwestern China; lower Lianmuqin.
HABITAT Rivers and lakes.
NOTES *Lonchognathosaurus acutirostris* and *Noripterus complicidens* are probably juveniles of this species. If so, then beak became increasingly upturned, originally pointed teeth more robust and rounded, and bottom of orbit more closed off, with maturity. Skeleton has not been redescribed in recent years.

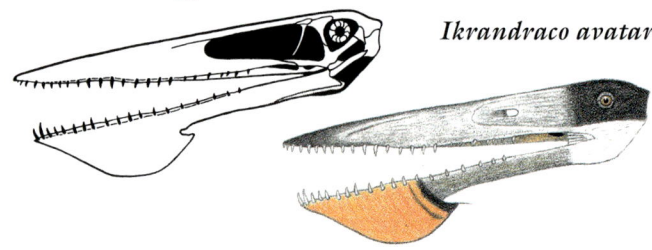

Ikrandraco avatar

NOTES Relationships with other eupterodactyloids uncertain. Shared its habitat with *Vesperopterylus, Forfexopterus, Hongshanopterus, Nurhachius, Istiodactylus sinensis, Sinopterus, Chaoyangopterus, Shenzhoupterus, Guidraco, Liaoningopterus.*

MIMODACTYLIDS

EUPTERODACTYLOIDS OF THE EARLY TO LATE CRETACEOUS OF EURASIA

ANATOMICAL CHARACTERISTICS Head fairly shallow, beak fairly long, teeth fairly widely spaced, subconical, subvertical, throat pouch long. Wings long. Legs short.
HABITAT Freshwater courses and coastal.
HABITS Probable thermal soarers, fishers.
NOTES Absence from other continents may reflect lack of sufficient sampling.

Haopterus gracilis (Wang and Lü 2001)
Adult size uncertain

FOSSIL REMAINS Majority of immature skull and skeleton.
ANATOMICAL CHARACTERISTICS Teeth along first two-thirds of jaws.
AGE Early Cretaceous, early Aptian.
DISTRIBUTION AND FORMATIONS Northeastern China; lower Yixian.
HABITAT Well-watered highland forests and lakes, winters chilly with some snow.
NOTES Jumbled remains preclude skeletal restoration and identification. Has been placed near base of eupterodactyloids, in istiodactylids, and ornithocheiroids. Shared its habitat with *Elanodactylus, Eosipterus, Beipiaopterus, Boreopterus, Moganopterus, Gegepterus, Pterofiltrus.*

Yixianopterus jingangshanensis (Lü et al. 2006)
2 m (6.5 ft) WS, 2 kg (4 lb)

FOSSIL REMAINS Majority and minority of two specimens.
ANATOMICAL CHARACTERISTICS Insufficient information.
AGE Early Cretaceous, early Aptian.
DISTRIBUTION AND FORMATIONS Northeastern China; upper Yixian.
HABITAT Well-watered highland forests and lakes, winters chilly with some snow.
NOTES Lack of most of skull precludes restoration. Shared its habitat with *Luchibang, Gladocephaloideus, Eopteranodon.*

Mimodactylus libanensis (Kellner et al. 2019a)
Adult size uncertain

FOSSIL REMAINS Partial skull and majority of skeleton, immature.
ANATOMICAL CHARACTERISTICS Teeth along first third of jaws.
AGE Late Cretaceous, late Cenomanian.
DISTRIBUTION AND FORMATIONS Lebanon; Sannine Limestone.
HABITAT Nearshore marine.

ISTIODACTYLIDS

LARGE EUPTERODACTYLOIDS OF THE EARLY CRETACEOUS OF EURASIA

ANATOMICAL CHARACTERISTICS Head long, shallow, subtriangular, moderately broad, beak fairly short and preorbital openings very long, snout tip rounded in top view, back of skull fairly deep, teeth along first one-fifth to half of jaws, teeth stout, triangular. Neck length moderate. Notarium present. Inner wings well developed, inner hand length moderate. Leg length moderate, much shorter than inner wings.
HABITAT Freshwater courses and coastal wetlands.
HABITS Probable vulture analogues, thermal soarers, scavengers, small-game predators.
NOTES Absence from other continents may reflect lack of sufficient sampling.

Hongshanopterus lacustris (Wang et al. 2008)
2 m (6.5 ft) WS, 2 kg (4 lb)

FOSSIL REMAINS Majority of skull.
ANATOMICAL CHARACTERISTICS Teeth along first half of jaws.
AGE Early Cretaceous, early or middle Aptian.
DISTRIBUTION AND FORMATIONS Northeastern China; Jiufotang.
HABITAT Well-watered highland forests and lakes, winters chilly with some snow.
NOTES Shared its habitat with *Forfexopterus, Hongshanopterus, Istiodactylus sinensis, Sinopterus, Chaoyangopterus, Shenzhoupterus, Guidraco, Liaoningopterus, Ikrandraco, Vesperopterylus, Nurhachius.*

Nurhachius luei (Zhou et al. 2019)
1.75 m (5.75 ft) WS, 1.8 kg (4 lb)

FOSSIL REMAINS Skull and minority of skeleton.
ANATOMICAL CHARACTERISTICS Head very large, teeth along first third of jaws, not tightly packed. Feet small.
AGE Early Cretaceous, early or middle Aptian.

DISTRIBUTION AND FORMATIONS Northeastern China; lower Jiufotang.
HABITAT Well-watered highland forests and lakes, winters chilly with some snow. May be direct ancestor of *N. ignaciobritoi.*

Nurhachius ignaciobritoi (Wang et al. 2005)
1.75 m (5.75 ft) WS, 0.68 m (2.2 ft) TL, 1.8 kg (4 lb)

FOSSIL REMAINS Majority of skull and skeleton.
ANATOMICAL CHARACTERISTICS Head very large, teeth along first third of jaws, not tightly packed. Wingspan moderate relative to mass. Feet small.
AGE Early Cretaceous, early or middle Aptian.
DISTRIBUTION AND FORMATIONS Northeastern China; upper Jiufotang.
HABITAT Well-watered highland forests and lakes, winters chilly with some snow.

Luchibang xingzhe (Hone et al. 2020)
Adult size uncertain

FOSSIL REMAINS Majority of skull and skeleton.
ANATOMICAL CHARACTERISTICS Tooth rows very short. Inner hands long. Feet fairly large.
AGE Early Cretaceous, early Aptian.
DISTRIBUTION AND FORMATIONS Northeastern China; upper Yixian.
HABITAT Well-watered highland forests and lakes, winters chilly with some snow.
NOTES Shared its habitat with *Yixianopterus, Gladocephaloideus, Eopteranodon.*

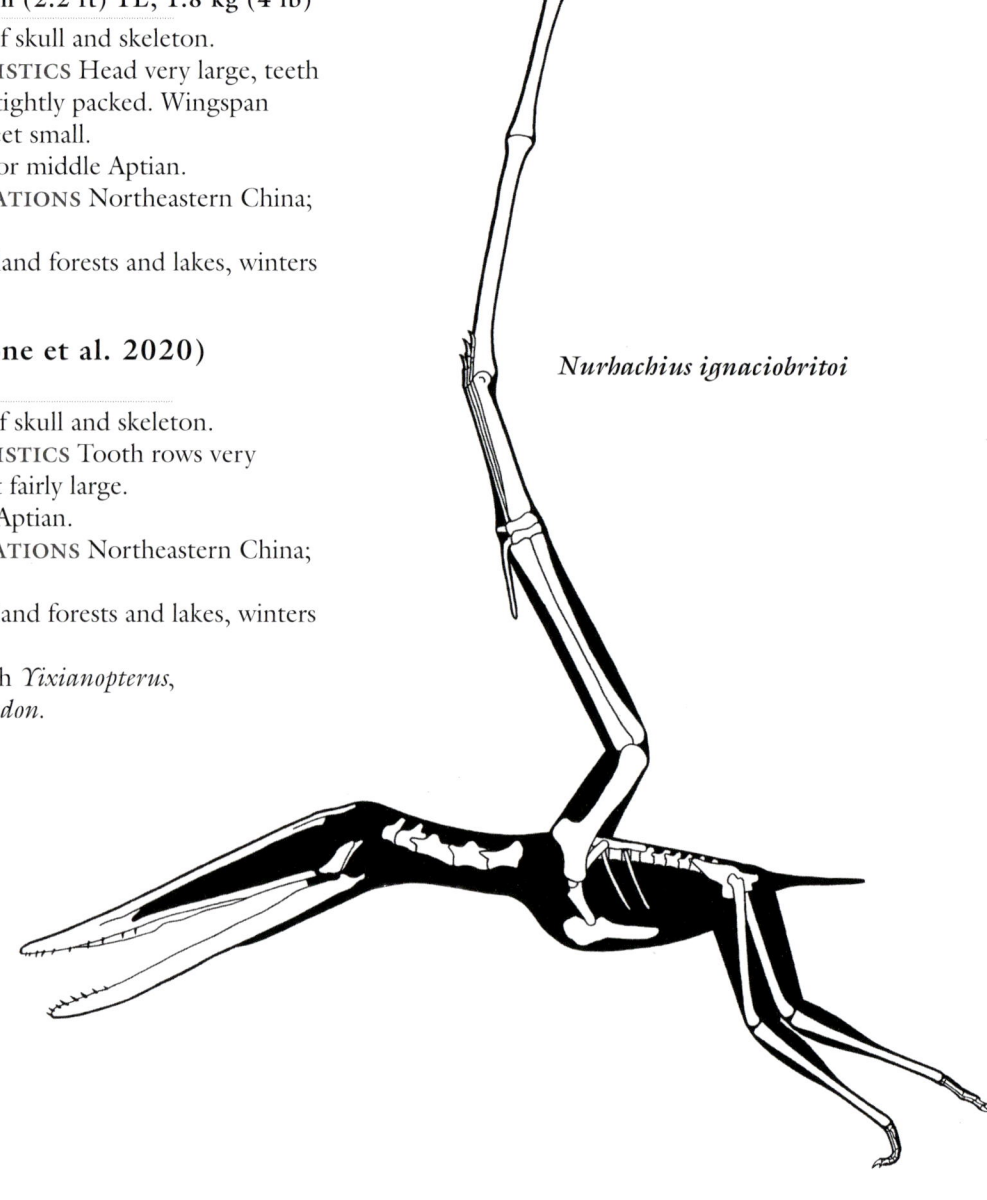

Nurhachius ignaciobritoi

Istiodactylus sinensis (Andres and Qiang 2006)
2.6 m (8.5 ft) WS, 0.75 m (2.5 ft) TL, 3 kg (7 lb)

FOSSIL REMAINS Majority of skull and skeleton.
ANATOMICAL CHARACTERISTICS Head length moderate, front teeth tightly packed to form interlocking shears. Wingspan very long relative to mass.
AGE Early Cretaceous, early or middle Aptian.
DISTRIBUTION AND FORMATIONS Northeastern China; Jiufotang.
HABITAT Well-watered highland forests and lakes, winters chilly with some snow.

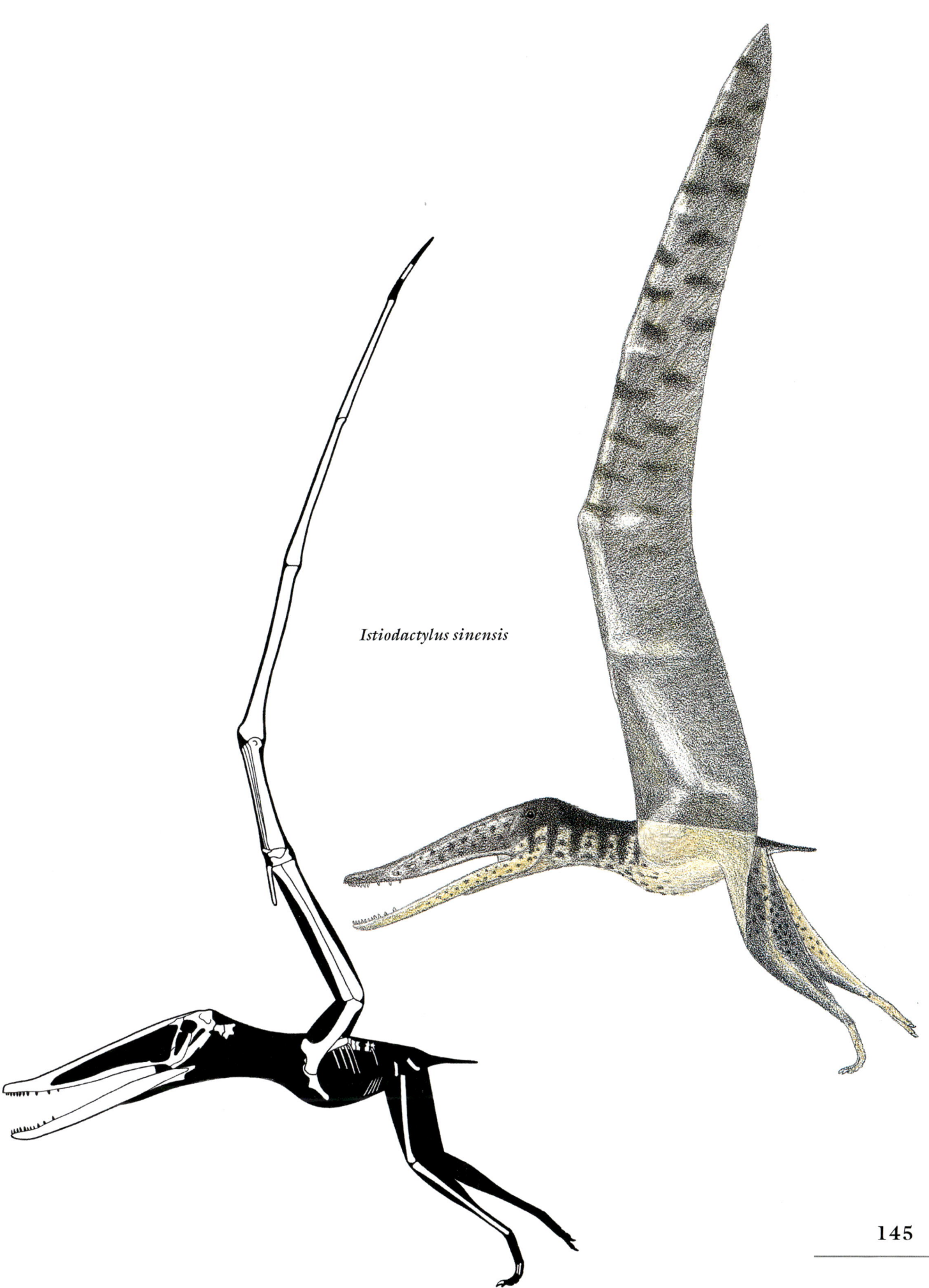

Istiodactylus sinensis

Istiodactylus sinensis

Istiodactylus latidens (Howse et al. 2001)
4 m (13 ft) WS, 10 kg (20 lb)

FOSSIL REMAINS Majority of skull and partial skeleton.
ANATOMICAL CHARACTERISTICS Teeth along first third of jaws, front teeth packed to form interlocking shears.
AGE Early Cretaceous, late Barremian.
DISTRIBUTION AND FORMATIONS Isle of Wight, England; Wessex.
HABITAT Coastal wetlands.
NOTES Fragmentary bones may indicate wingspans up to 8 m (25 ft).

AZHDARCHOIDS

MEDIUM-SIZED TO GIGANTIC EUPTERODACTYLOIDS OF THE EARLY CRETACEOUS TO THE END OF THE MESOZOIC ON MOST OR ALL CONTINENTS

ANATOMICAL CHARACTERISTICS Preorbital openings very large, orbits modest sized and low on head, toothless. Inner hands long but overall wings rather short, wing finger cross section shallow inverted T shape. Legs at least fairly long, more erect in at least some examples.
HABITS Terrestrial and shoreline, not strongly aquatic feeders.
NOTES At what stage of azhdarchoid evolution a more erect leg posture evolved is not certain.

TAPEJARIDS

MEDIUM-SIZED TO GIGANTIC AZHDARCHOIDS OF THE EARLY TO THE LATE CRETACEOUS OF THE NORTHERN AND SOUTHERN HEMISPHERES

ANATOMICAL CHARACTERISTICS Head fairly large, deep, subrectangular, back of head moderately broad, deep beak parrot-like, preorbital opening very large, long, and deep, deep bony crest atop snout, tapering bony crest atop back of head anchored very large soft tissue crest, lower jaw fairly robust, subtriangular bony crest projects below front lower jaw, jaw muscles fairly powerful in deeper-headed examples. Neck at least moderately long, slender. Inner arms fairly well developed. Toe bones arced in juveniles. Inner brachiopatagium attached near ankle.
GROWTH AND HABITS Curved toes of juveniles indicate they were more arboreal than adults. Flight performance moderate. Fruit eaters, possibly other vegetation and small-game omnivory.
NOTES Only pterosaur group that seems to have been markedly or predominantly herbivorous. Absence from at least some other continents probably reflects lack of sufficient sampling.

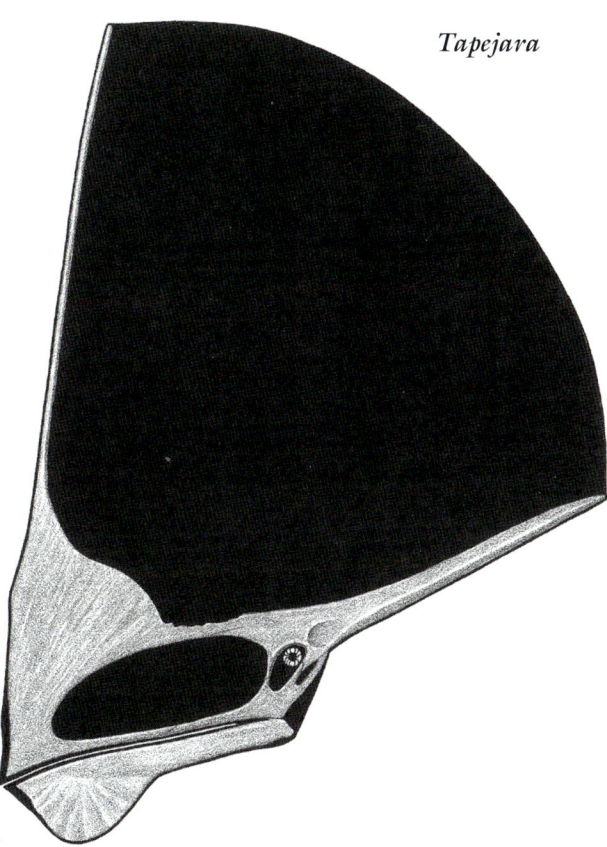

Tapejara

Sinopterus dongi (Wang and Zhou 2003a)
1.9 m (6 ft) WS, 0.75 m (2.5 ft) TL, 4 kg (9 lb)

FOSSIL REMAINS A number of skulls and skeletons, adult and juvenile, some complete or nearly so.

ANATOMICAL CHARACTERISTICS Bony crest on snout modest in size, bony crest atop back of skull subvertical in adults that are possible males, crest may be shorter and less vertical in females, upper edge of lower jaw not kinked. Neck moderately long. Wingspan rather short relative to mass.

AGE Early Cretaceous, early or middle Aptian.

DISTRIBUTION AND FORMATIONS Northeastern China; Jiufotang.

HABITAT Well-watered highland forests and lakes, winters chilly with some snow.

Europejara olcadesorum (Vullo et al. 2012)
2 m (6.5 ft) WS, 5 kg (10 lb)

FOSSIL REMAINS Minority of skull.

ANATOMICAL CHARACTERISTICS Deep bony crest below front of lower jaw.

AGE Early Cretaceous, early Barremian.

DISTRIBUTION AND FORMATIONS Spain; La Huérguina.

HABITAT Lake deposits.

Eopteranodon lii (Lü and Zhang 2005)
1.1 m (3.5 ft) WS, 0.7 kg (1.5 lb)

FOSSIL REMAINS Poorly preserved skull and majority of at least one skeleton.

ANATOMICAL CHARACTERISTICS Insufficient information.

AGE Early Cretaceous, early Aptian.

DISTRIBUTION AND FORMATIONS Northeastern China; upper Yixian.

HABITAT Well-watered highland forests and lakes, winters chilly with some snow.

NOTES Not a close relative of *Pteranodon*. Has not been adequately described. Shared its habitat with *Yixianopterus, Gladocephaloideus, Luchibang.*

Sinopterus dongi

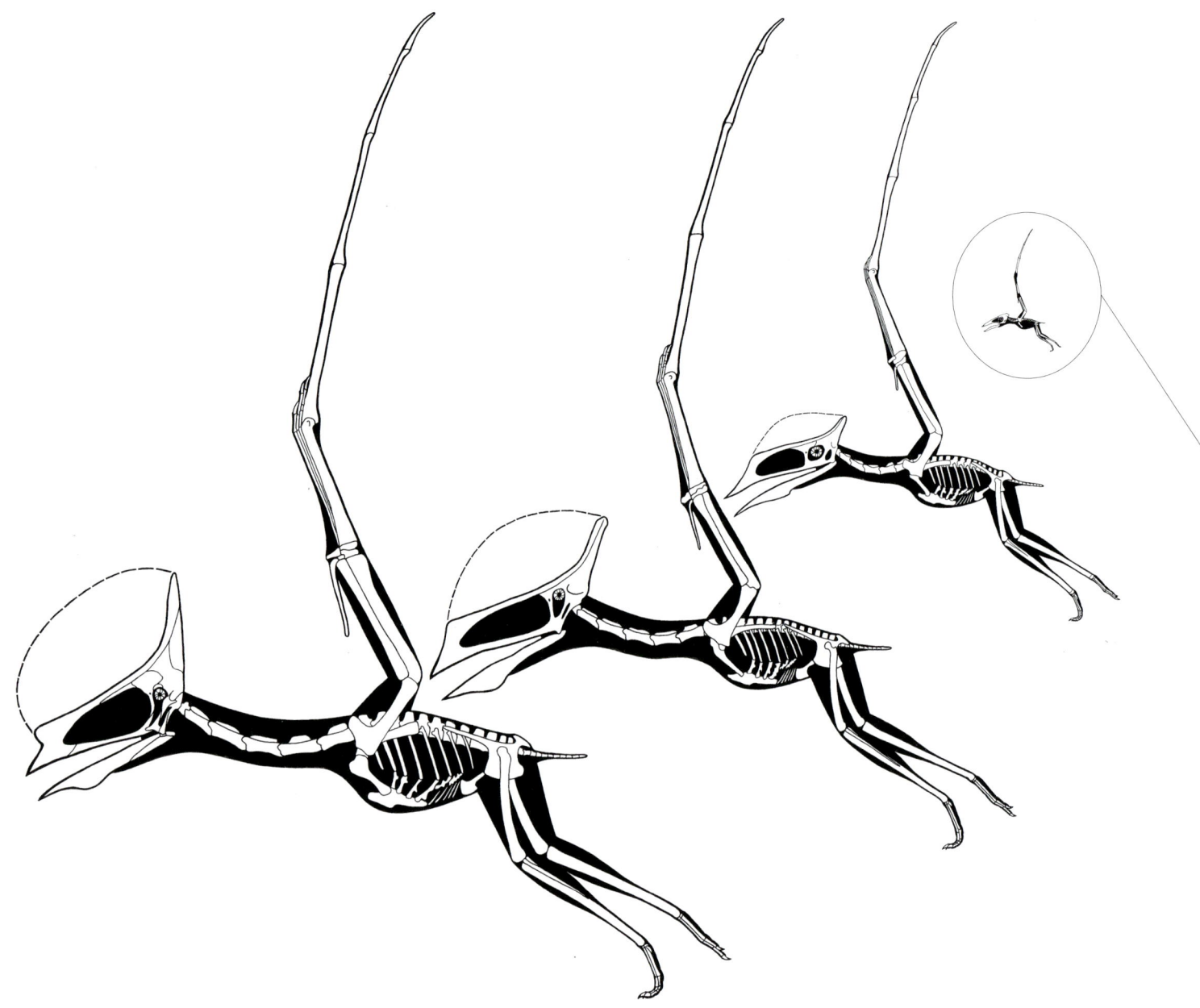

Sinopterus dongi, from left: adult, immature, large juvenile, and hatchling

GROWTH AND HABITS Presence of fossil hatchlings far away from nests indicates they were active independent fliers upon hatching. Seeds in abdominal cavity affirm frugivory.

NOTES Jiufotang specimens placed in a number of species of *Sinopterus*, *Huaxiapterus*, *Nemicolopterus* probably represent growth stages and possibly sexes of *S. dongi*. Fruit seeds found in abdominal region of a skeleton support frugivory of tapejaromorphs. Shared its habitat with *Vesperopterylus*, *Forfexopterus*, *Hongshanopterus*, *Nurhachius*, *Istiodactylus sinensis*, *Chaoyangopterus*, *Shenzhoupterus*, *Guidraco*, *Liaoningopterus*.

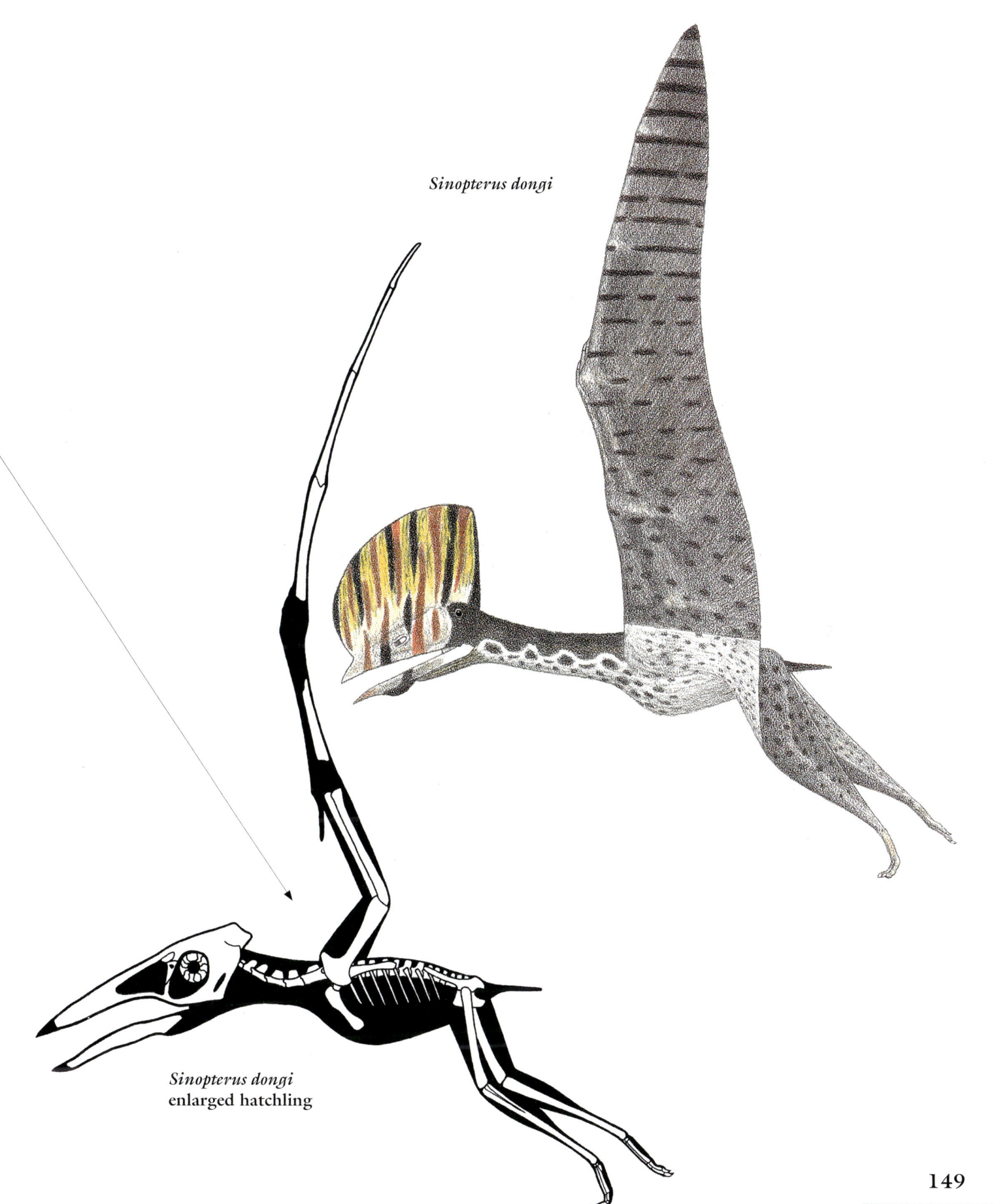

Sinopterus dongi

Sinopterus dongi
enlarged hatchling

Tapejara (or Tupandactylus) imperator (Pinheiro et al. 2011)
7–8 m (23–25 ft) WS, 150–200 kg (330–441 lb)

FOSSIL REMAINS A few skulls, none entirely complete, with partial soft tissue crests.
ANATOMICAL CHARACTERISTICS Bony and soft tissue crest forms an enormous, tall subtriangle with rounded back edge in adult males, crest tall but shorter in females, upper edge of lower jaw not kinked.
AGE Early Cretaceous, early Aptian?
DISTRIBUTION AND FORMATIONS Northeastern Brazil; Crato.
HABITAT Nearshore marine.
GROWTH AND HABITS Adults too large for climbing.
NOTES *Tupandactylus navigans* may be a female of this species. Size of this species has been underappreciated, earliest known gigantic pterosaur, may be heaviest known nonazhdarchid pterosaur. Shared its habitat with *Lacusovagus, Brasileodactylus, Arthurdactylus.*

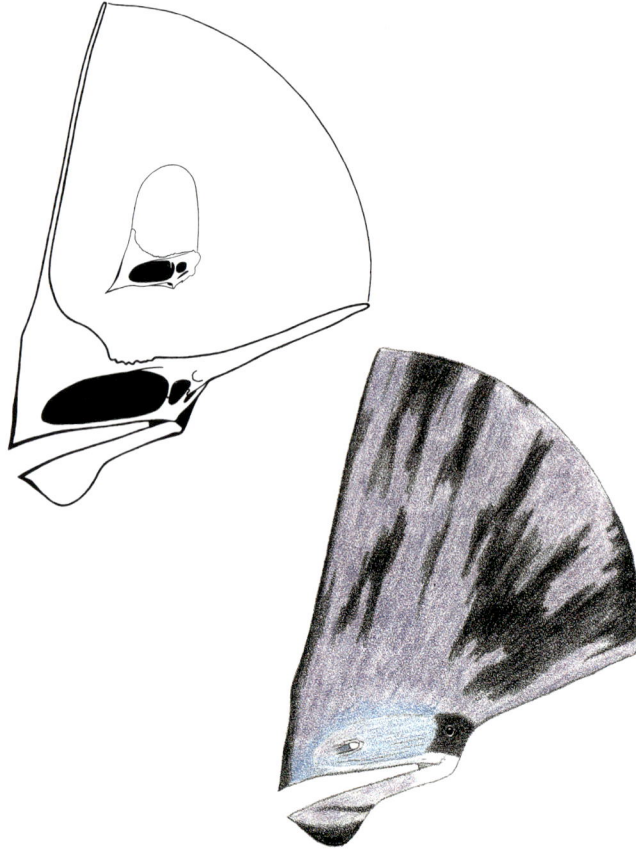

Tapejara imperator

Tapejara wellnhoferi
Adult size uncertain

FOSSIL REMAINS A few immature skulls and partial skeleton.
ANATOMICAL CHARACTERISTICS Shallow upward kink near front of lower jaw.
AGE Early Cretaceous, late Aptian.
DISTRIBUTION AND FORMATIONS Northeastern Brazil; Romualdo.
HABITAT Nearshore marine.
HABITS Configuration of bony eye rings may indicate flexible diurnal/nocturnal activity in this genus.
NOTES May be direct descendant of *T. imperator*, probably not as large as latter. Shared its habitat with *Tupuxuara, Anhanguera, Caupedactylus.*

Caupedactylus ybaka (Kellner 2013)
3 m (10 ft) WS, 15 kg (30 lb)

FOSSIL REMAINS Majority of skull and partial skeleton.
ANATOMICAL CHARACTERISTICS Insufficient information.
AGE Early Cretaceous, late Aptian.
DISTRIBUTION AND FORMATIONS Northeastern Brazil; Romualdo.
HABITAT Nearshore marine.

Caiuajara dobruskii (Manzig et al. 2014)
2.5 m (8 ft) WS, 8 kg (18 lb)

FOSSIL REMAINS Dozens of partial specimens, mainly juveniles to a few adults.
ANATOMICAL CHARACTERISTICS Bony snout crest subvertical, shallow upward kink near front upper edge of lower jaw.
AGE Late Cretaceous, Turonian.
DISTRIBUTION AND FORMATIONS Southern Brazil; Goio-Erê.
HABITAT Dune desert lake.
NOTES Shared its habitat with *Keresdrakon.*

Montanazhdarcho minor (McGowen et al. 2002)
2.5 m (8 ft) WS, 8 kg (18 lb)

FOSSIL REMAINS Minority of skeleton.
ANATOMICAL CHARACTERISTICS Insufficient information.
AGE Late Cretaceous, late Campanian.
DISTRIBUTION AND FORMATIONS Montana; upper Two Medicine.
HABITAT Well-watered, forested floodplain with coastal swamps and marshes, cool winters, drier uplands.
NOTES If a tapejarid, extends survival of group into late Late Cretaceous.

Javelinadactylus sagebieli (Campos 2021)
5 m (15 ft) WS, 50 kg (100 lb)

FOSSIL REMAINS Partial skull and possibly minority of skeleton.

ANATOMICAL CHARACTERISTICS No bony crest over snout. Neck may be very long.

AGE Late Cretaceous, middle Maastrichtian.

DISTRIBUTION AND FORMATIONS Southern Texas; lower Javelina.

HABITAT Seasonally dry coastal plain.

NOTES If a tapejarid rather than an azhdarchid, then extends survival of former group into very late and possibly to end of Late Cretaceous.

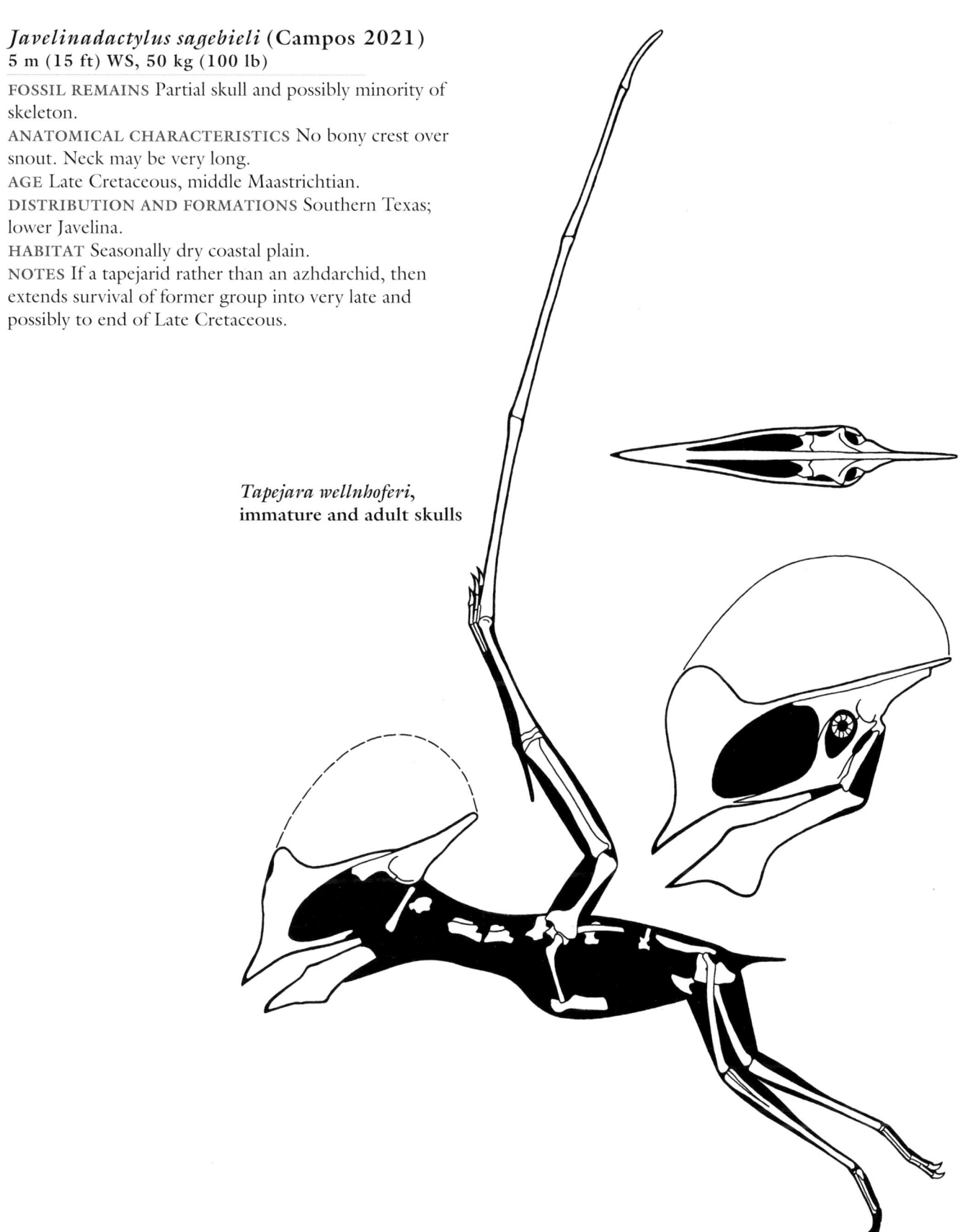

Tapejara wellnhoferi,
immature and adult skulls

NEO-AZHDARCHIANS

MEDIUM-SIZED TO GIGANTIC AZHDARCHOIDS OF THE EARLY CRETACEOUS TO THE END OF THE MESOZOIC ON MOST OR ALL CONTINENTS
ANATOMICAL CHARACTERISTICS Head large, at least fairly deep at back, subtriangular, preorbital openings large, bony crest atop back of head, lower jaws shallow. Inner wings at least fairly well developed. Legs long.
HABITS Flight performance modest to mediocre, some burst fliers. Good ground walkers, strong bite force, terrestrial and wading small-game predators and scavengers, not highly adapted for scavenging large carcasses. These relatively delicately built pterosaurs were vulnerable to attack by larger, more strongly built, blade-toothed, predaceous theropods, probably avoided proximity to large dinosaurian competitors for carcasses.
NOTES Straight-tipped beaks were not highly adapted for scavenging large carcasses.

NEOAZHDARCHIAN MISCELLANEA

Cretornis hlavaci (Averianov and Ekrt 2015)
1.5 m (5 ft) WS, 1.5 kg (3 lb)

FOSSIL REMAINS Majority of wing.
ANATOMICAL CHARACTERISTICS Insufficient information.
AGE Late Cretaceous, Turonian.
DISTRIBUTION AND FORMATIONS Czech Republic; Middle Iser Shales.
HABITAT Shallow interior seaway.

Ornithostoma sedgwicki (Smith et al. 2020)
Size uncertain

FOSSIL REMAINS Numerous fragments.
ANATOMICAL CHARACTERISTICS Insufficient information.
AGE Early Cretaceous, late Albian.
DISTRIBUTION AND FORMATIONS Southeastern England; Cambridge Greensand.
HABITAT Shallow interior seaway.
NOTES Shared its habitat with *Ornithocheirus*.

Chaoyangopterus zhangi

CHAOYANGOPTERIDS

MEDIUM-SIZED TO LARGE NEOAZHDARCHIANS OF THE EARLY/LATE CRETACEOUS OF THE NORTHERN AND SOUTHERN HEMISPHERES

ANATOMICAL CHARACTERISTICS Head deep at back, fairly deep beak sharp tipped, bony crest atop back of head moderate sized. Neck moderately long, slender. Inner wings fairly well developed, wingspan moderate relative to mass.

NOTES Absence from at least some other continents probably reflects lack of sufficient sampling.

Chaoyangopterus zhangi
(Wang and Zhou 2003b)
1.3 m (4 ft) WS, 0.7 m (2.2 ft) TL, 1.15 kg (2.5 lb)

FOSSIL REMAINS Two partial skulls with majority of skeletons.

ANATOMICAL CHARACTERISTICS As for group.

AGE Early Cretaceous, early or middle Aptian.

DISTRIBUTION AND FORMATIONS Northeastern China; Jiufotang.

HABITAT Well-watered highland forests and lakes, winters chilly with some snow.

HABITS At least some may have been adapted for feeding on hard-bodied invertebrates.

NOTES Probably includes *Jidapterus edentus*, may include *Shenzhoupterus chaoyangensis*. Shared its habitat with *Vesperopterylus*, *Forfexopterus*, *Hongshanopterus*, *Nurhachius*, *Istiodactylus sinensis*, *Sinopterus*, *Guidraco*, *Liaoningopterus*, *Shenzhoupterus*.

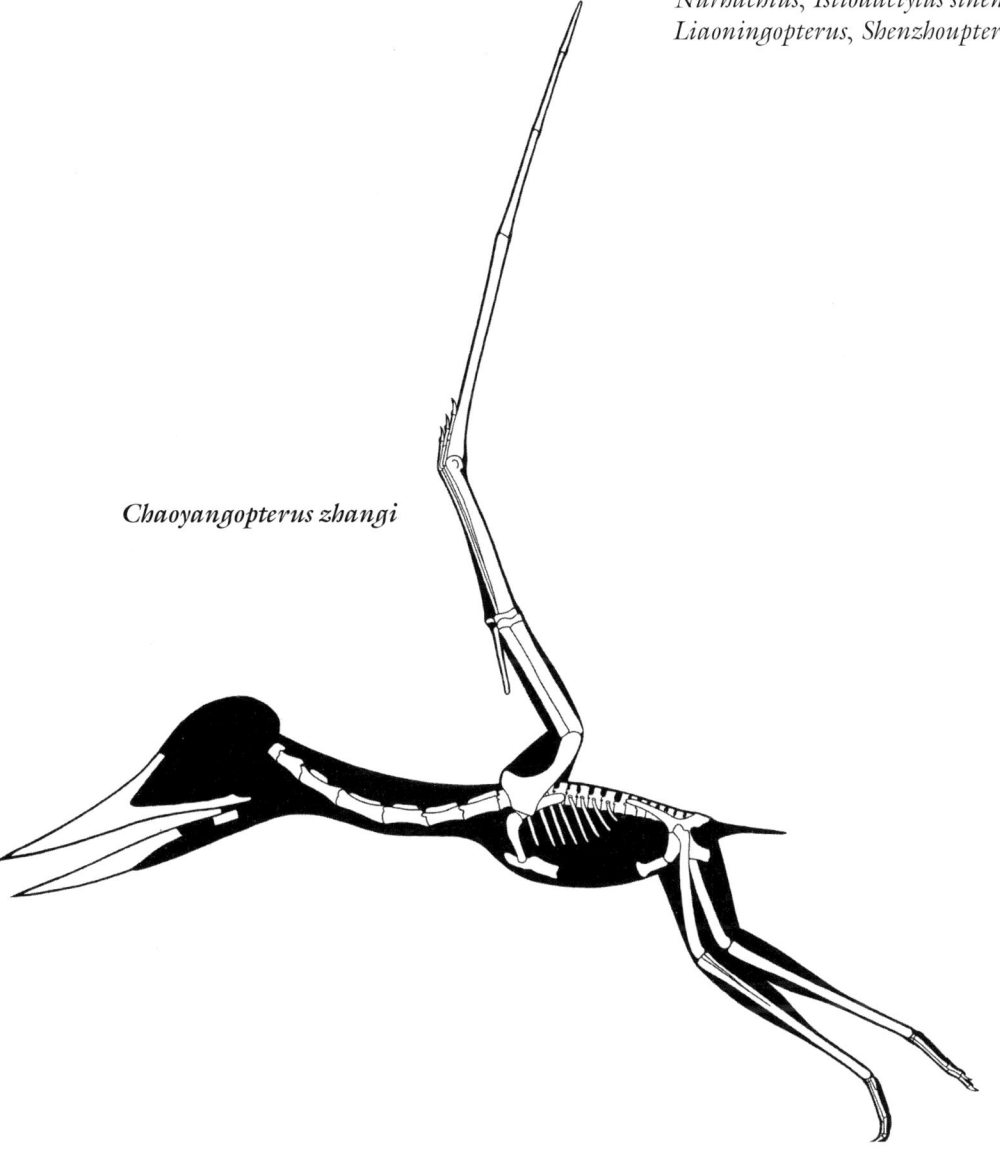

Chaoyangopterus zhangi

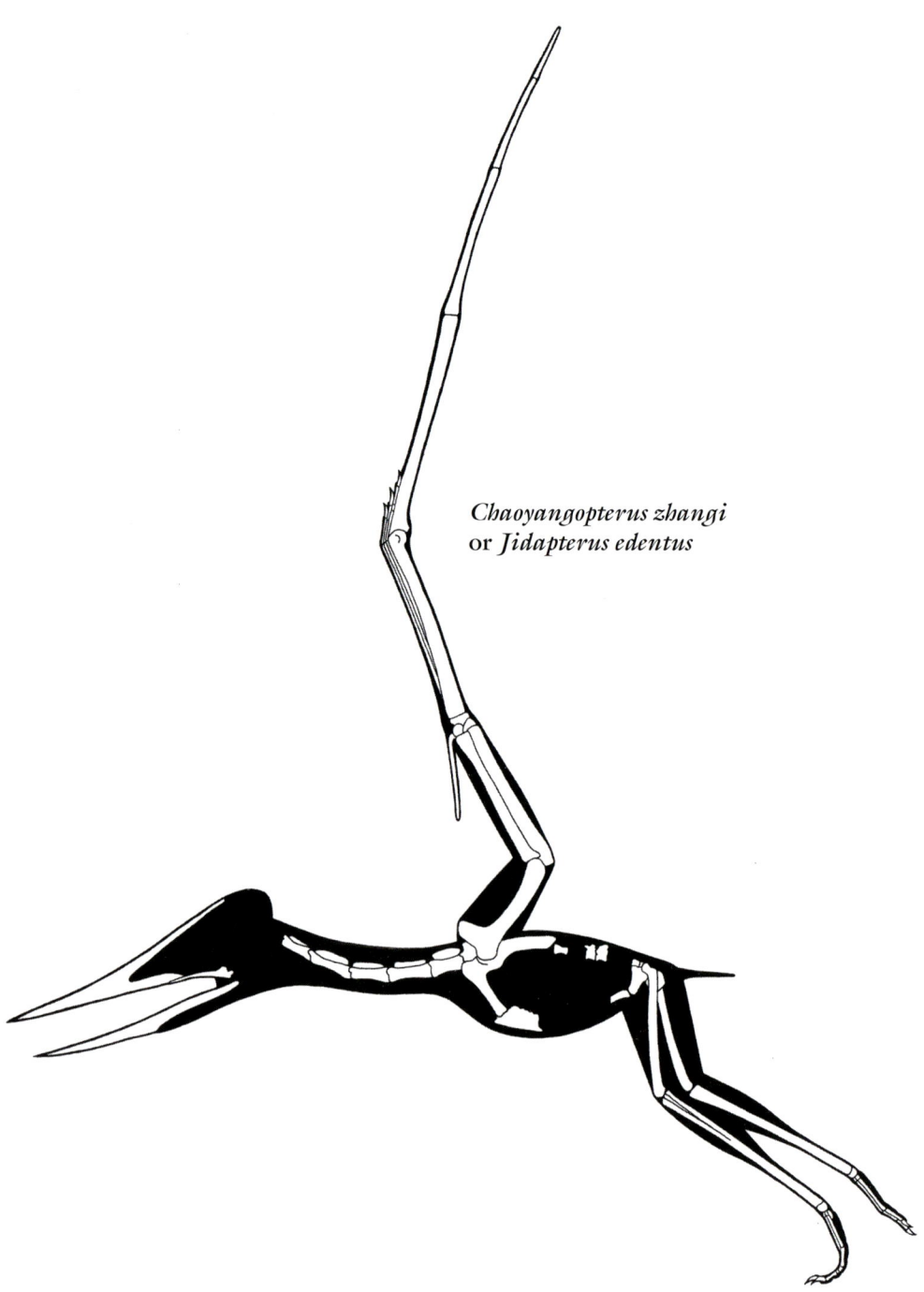

Chaoyangopterus zhangi
or *Jidapterus edentus*

Shenzhoupterus chaoyangensis?
(Lü et al. 2008)
1.3 m (4 ft) WS, 0.7 m (2.2 ft) TL, 1.1 kg (2.5 lb)

FOSSIL REMAINS Poorly preserved skull with majority of skeleton.

ANATOMICAL CHARACTERISTICS Bony head crest appears to be sharp tipped.

AGE Early Cretaceous, early or middle Aptian.

DISTRIBUTION AND FORMATIONS Northeastern China; Jiufotang.

HABITAT Well-watered highland forests and lakes, winters chilly with some snow.

NOTES. Badly damaged skull hinders understanding of its configuration compared to *Chaoyangopterus zhangi*, may be male of latter.

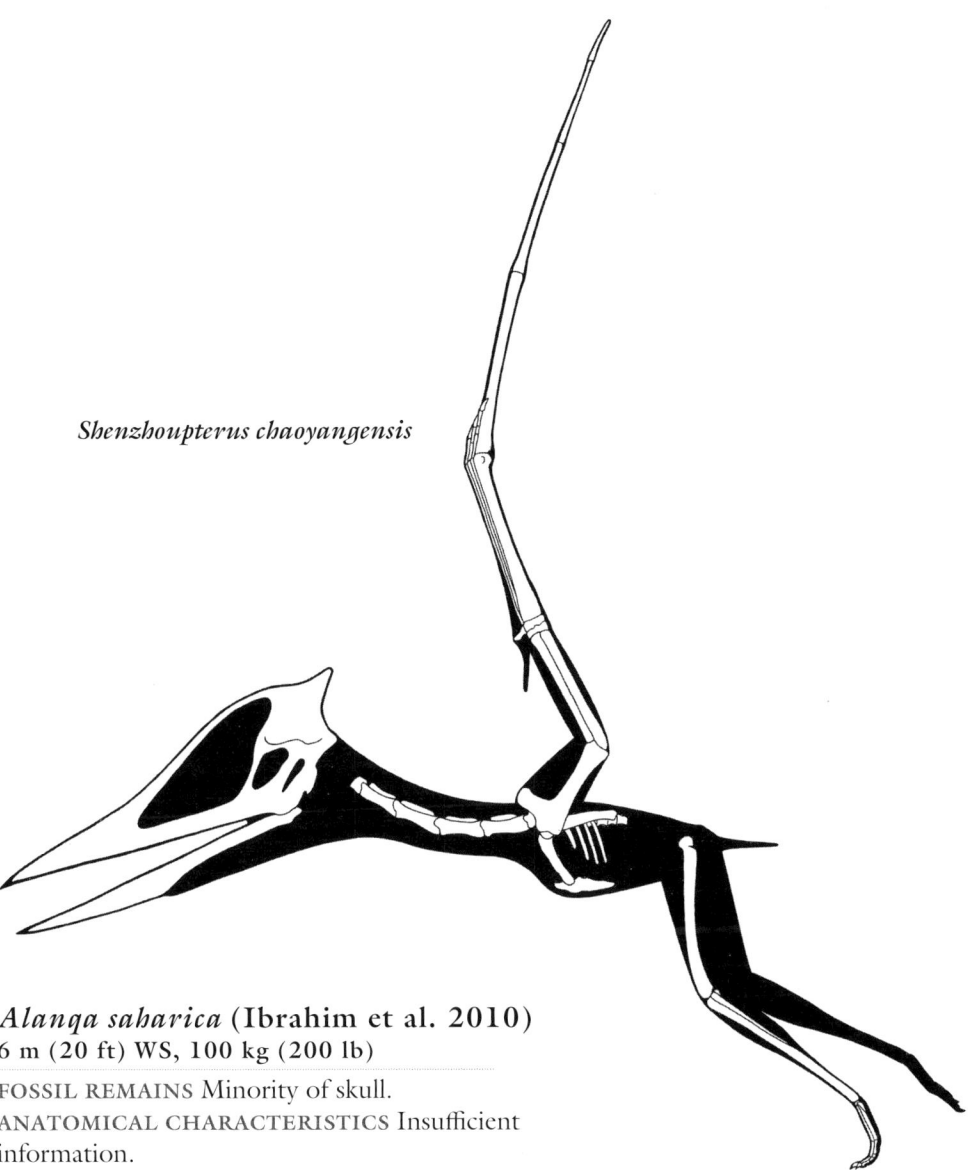

Shenzhoupterus chaoyangensis

Alanqa saharica (Ibrahim et al. 2010)
6 m (20 ft) WS, 100 kg (200 lb)

FOSSIL REMAINS Minority of skull.
ANATOMICAL CHARACTERISTICS Insufficient information.
AGE Late Cretaceous, early Cenomanian.
DISTRIBUTION AND FORMATIONS Morocco; Kem.
HABITAT Coastal mangroves.
NOTES Has been considered an azhdarchid. Shared its habitat with *Siroccopteryx*.

Lacusovagus magnificens (Witton 2013)
4 m (13 ft) WS, 30 kg (60 lb)

FOSSIL REMAINS Partial skull.
ANATOMICAL CHARACTERISTICS Beak lightly constructed.
AGE Early Cretaceous, early Aptian.
DISTRIBUTION AND FORMATIONS Northeastern Brazil; Crato.
HABITAT Shallow interior seaway.
NOTES May be a thalassodromid. Shared its habitat with *Tapejara imperator*, *Brasileodactylus*, *Arthurdactylus*.

THALASSODROMIDS

GIGANTIC NEOAZHDARCHIANS OF THE EARLY TO LATE CRETACEOUS OF SOUTH AMERICA

ANATOMICAL CHARACTERISTICS Head deep at back, bony head crest atop back of head large, deep. Neck slender. Notarium present.
NOTES Suggestions that these were surface-skimming fishers have been discredited. Absence from at least some other continents probably reflects lack of sufficient sampling.

Tupuxuara longicristatus (Witton 2013)
4.7 m (15 ft) WS, 2 m (6.5 ft) TL, 25 kg (55 lb)

FOSSIL REMAINS A few skulls and partial skeletons.
ANATOMICAL CHARACTERISTICS Crest along back of head deep and projecting fairly far backward, very large in males. Wings long relative to mass.
AGE Early Cretaceous, late Aptian.
DISTRIBUTION AND FORMATIONS Northeastern Brazil; Romualdo.
HABITAT Nearshore marine.
HABITS Configuration of bony eye rings may indicate diurnal activity for this genus and perhaps group.
NOTES A number of thalassodromid remains from the Romualdo Formation have been assigned to a large number of taxa, including *Thalassodromeus* *sethi, Tupuxuara leonardii, T. deliradamus, Banguela oberlii*. Some of these may be distinct taxa, either contemporaneous or stratigraphic—these possibilities are difficult to examine because many remains are incomplete, and there are no detailed stratigraphic data. But anatomical variation is limited largely to crests, the variation of which is broadly similar to that proposed to be present within species of *Pteranodon*. All but largest specimens are immature, so these are all presented herein as growth stages, individual variants, different sexes, and subspecies of a single gigantic species. The full extent of the bony crest, the border of which tends to be irregular and potentially broken, is often uncertain. The complete immature skeletal is a composite. Shared its habitat with *Tapejara wellnhoferi, Caupedactylus, Anhanguera.*

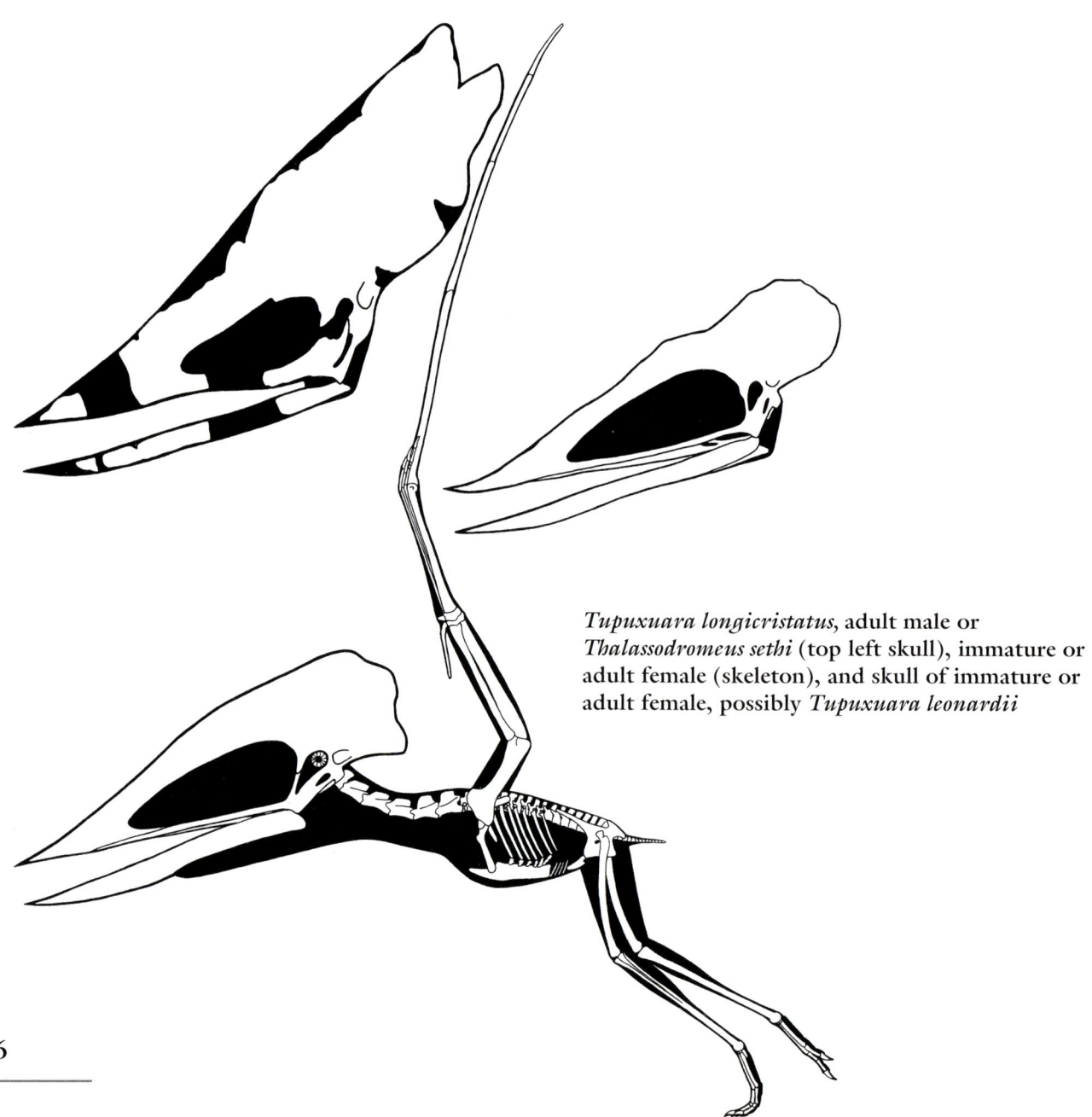

Tupuxuara longicristatus, adult male or *Thalassodromeus sethi* (top left skull), immature or adult female (skeleton), and skull of immature or adult female, possibly *Tupuxuara leonardii*

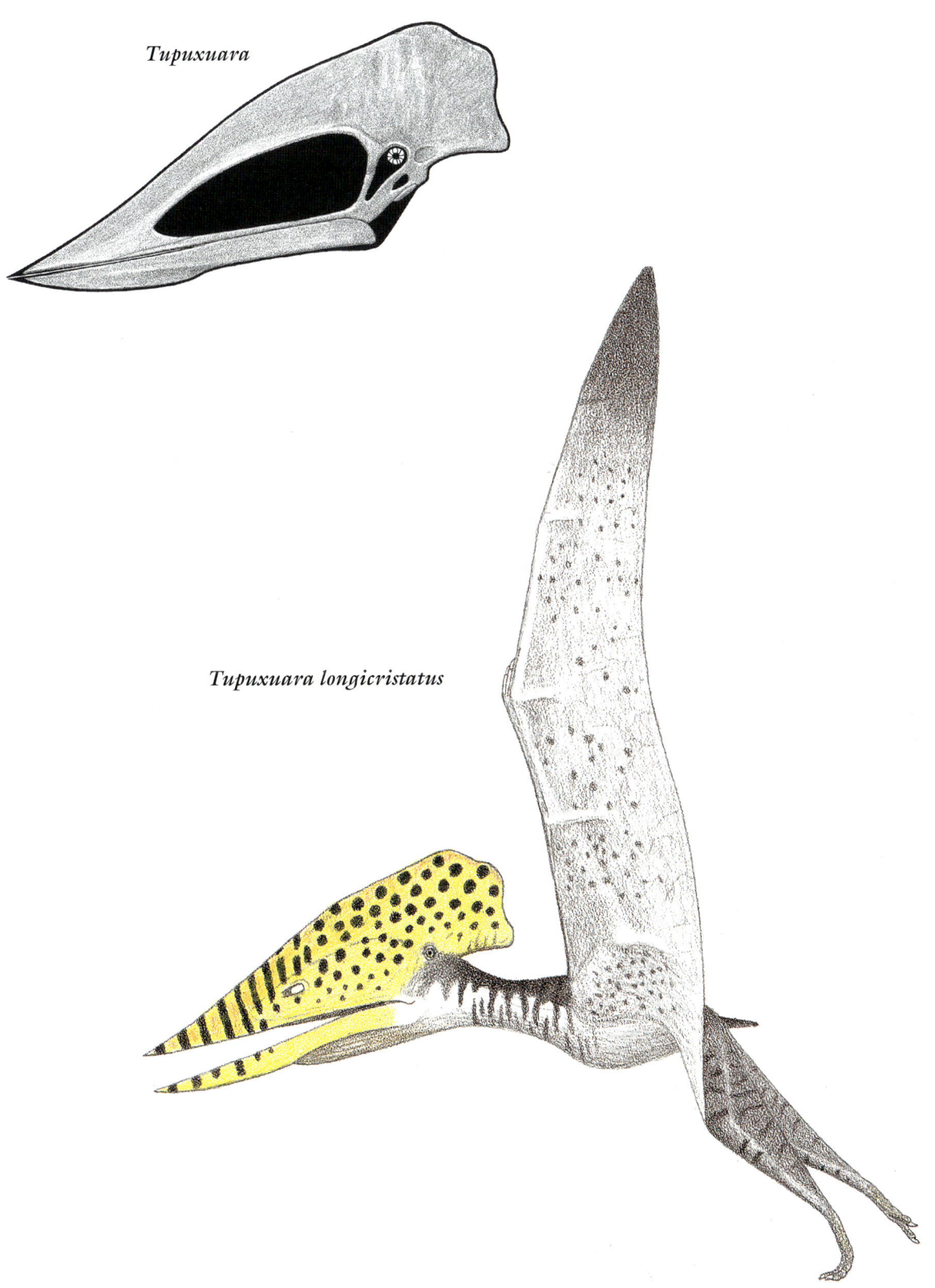

Tupuxuara

Tupuxuara longicristatus

Keresdrakon vilsoni (Kellner et al. 2019b)
Adult size uncertain

FOSSIL REMAINS Partial skulls and skeletons, immature.
ANATOMICAL CHARACTERISTICS Insufficient information.
AGE Late Cretaceous, Turonian.
DISTRIBUTION AND FORMATIONS Southern Brazil; Goio-Erê.
HABITAT Dune desert lake.
NOTES Exact position not certain. Shared its habitat with *Caiuajara*.

AZHDARCHIDS

LARGE TO GIGANTIC NEOAZHDARCHIANS OF THE LATE LATE CRETACEOUS OF THE NORTHERN AND SOUTHERN HEMISPHERES

ANATOMICAL CHARACTERISTICS Head not very deep, orbits located very low, rather small. Notarium present. Hip sockets face strongly downward so legs erect and narrow gauged, legs much shorter than inner wings, foot size moderate.
HABITAT Continental, including coastal.
HABITS May have been short-range burst fliers, or flap/glide fliers with more extended range.
NOTES Fragmentary remains suggest presence of group in early Early Cretaceous. Absence from at least some other continents probably reflects lack of sufficient sampling. These are the only pterosaurs known to have exceeded 200 kg (400 lb), derived members include the largest-known flying animals. Development of the most erect legs among pterosaurs before or within these azhdarchoids may have rendered such gigantism possible. No head crest is well preserved.

Aralazhdraco bostobensis
Adult size uncertain

FOSSIL REMAINS Minority of skeleton, possibly juvenile.
ANATOMICAL CHARACTERISTICS Insufficient information.
AGE Late Cretaceous, late Santonian or early Campanian.
DISTRIBUTION AND FORMATIONS Kazakhstan; Bostobe Svita.
HABITAT Coastal.

Volgadraco bogolubovi (Averianov et al. 2008)
Adult size uncertain

FOSSIL REMAINS Minority of skull and skeleton, possibly immature.
ANATOMICAL CHARACTERISTICS Insufficient information.
AGE Late Cretaceous, early Campanian.
DISTRIBUTION AND FORMATIONS Southwestern Russia; Rybushka.
NOTES *Bogolubovia orientalis* may be the same taxon. May be a pteranodontid.

Zhejiangopterus linhaiensis (Cai and Wei 1994)
2.75 m (9 ft) WS, 1.7 m (5.5 ft) TL, 6 kg (13 lb)

FOSSIL REMAINS Complete skull and a few skeletons of varying completeness, juvenile to adult.
ANATOMICAL CHARACTERISTICS Beak fairly long, small and rounded bony crest atop back of head at least in juveniles. Neck very long. Wingspan fairly long relative to mass.
AGE Late Cretaceous, early Campanian.
DISTRIBUTION AND FORMATIONS Northeastern China; Tangshang.
NOTES Although this is the most completely preserved azhdarchid, because the preserved skull is a juvenile, the form of the adult head is not certain.

Cryodrakon boreas (Hone et al. 2019)
10 m (33 ft) WS

FOSSIL REMAINS Minority of skeleton.
ANATOMICAL CHARACTERISTICS Neck long.
AGE Late Cretaceous, late Campanian.
DISTRIBUTION AND FORMATIONS Alberta; lower Dinosaur Park.
HABITAT Coastal.
NOTES Remains tentatively assigned to this species are known from higher in the Dinosaur Park Formation.

Aerotitan sudamericanus (Novas et al. 2012)
5 m (16 ft) WS

FOSSIL REMAINS Minority of skull.
ANATOMICAL CHARACTERISTICS Insufficient information.
AGE Late Cretaceous, early Maastrichtian.
DISTRIBUTION AND FORMATIONS Southern Argentina; upper Allen.
HABITAT Desert coastline.

OPPOSITE: *Tupuxuara longicristatus*

Unnamed genus and species
10–14 m (33–45 ft) WS

FOSSIL REMAINS Minority of skeleton.
ANATOMICAL CHARACTERISTICS Insufficient information.
AGE Late Cretaceous, Maastrichtian.
DISTRIBUTION AND FORMATIONS Transylvania; Sebes.
HABITAT Large forested island.
NOTES May have shared its habitat with *Eurazhdarcho*. Appears to be a contender for the largest known flying animal along with *Hatzegopteryx* and *Quetzalcoatlus northropi*.

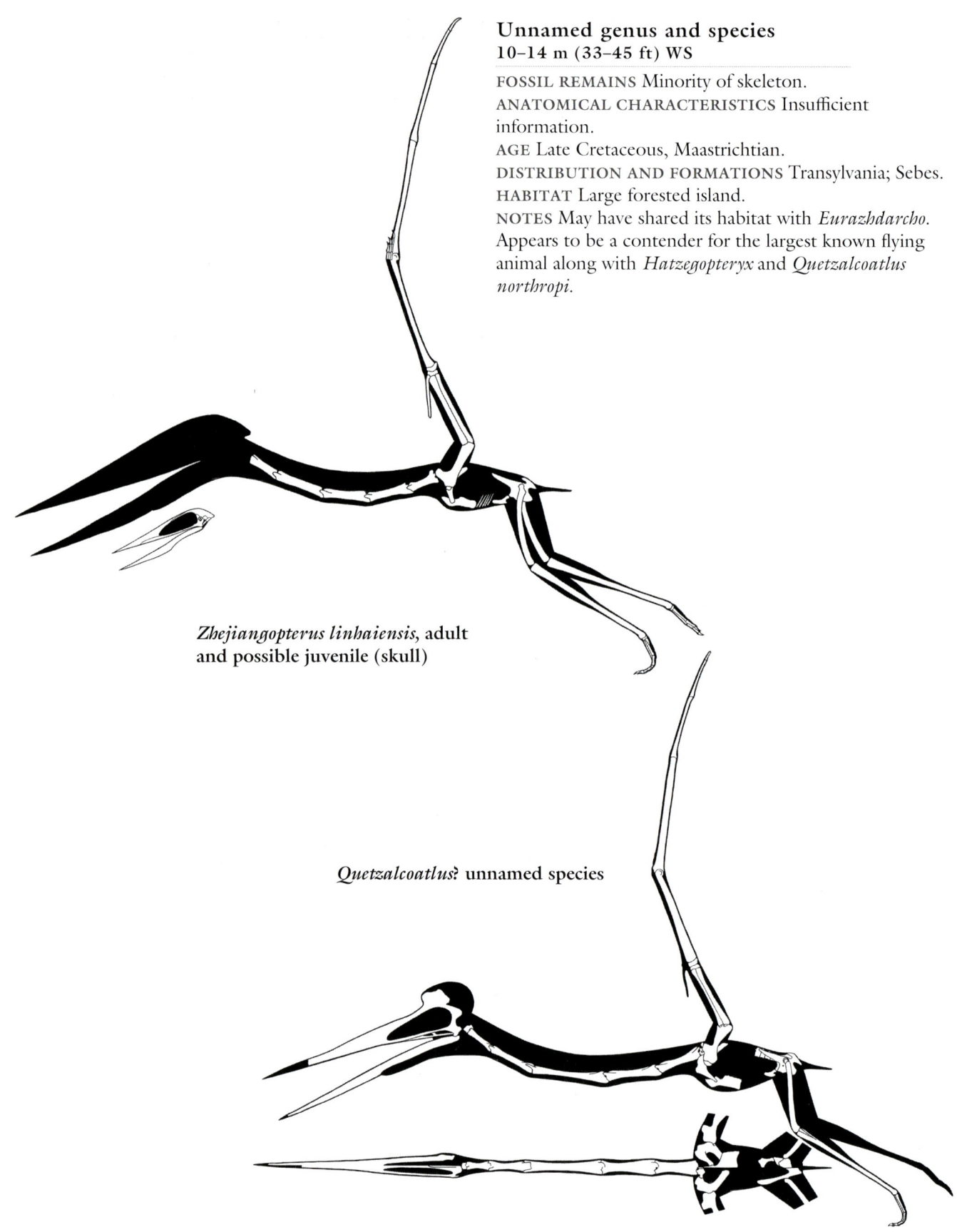

Zhejiangopterus linhaiensis, adult
and possible juvenile (skull)

Quetzalcoatlus? unnamed species

Eurazhdarcho langendorfensis
(Vremir et al. 2013)
3 m (10 ft) WS

FOSSIL REMAINS Minority of skeleton.
ANATOMICAL CHARACTERISTICS Neck long.
AGE Late Cretaceous, early Maastrichtian.
DISTRIBUTION AND FORMATIONS Transylvania; lower Sebes.
HABITAT Large forested island.

Arambourgiania philadelphiae
(Martill et al. 1998)
8–9 m (26–30 ft) WS

FOSSIL REMAINS Neck vertebra.
ANATOMICAL CHARACTERISTICS Neck very long.
AGE Late Cretaceous, Maastrichtian.

DISTRIBUTION AND FORMATIONS Jordan, Morocco? Phosphorite Unit, unnamed?
HABITAT Coastal.
NOTES Originally *Titanopteryx*, that name preoccupied by another organism. A partial vertebra from the late Campanian of Tennessee is unlikely to belong to this later species.

Quetzalcoatlus? unnamed species
5 m (16 ft) WS, 3.5 m (11 ft) TL, 65 kg (140 lb)

FOSSIL REMAINS Majority of skulls and skeletons.
ANATOMICAL CHARACTERISTICS Beak very long, bony crest atop back of head. Neck very long. Wingspan rather short relative to mass.
AGE Late Cretaceous, latest Maastrichtian.
DISTRIBUTION AND FORMATIONS Southern Texas; upper Javelina and lower Black Peaks.

Quetzalcoatlus

HABITAT Seasonally dry coastal plain.

NOTES Often labeled *Quetzalcoatlus* sp., this is not the juvenile of the much larger *Q. northropi*, whether these are the same genus is not certain.

Quetzalcoatlus northropi
10–11 m (33–36 ft) WS, 450–650 kg (1,000–1,400 lb)

FOSSIL REMAINS Minority of a few skeletons.

ANATOMICAL CHARACTERISTICS Inner arms very robust, wingspan probably rather short, or short relative to mass.

AGE Late Cretaceous, latest Maastrichtian.

DISTRIBUTION AND FORMATIONS Southern Texas; upper Javelina and lower Black Peaks.

HABITAT Seasonally dry coastal plain.

NOTES One of the iconic pterosaurs despite being a fragmentary, poorly known species. The common assumption that this giant possessed the same very elongated, shallow head of the much smaller azhdarchid from the same sediments is not certain. A contender for the largest-known flying animal. A redescription of this and prior two species is underway.

Phosphatodraco mauritanicus
(Longrich et al. 2018)
5 m (16 ft) WS

FOSSIL REMAINS Majority of necks.

ANATOMICAL CHARACTERISTICS Neck not exceptionally long.

AGE Late Cretaceous, late Maastrichtian.

DISTRIBUTION AND FORMATIONS Morocco; unnamed.

HABITAT Coastal.

NOTES Shared its habitat with *Tethydraco, Alcione*.

Hatzegopteryx thambema
(Naish and Witton 2017)
10–12 m (33–40 ft) WS

FOSSIL REMAINS Minority of a few skulls and skeletons.

ANATOMICAL CHARACTERISTICS Robustly built. Back of head broad. Neck not exceptionally long.

AGE Late Cretaceous, latest Maastrichtian.

DISTRIBUTION AND FORMATIONS Transylvania; middle Densus Ciula.

HABITAT Large forested island.

HABITS Most powerful pterosaur predator.

NOTES Preserved neck vertebrae are much shorter than those of other azhdarchids, but their position is uncertain, so neck length is not entirely certain. *Albadraco tharmisensis* may be a large juvenile of this species. A contender for the largest known flying animal.

ORNITHO-CHEIROIDS

MEDIUM-SIZED TO GIGANTIC EUPTERO-DACTYLOIDS OF THE EARLY CRETACEOUS TO THE END OF THE MESOZOIC ON ALL OCEANS AND CONTINENTS

ANATOMICAL CHARACTERISTICS Head long, shallow, preorbital openings small or modest sized. Neck length at most moderate, fairly robust. Trunk small, notarium present. Wingspan long relative to mass, free fingers small if present. Hindlimb length moderate and much shorter than inner wings, feet small.

HABITAT Freshwaters and oceans.

HABITS Emphasized aerial over ground locomotion, superb soarers. Fishers, predominantly on the wing, perhaps also while floating on shallow waters especially in some cases, scavenged floating and beached carcasses, stole food.

BOREOPTERIDS

LARGE ORNITHOCHEIROIDS OF THE EARLY CRETACEOUS OF ASIA

ANATOMICAL CHARACTERISTICS Head shallow, subrectangular, lower jaw shallow, teeth fairly slender, long at front, short at back. Inner wings not powerfully developed, inner hands fairly long.

HABITS Not powerful flapping fliers. Fishing, including shallow-water floating, small-game hunting.

NOTES Absence from at least some other continents may reflect lack of sufficient sampling.

Boreopterus cuiae (Lü and Ji 2005)
3.2 m (12 ft) WS, 1.3 m (4.3 ft) TL, 6 kg (13 lb)

FOSSIL REMAINS Two skulls and skeletons.

ANATOMICAL CHARACTERISTICS Head very large, shallow, snout straight, subrectangular bony crest above middle third of head, teeth along nearly entire jaws, project somewhat forward. Wingspan very long relative to mass. Feet very small.

AGE Early Cretaceous, early Aptian.

DISTRIBUTION AND FORMATIONS Northeastern China; lower Yixian.

HABITAT Well-watered highland forests and lakes, winters chilly with some snow.

NOTES *Zhenyuanopterus longirostris* is probably an adult of this species. Shared its habitat with *Elanodactylus, Eosipterus, Beipiaopterus, Haopterus, Moganopterus, Gegepterus, Pterofiltrus*.

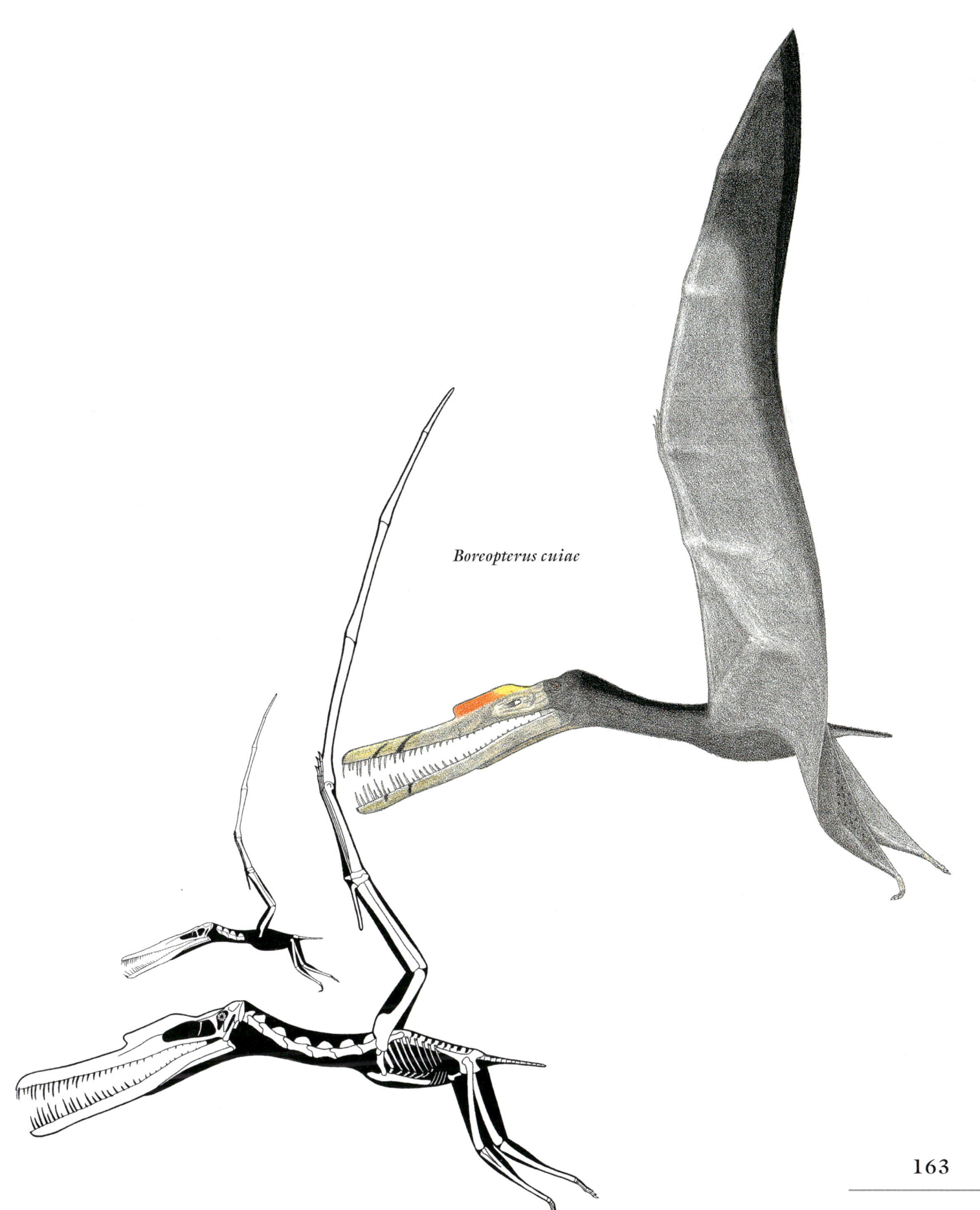

Boreopterus cuiae

Guidraco venator (Wang et al. 2012)
4.5 m (15 ft) WS, 15 kg (30 lb)

FOSSIL REMAINS Skull and minority of skeleton.
ANATOMICAL CHARACTERISTICS Fairly short, deep snout a little upcurved, rounded bony crest atop back of head, teeth along three-quarters of jaws, frontmost teeth very long, project strongly forward.
AGE Early Cretaceous, early or middle Aptian.
DISTRIBUTION AND FORMATIONS Northeastern China; Jiufotang.
HABITAT Well-watered highland forests and lakes, winters chilly with some snow.
NOTES Shared its habitat with *Vesperopterylus, Forfexopterus, Hongshanopterus, Nurhachius, Istiodactylus sinensis, Sinopterus, Chaoyangopterus, Shenzhoupterus, Liaoningopterus.*

Guidraco venator

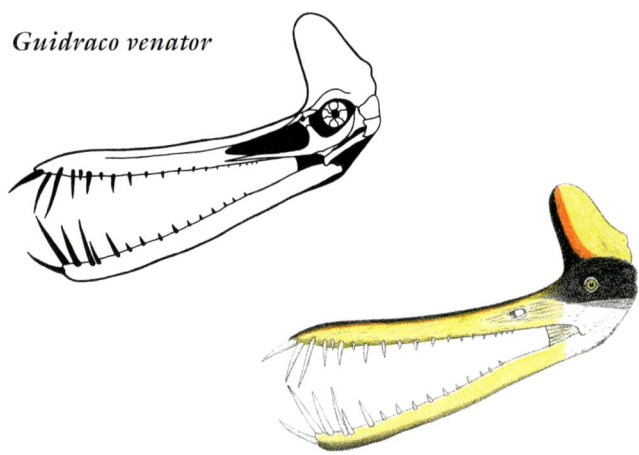

LONCHODECTIDS

LARGE TO GIGANTIC ORNITHOCHEIROIDS OF THE EARLY TO LATE CRETACEOUS

ANATOMICAL CHARACTERISTICS Beak large, toothed.
HABITAT Mostly marine.
HABITS Probably fishers, also consumed some invertebrates.
NOTES A very poorly known group, known largely and often only from fragments of jaws from Europe and South America labeled *Lonchodectes, Lonchodraco, Serradraco, Palaeornis, Prejanopterus, Unwindia,* none are sufficiently known, and relationships among them and with other pterodactyloids are very uncertain.

ORNITHO-CHEIRANS

GIGANTIC ORNITHOCHEIROIDS OF THE EARLY CRETACEOUS TO THE END OF THE MESOZOIC, ALL OCEANS

ANATOMICAL CHARACTERISTICS Head subtriangular, teeth small or absent, jaw muscles not powerful. Wings very long.
HABITS Good flapping fliers as well as excellent soarers. Highly marine. Nested in isolated locations lacking nonaerial predators, especially oceanic islands. Modest bite force suitable for snatching weak prey. Probably tracked large predaceous sea reptiles, sharks, and fish in order to dip feed on small fish and cephalopods disturbed, injured, and cut up by undersea hunters.

HAMIPTERIDS

LARGE ORNITHOCHEIRANS OF THE EARLY CRETACEOUS, NORTHERN OCEANS

ANATOMICAL CHARACTERISTICS Long, shallow, bony crest atop most of head, larger crests probably on males, teeth widely spaced, stout, large at front.
HABITAT Freshwater courses and coastal.
NOTES Absence from southern oceans may reflect lack of sufficient sampling.

Hamipterus tianshanensis (Wang et al. 2015)
3.5 m (11 ft) WS, 10 kg (20 lb)

FOSSIL REMAINS Skulls, partial skeletons, hundreds of eggs, some containing embryos.
ANATOMICAL CHARACTERISTICS Head long, shallow, snout long, preorbital opening size modest, teeth along over first half of jaws, project forward.
AGE Early Cretaceous.
DISTRIBUTION AND FORMATIONS Northeastern China; Shalidatun.
HABITAT Well-watered forests and lakes, winters chilly with some snow.
GROWTH AND HABITS Poor development of embryos may indicate hatchlings were not flight capable.
NOTES Some paleozoologists contend that embryos were not yet hatchlings and therefore do not indicate initial flight capability. Skeletons have not yet been detailed.

Hamipterus tianshanensis

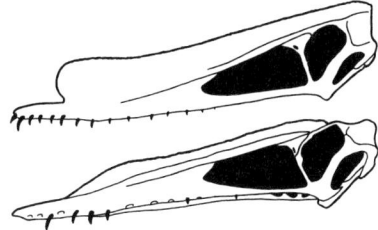

Iberodactylus andreui (Holgado et al. 2019)
4 m (13 ft) WS, 15 kg (30 lb)

FOSSIL REMAINS Partial skull.
ANATOMICAL CHARACTERISTICS Insufficient information.
AGE Early Cretaceous, middle Barremian.
DISTRIBUTION AND FORMATIONS Northeastern Spain; middle Blesa.
HABITAT Nearshore marine.

ORNITHOCHEIRIDS

GIGANTIC ORNITHOCHEIRANS OF THE EARLY TO LATE CRETACEOUS, ALL OCEANS

ANATOMICAL CHARACTERISTICS Snout tip not pointed, lower jaw shallow, teeth fairly robust, long at front, short at back. Wingspan very long relative to mass even though inner hand length moderate.
HABITAT Fresh and salt water.
NOTES Same family as anhanguerids. May have made it into earliest Late Cretaceous. Marine examples presumably possessed salt glands to deal with high sodium intake.

Liaoningopterus gui (Wang and Zhou 2003b)
5 m (16 ft) WS, 20 kg (45 lb)

FOSSIL REMAINS Majority of skull and minority of skeleton.
ANATOMICAL CHARACTERISTICS Snout crest shallow, if present.
AGE Early Cretaceous, early or middle Aptian.
DISTRIBUTION AND FORMATIONS Northeastern China; Jiufotang.
HABITAT Well-watered highland forests and lakes, winters chilly with some snow.
NOTES Skull so badly damaged it is difficult to restore. Shared its habitat with *Vesperopterylus*, *Forfexopterus*, *Hongshanopterus*, *Nurhachius*, *Istiodactylus sinensis*, *Sinopterus*, *Chaoyangopterus*, *Shenzhoupterus*, *Guidraco*.

Mythunga camara (Pentland and Poropat 2018)
5 m (16 ft) WS, 20 kg (45 lb)

FOSSIL REMAINS Minority of skull.
ANATOMICAL CHARACTERISTICS Snout crest large, subtriangular, teeth subvertical.
AGE Early Cretaceous, middle Albian.
DISTRIBUTION AND FORMATIONS Northeastern Australia; Toolebuc.
HABITAT Shallow interior seaway.

Ferrodraco lentoni (Pentland et al. 2019)
4 m (13 ft) WS, 10 kg (20 lb)

FOSSIL REMAINS Minority of skull and skeleton.
ANATOMICAL CHARACTERISTICS Snout crest subtriangular, apex a little aft.
AGE Late Cretaceous, late Cenomanian or early Turonian.
DISTRIBUTION AND FORMATIONS Northeastern Australia; Winton.
HABITAT River and estuary deposits, largely terrestrial.
HABITS Appears to have been less marine than other ornithocheirids.
NOTES May be the last known ornithocheirid.

Brasileodactylus araripensis
4 m (13 ft) WS, 10 kg (20 lb)

FOSSIL REMAINS Two skulls, adult and juvenile, latter with minority of skeleton.
ANATOMICAL CHARACTERISTICS No bony crest on snout, bony crest atop back of head.
AGE Early Cretaceous, late Aptian and/or Albian.
DISTRIBUTION AND FORMATIONS Northeastern Brazil; Crato.
HABITAT Nearshore marine.
NOTES *Ludodactylus sibbicki* may be adult of this species. Whether a large, stiff plant frond found between right and left mandibles was cause of adult specimen's death is uncertain. Shared its habitat with *Tapejara imperator*, *Lacusovagus*, *Arthurdactylus*.

Brasileodactylus araripensis

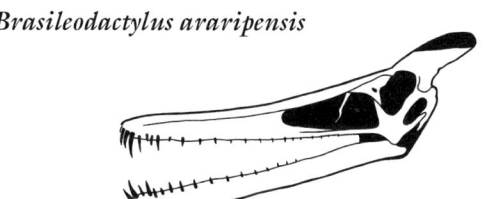

Arthurdactylus? conandoylei
(Frey and Martill 1994)
4.5 m (15 ft) WS, 15 kg (30 lb)

FOSSIL REMAINS Majority of skeleton.
ANATOMICAL CHARACTERISTICS Insufficient
information.
AGE Early Cretaceous, late Aptian and/or Albian.
DISTRIBUTION AND FORMATIONS Northeastern Brazil;
Crato.
HABITAT Nearshore marine.
NOTES May be same genus as *Anhanguera* and/or
Ornithocheirus. May be direct ancestor of *A. blittersdorfi*.

Anhanguera blittersdorfi
(Pinheiro and Rodrigues 2017)
9.5 m (30 ft) WS, 100+ kg (200+ lb)

FOSSIL REMAINS A number of skulls and partial
skeletons.
ANATOMICAL CHARACTERISTICS Curved bony crest
atop and beneath snout expands forward and becomes
much deeper and blunter tipped with increasing size
until upper crest is half-moon shaped, perhaps especially
on males, front teeth change from large and sharp to
relatively smaller and less sharp with increasing size.
AGE Early Cretaceous, late Aptian?

Anhanguera blittersdorfi

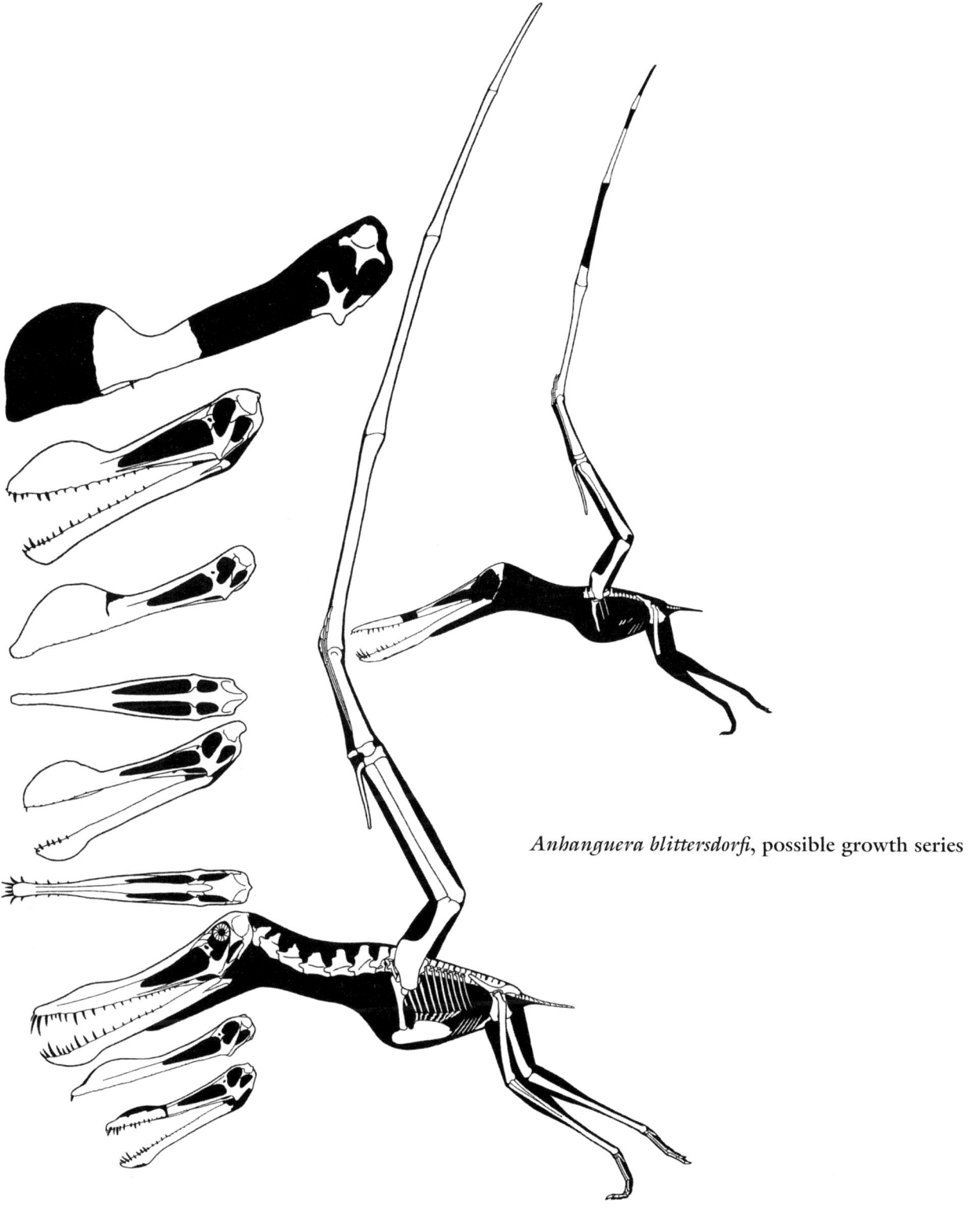

Anhanguera blittersdorfi, possible growth series

DISTRIBUTION AND FORMATIONS Northeastern Brazil; Romualdo.
HABITAT Nearshore marine.
GROWTH AND HABITS Strong sexual dimorphism indicates males probably fought each other to establish breeding harems. Did not become oceanic until wingspan was 4 m (12 ft) and mass 10 kg (20 lb), then continued to mature to adult size over a few years, and then returned to isolated locations to reproduce. Juveniles may have grown rapidly to marine size under parental care over some months.

NOTES A number of ornithocheirid remains from Romualdo Formation have been assigned to a large number of taxa including, in general order of decreasing size, *Tropeognathus* (or *Ornithocheirus*) *mesembrinus*, *Anhanguera piscator*, *A. spielbergi*, *A. robustus*, *A.* (or *Araripesaurus*) *santanae*, *Cearadactylus atrox*, *Barbosania gracilirostris*, *B.* (or *Santanadactylus*) *araripensis*. Some of these may be distinct taxa, either contemporaneous or stratigraphic. These possibilities are difficult to examine because many remains are incomplete, and there are no detailed stratigraphic data. But anatomical variation is limited largely to snout crests, which tend to become deeper and more forward expanded with size, and teeth that tend to become relatively smaller and less sharp. Variation is broadly similar to that proposed to be present within species of *Pteranodon*, and all but largest specimens are immature, so these are all presented herein as growth stages, individual variants, different sexes, and subspecies of a single gigantic species. If *Anhanguera* includes blunt-snouted *Tropeognathus*, then this may be the same genus as *Ornithocheirus*. Complete skeletal is a composite. Largest specimen is a partial humerus. May have been the largest-spanned nonazhdarchid pterosaur. Second most lightly built known pterosaur relative to wingspan. Shared its habitat with *Tapejara wellnhoferi*, *Caupedactylus*, *Tupuxuara*.

Ornithocheirus simus
(Rodrigues and Kellner 2013)
5 m (16 ft) WS, 20 kg (40 lb)

FOSSIL REMAINS A number of jaw tips.
ANATOMICAL CHARACTERISTICS Snout crest half-moon shaped.
AGE Early Cretaceous, late Albian.
DISTRIBUTION AND FORMATIONS Southern England; Cambridge Greensand.
HABITAT Shallow interior seaway.
HABITS Although ornithocheirids are generally seen as fishers, tooth microwear patterns of some examples with deep jaw tips suggest a broader diet including soft invertebrates and small game.
NOTES Some Early Cretaceous specimens from England with deep jaw tips such as *Coloborhynchus* may or may not be in this genus. Shared its habitat with *Ornithostoma*.

Anhanguera blittersdorfi

PTERANODONTS

MEDIUM-SIZED TO GIGANTIC ORNITHO-
CHEIRANS OF THE LATE CRETACEOUS,
MOST OR ALL OCEANS

ANATOMICAL CHARACTERISTICS Head large, bony
crest restricted to back of head, large bony crest probably
on males only, lower jaw fairly deep at midlength,
toothless. Wingspan very long relative to mass in part
because inner hands very long and inner wings well
developed.
HABITAT Do not appear to have soared over polar waters
north of US-Canadian border.
GROWTH Head crests did not become large until near
full adult body size.
NOTES Shortest-necked pterodactyloids.

Pteranodon

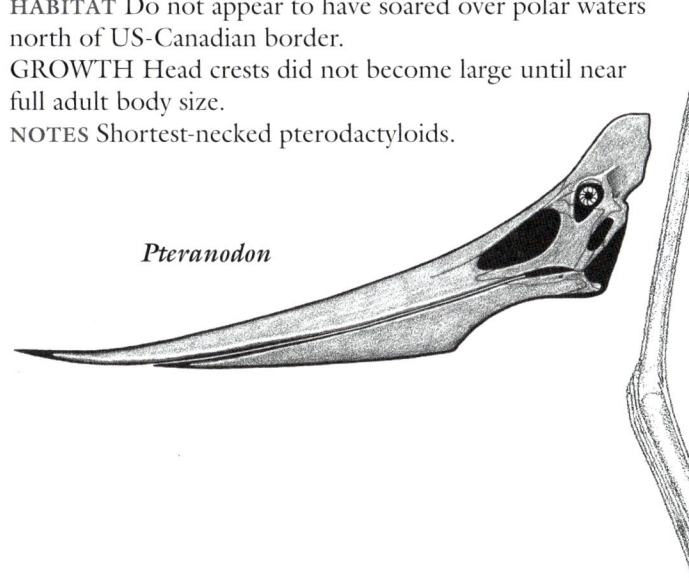

Pteranodon muscle study

PTERANODONTIDS

GIGANTIC PTERANODONTS OF THE LATE
CRETACEOUS, MOST OR ALL OCEANS

ANATOMICAL CHARACTERISTICS Orbits set high
and rather small, head crest moderately tall or long at
least in males, lower jaw shorter than upper. Notarium
well developed. Trunk compact. Tail ends as long rod.
Probable males much larger than probable females.
GROWTH AND HABITS Strong sexual dimorphism
indicates males probably fought each other to establish
breeding harems. *Pteranodon* did not become oceanic
until wingspan was 3 m (10 ft) and mass 5 kg (10 lb),
then continued to mature to adult size
over a few years, and then returned
to isolated locations to reproduce.
Juveniles may have grown rapidly
to marine size under parental
care over some months—up to
a year as in large albatrosses—
or they may have left their
nests soon after hatching
and led nonmarine, coastal
lifestyles over some years
until sufficiently mature
to become fully oceanic
as large juveniles. Latter
appears to be supported
by presence of a 1.75 m
(6 ft) pteranodontid wing
in Niobrara Formation,
but that fossil may belong
to a medium-sized
pteranodontid
species that

normally dwelled elsewhere than the Niobrara seaway,
and the scarcity of pteranodontid remains in Late
Cretaceous coastal deposits compared to neoazhdarchian
pterodactyloids indicates that pteranodonts were
inherently highly marine.
NOTES May be most extreme known open-ocean
pterosaurs. Although *Pteranodon* is known from a very
large number of specimens, none are entirely complete.
Whether the long tail rod supported wing membranes,
a separate elevator or rudder surface, or other soft tissue
structure is not certain.

Pteranodon sternbergi (Martin-Silverstone et al. 2017)
5 m (17 ft) WS, 2.3 m (7.5 ft) TL,
15 kg (30 lb)

FOSSIL REMAINS Dozens of specimens
of varying completeness, adult and large
juvenile.
ANATOMICAL CHARACTERISTICS Beak
upcurved, very long especially in adult
males relative to main body of skull and
skeleton, lower jaw much shorter than
upper, broad head crest subvertical, leading
edge strongly convex, trailing edge straight.
AGE Late Cretaceous, middle Coniacian to
early Santonian.
DISTRIBUTION AND FORMATIONS Kansas;
middle Niobrara.
HABITAT Shallow interior seaway at a maximum
in breadth and depth.
HABITS *P. sternbergi* fossils are abundant in

Pteranodon sternbergi, juvenile (left), immature
male? (centre), and adult male (right)

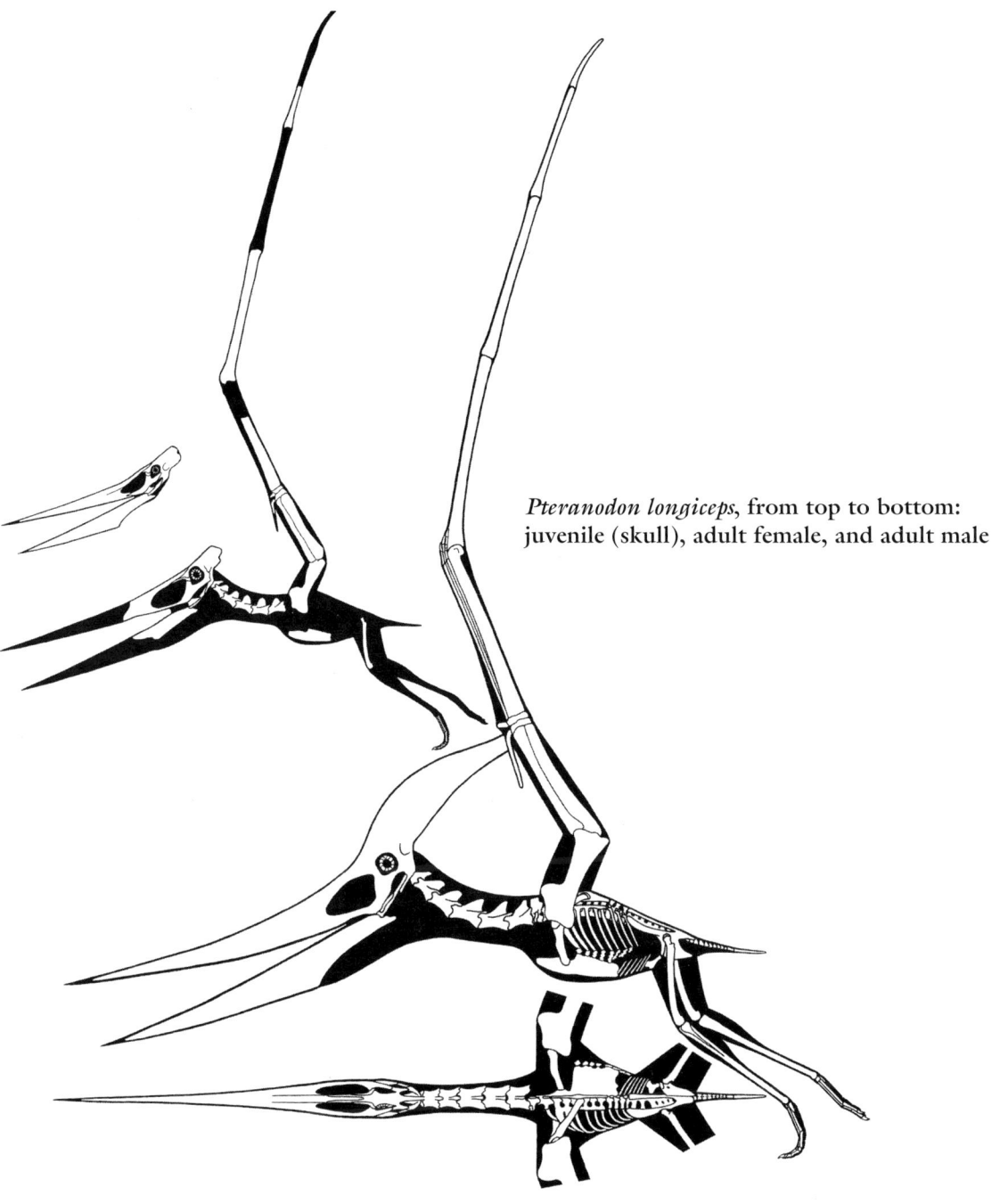

Pteranodon longiceps, from top to bottom: juvenile (skull), adult female, and adult male

the seaway when it was exceptionally deep and broad, indicating that at least some pteranodonts were not coastal pterosaurs.

NOTES Not sufficiently different from *P. longiceps* to be the distinct genera *Geosternbergia* and/or *Dawndraco*, may consist of two species, the other being *P.* (= *Dawndraco*) *kanzai*. Female skull and skeleton illustrated herein represent the most complete known *Pteranodon* specimen. May be smallest known *Pteranodon* species. May be direct ancestor of *P. longiceps*.

Pteranodon longiceps (Bennett 2001)
6.5 m (21 ft) WS, 2.6 m (8.5 ft) TL, 50 kg (110 lb)

FOSSIL REMAINS Dozens of specimens of varying completeness, adult and large juvenile.

ANATOMICAL CHARACTERISTICS Beak nearly straight, length moderate relative to main body of skull and skeleton, lower jaw only a little shorter than upper, very variable shallow, sword-like head crest strongly swept and slightly arced backward, sometimes with gentle S curve at end.

AGE Late Cretaceous, middle Santonian to earliest
Campanian.
DISTRIBUTION AND FORMATIONS Kansas; upper
Niobrara.
HABITAT Shallow interior seaway not as broad or deep.
NOTES The iconic pterosaur. Complete adult male
skeletal is a composite of the largest known skull and
partial skeleton. Habitat may have overlapped at least
somewhat with that of more-nearshore *Nyctosaurus
gracilis.*

Pteranodon maysei
7.0 m (23 ft) WS, 60 kg (130 lb)

FOSSIL REMAINS A few specimens of varying
completeness.
ANATOMICAL CHARACTERISTICS Insufficient
information.
AGE Late Cretaceous, early Campanian.
DISTRIBUTION AND FORMATIONS Kansas; lower Pierre
Shale.
HABITAT Shallow interior seaway less broad and deep.
NOTES No specimen with a mature male crest known.
Apparently not distinct enough to be distinct genus
Geosternbergia, the type species of which appears
much earlier in any case, may be *P. longiceps* or direct
descendant. May be largest known *Pteranodon* species.

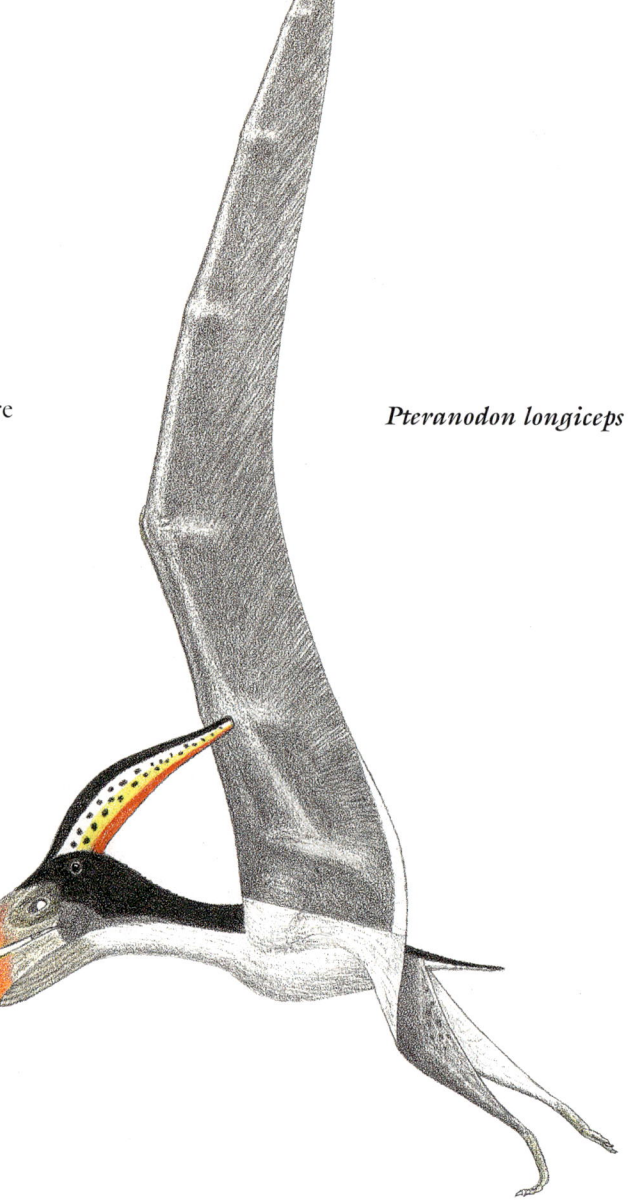

Pteranodon longiceps

Tethydraco regalis (Longrich et al. 2018)
5.0 m (15 ft) WS, 15 kg (30 lb)

FOSSIL REMAINS Minority of several specimens, poorly
preserved.
ANATOMICAL CHARACTERISTICS Insufficient
information.

AGE Late Cretaceous, late Maastrichtian.
DISTRIBUTION AND FORMATIONS Morocco; unnamed.
HABITAT Coastal marine.
NOTES Has also been placed among azhdarchids. Shared
its habitat with *Alcione, Phosphatodraco.*

Pteranodon longiceps

NYCTOSAURIDS

MEDIUM-SIZED TO LARGE PTERANODONTS OF THE LATE CRETACEOUS, WESTERN HEMISPHERE

ANATOMICAL CHARACTERISTICS Action of shoulder joints limited, some wing tendons ossified, pteroids stout, three inner fingers absent and their metacarpals reduced, fourth wing fingertip bones reduced or absent.

HABITS Long and unusually stiff wings indicate exceptionally well-developed soaring performance, but large pectoral crest and sternum indicate strong power stroke. Probably more nearshore than pteranodontids. Because of extreme inner arm/leg length ratio, were exceptionally awkward on land.

NOTES All nyctosaurs may have had just three wing finger bones, an extreme streamlining feature otherwise found only in *Anurognathus*. Absence from Eastern Hemisphere may reflect lack of sufficient sampling.

Muzquizopteryx coahuilensis (Frey et al. 2006)
1.7 m (5.5 ft) WS, 0.6 m (2 ft) TL, 1.4 kg (3 lb)

FOSSIL REMAINS Partial skull and skeleton, partial wing.
ANATOMICAL CHARACTERISTICS Bony head crest may
be small. Wingspan long relative to mass. Feet very small.
AGE Late Cretaceous, early Coniacian.
DISTRIBUTION AND FORMATIONS Northern Mexico;
Austin Group.
HABITAT Continental shelf.
NOTES Possibility that one known skull is not a mature
male precludes full determination of form of crest.

Nyctosaurus gracilis (Bennett 2003)
2.1 m (6.9 ft) WS, 0.63 m (2.1 ft) TL, 0.9 kg (2 lb)

FOSSIL REMAINS Several skulls and skeletons.
ANATOMICAL CHARACTERISTICS Jaw tip very slender
and needle-like, bony crest of at least mature males
consists of extremely long, very thin struts forming
a backward-directed Y. Wing finger reduced to three
elements, wingspan and inner arm/leg length longest
relative to mass and leg length among known pterosaurs,
even though leg including foot is not exceptionally short.
Males not larger than females.
AGE Late Cretaceous, late Santonian.
DISTRIBUTION AND FORMATIONS Kansas; upper
Niobrara.
HABITAT Shallow interior seaway.
HABITS AND GROWTH Most extremely adapted and
highest-performing known glider among pterosaurs and
perhaps flying vertebrates. Similar size of sexes indicates
breeding males not strongly competitive with each other,

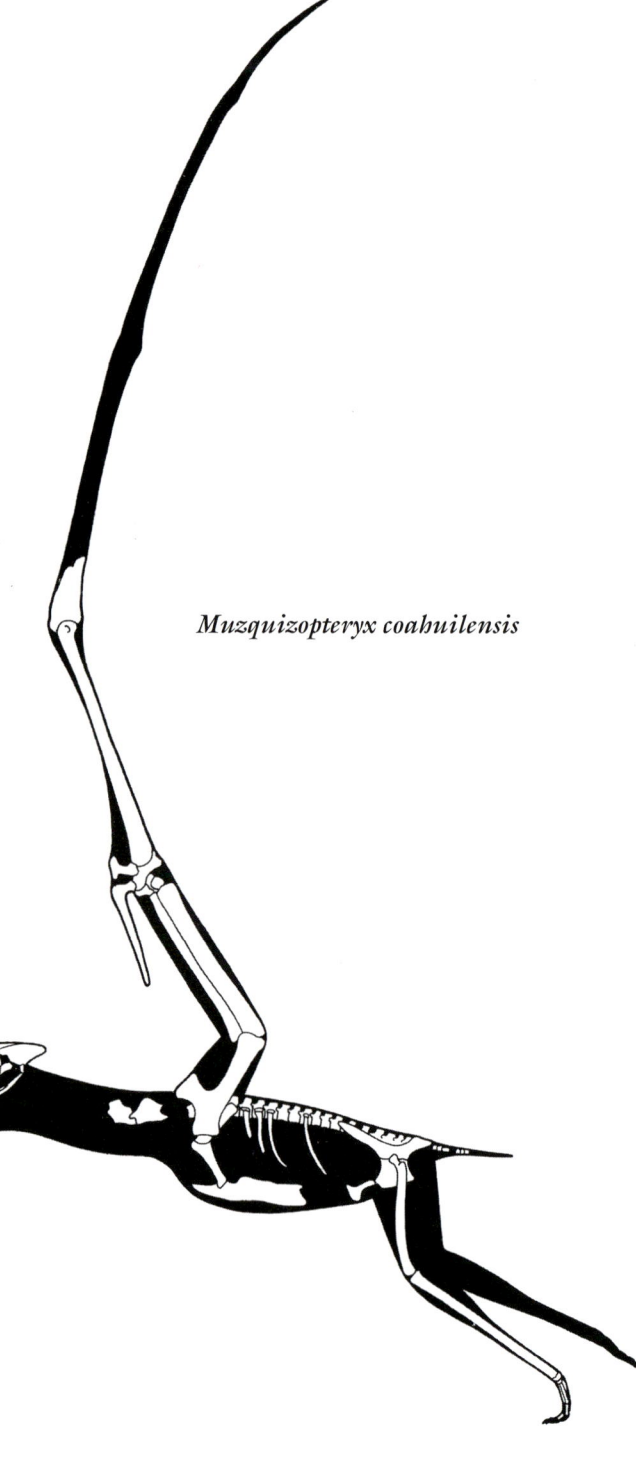

Muzquizopteryx coahuilensis

crests may have been used to impress females. Juveniles
apparently grew rapidly to 1.5 m (5 ft) wingspan and
0.3 kg (0.7 lb) over some months before becoming
oceanic, then continued to mature to adult size over a few
years before returning to isolated locations to reproduce.
NOTES The area of male crest being greatly expanded by
soft tissues cannot be ruled out, shape of female crests
not known. Complete skeletal is a composite combining
best male skull and most complete skeleton. Lightest
known pterosaur relative to wingspan. Apparently the
genus existed for only a short period, although absence
from middle Niobrara may be because those deposits
were laid in a broader, deeper seaway less suitable for this
more-nearshore pterosaur, which may not have strongly
overlapped habitats with more open-sea *Pteranodon
longiceps*. Although much rarer than latter, specimens of
this species tend to be more complete, perhaps because of
their smaller size.

Alcione elainus (Longrich et al. 2018)
4 m (13 ft) WS, 4 kg (9 lb)

FOSSIL REMAINS Minority of several skeletons, juvenile to adult.

ANATOMICAL CHARACTERISTICS Wings not exceptionally long.

AGE Late Cretaceous, late Maastrichtian.

DISTRIBUTION AND FORMATIONS Morocco; unnamed.

HABITAT Coastal marine.

HABITS Least extreme adaptations for soaring of known nyctosaurs.

NOTES *Simurghia robusta* and *Barbaridactylus grandis* may be more mature members of this species, or distinct taxa. Shared its habitat with *Tethydraco*, *Phosphatodraco*.

Nyctosaurus gracilis

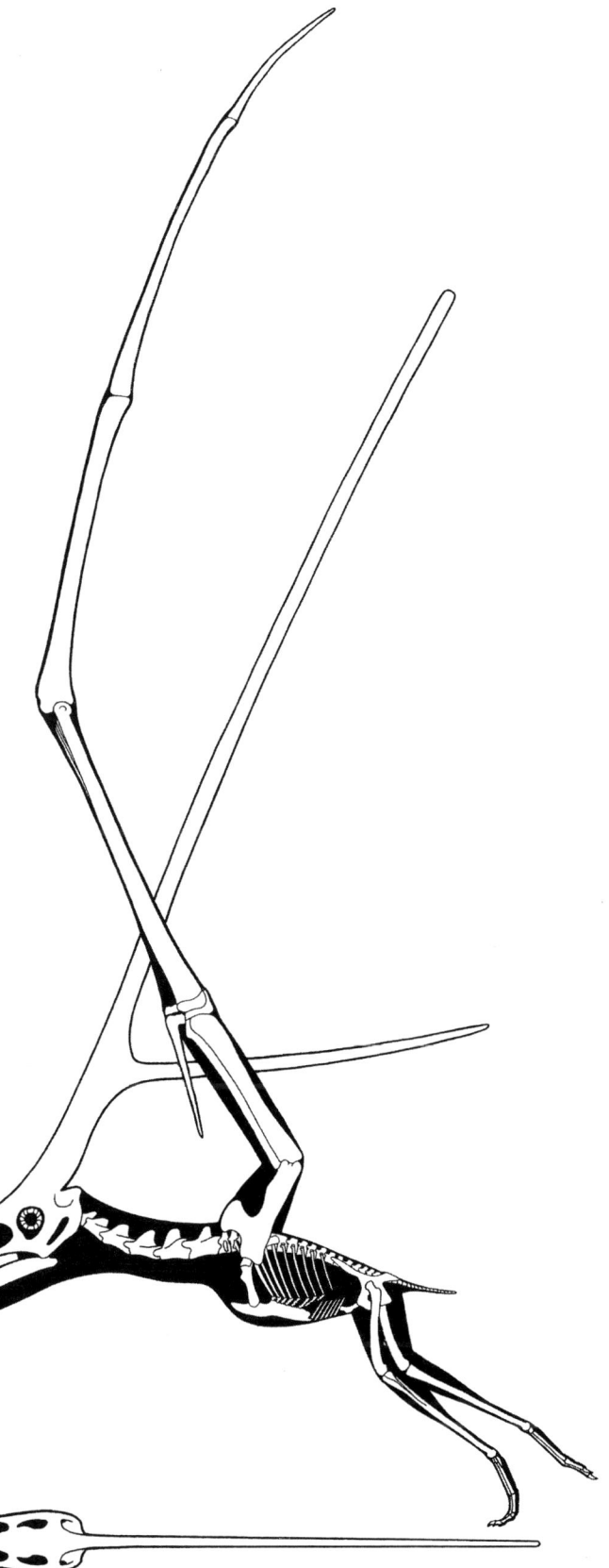

Nyctosaurus gracilis

REFERENCES

Aires, A. S., et al. 2020. Development and evolution of the notarium in Pterosauria. *Journal of Anatomy* 238:400–415.

Andres, B., J. M. Clark, and X. Xu. 2010. A new rhamphorhynchoid pterosaur from the Upper Jurassic of Xinjiang, and the phylogenetic relationships of basal pterosaurs. *Journal of Vertebrate Paleontology* 30:163–87.

———. 2014. The earliest pterodactyloid and the origin of the group. *Current Biology* 24:1011–1016.

Andres, B., and Q. Ji. 2008. A new pterosaur from the Liaoning Province of China, the phylogeny of the Pterodactyloidea, and convergence in their cervical vertebrae. *Palaeontology* 51:453–69.

Andres, B., and J. Qiang. 2006. A new species of *Istiodactylus* (Pterosauria, Pterodactyloidea) from the Lower Cretaceous of Liaoning, China. *Journal of Vertebrate Paleontology* 26:70–78.

Averianov, A. O., et al. 2008. A new Late Cretaceous azhdarchid from the Volga region. *Paleontological Journal* 42:634–42.

Averianov, A., and B. Ekrt. 2015. *Cretornis hlavaci* Frič, 1881 from the Upper Cretaceous of Czech Republic. *Cretaceous Research* 55:164–75.

Baron, M. G. 2020. Testing pterosaur ingroup relationships through broader sampling of avemetatarsalian taxa and characters and a range of phylogenetic analysis techniques. *PeerJ* 8:e9604.

Bennett, S. C. 1997. Terrestrial locomotion of pterosaurs: A reconstruction based on *Pteraichnus* trackways. *Journal of Vertebrate Paleontology* 17:104–13.

———. 2000. Pterosaur flight: The role of actinofibrils in wing function. *Historical Biology* 14:255–84.

———. 2001. The osteology and functional morphology of the Late Cretaceous pterosaur *Pteranodon*. *Palaeontographica A* 260:1–153.

———. 2003. New crested specimens of the Late Cretaceous pterosaur *Nyctosaurus*. *Paläontologische Zeitschrift* 77:61–75.

———. 2006. Juvenile specimens of the pterosaur *Germanodactylus cristatus*, with a revision of the genus. *Journal of Vertebrate Paleontology* 26:872–78.

———. 2007a. A review of the pterosaur *Ctenochasma*: Taxonomy and ontogeny. *Neues Jahrbuch für Geologie und Paläontologie–Abhandlungen* 245:23–31.

———. 2007b. A second specimen of the pterosaur *Anurognathus ammoni*. *Paläontologische Zeitschrift* 81:376–98.

———. 2013. The morphology and taxonomy of the pterosaur *Cycnorhamphus*. *Neues Jahrbuch für Geologie und Paläontologie–Abhandlungen* 267:23–41.

———. 2017. New smallest specimen of the pterosaur *Pteranodon* and ontogenetic niches in pterosaurs. *Journal of Paleontology* 92:254–71.

———. 2020. Reassessment of the Triassic archosauriform *Scleromochlus taylori*: Neither runner nor biped, but hopper. *PeerJ* 8:e8418.

Bestwick. J., et al. 2020. Dietary diversity and evolution of the earliest flying vertebrates revealed by dental microwear texture analysis. *Nature Communications* 11:5293.

Britt, B. B., et al. 2018. *Caelestiventus hanseni* gen. et sp. nov. extends the desert-dwelling pterosaur record back 65 million years. *Nature Ecology & Evolution* 2:1386–92.

Brusatte, S. L., et al. 2014-15. The extinction of the dinosaurs. *Biological Reviews* 90:628–42.

Cai, Z., and F. Wei. 1994. On a new pterosaur (*Zhejiangopterus linhaiensis* gen. et sp. nov.) from Upper Cretaceous in Linhai, Zhejiang, China. *Vertebrata PalAsiatica* 32:181–94.

Campos, H. B. N. 2021. A new azhdarchoid pterosaur from the Late Cretaceous Javelina Formation of Texas. *Biologia*.

Carpenter, K., et al. 2003. A new scaphognathine pterosaur from the Upper Jurassic Formation of Wyoming, USA. *Geological Society of London, Special Publications* 217:45–54.

Chen, H., et al. 2020. New anatomical information on *Dsungaripterus weii* Young, 1964 with focus on the palatal region. *PeerJ* 8:e8741.

Cheng, X., et al. 2012. A new scaphognathid pterosaur from western Liaoning, China. *Historical Biology* 24:101–11.

Czerkas, S. A., and Q. Ji. 2002. A new rhamphorhynchoid with a headcrest and complex integumentary structures. In *Feathered Dinosaurs and the Origin of Flight*, edited by S. J. Czerkas, 15–41. Blanding, UT: The Dinosaur Museum.

Dalla Vecchia, F. M. 1998. New observations on the osteology and taxonomic status of *Preondactylus buffarinii* Wild, 1984. *Bollettino della Società Paleontologica Italiana* 36:355–66.

———. 2019. *Seazzadactylus venieri* gen. et sp. nov., a new pterosaur from the Upper Triassic (Norian) of northeastern Italy. *PeerJ* 7:e7363.

Dalla Vecchia, F. M., R. Wild, H. Hopf, and J. Reitner. 2002. A crested rhamphorhynchid pterosaur from the Late Triassic of Austria. *Journal of Vertebrate Paleontology* 22:196–99.

Erickson, G. M., et al. 2017. Dinosaur incubation periods directly determined from growth-line counts in embryonic teeth show reptilian-grade development. *PNAS* 114:540–45.

Frey, E., and D. M. Martill. 1994. A new pterosaur from the Crato Formation of Brazil. *Neues Jahrbuch für Geologie und Paläontologie, Abhandlungen* 194:379–412.

Frey, E., et al. 2006. *Muzquizopteryx coahuilensis* n.g., n. sp., a nyctosaurid pterosaur with soft tissue preservation from the Coniacian (Late Cretaceous) of northeastern Mexico (Coahuila). *Oryctos* 6:19–39.

Fröbisch, N. B., and J. Fröbisch. 2006. A new basal pterosaur genus from the Upper Triassic of the Northern Calcareous Alps of Switzerland. *Palaeontology* 49:1081–90.

Gasparini, Z., M. Fernández, and M. Fuente. 2004. A new pterosaur from the Jurassic of Cuba. *Palaeontology* 47:919–27.

Goto, Y., et al. 2020. Soaring styles of extinct giant birds and pterosaurs. bioRxiv.org.

Habib, M. B. 2008. Comparative evidence for quadrupedal launch in pterosaurs. *Zitteliana* B28:161–68.

———. 2010. Maximum range and soaring efficiency of azhdarchid pterosaurs. *Journal of Paleontology* 30:99A–100A.

Harris, J. D., and K. Carpenter. 1996. A large pterodactyloid from the Morrison Formation (Late Jurassic) of Garden Park, Colorado. *Neues Jahrbuch für Geologie und Paläontologie–Monatshefte* 1996:473–84.

Hoffmann, R., et al. 2020. Pterosaurs ate soft-bodied cephalopods. *Scientific Reports* 10:1230.

Holgado, B., et al. 2019. On a new crested pterodactyloid from the Early Cretaceous of the Iberian Peninsula and the radiation of the clade Anhangueria. *Scientific Reports* 9:4940.

REFERENCES

Hone, D.W.E. 2020. A review of the taxonomy and palaeoecology of the Anurognathidae. *Acta Geologica Sinica* 94:1676–92.

Hone, D., M. Habib, and F. Therrien. 2019. *Cryodrakon boreas*, gen. et sp. nov., a Late Cretaceous Canadian azhdarchid pterosaur. *Journal of Vertebrate Paleontology* 39:e1649681.

Hone, D.W.E., M. K. Rooijen, and M. B. Habib. 2015. The wingtips of the pterosaurs: Anatomy, aeronautical function and ecological implications. *Palaeogeography, Palaeoclimatology, Palaeoecology* 440:431–39.

Hone, D.W.E., et al. 2012. A new non-pterodactyloid pterosaur from the Late Jurassic of southern Germany. *PLoS ONE* 7:e39312.

Hone, D.W.E., et al. 2020a. Unique near isometric ontogeny in the pterosaur *Rhamphorhynchus* suggests hatchlings could fly. *Lethaia* 54:106–12.

Hone, D.W.E., et al. 2020b. An unusual new genus of istiodactylid pterosaur from China based on a near complete specimen. *Palaeontologia Electronica* 23:a09.

Howse, S.C.B., A. R. Milner, and D. M. Martill. 2001. Pterosaurs. In *Dinosaurs of the Isle of Wight*, edited by D. M. Martill and D. Naish, 324–35. Guide 10, Field Guides to Fossils. London: The Palaeontological Association.

Hwang, K.-G., et al. 2002. New pterosaur tracks from the Late Cretaceous Uhangri Formation, southwestern Korea. *Geological Magazine* 139:421–35.

Ibrahim, N., et al. 2010. A new pterosaur from the Upper Cretaceous of Morocco. *PLoS ONE* 5:e10875.

Jenkins, F. A., Jr., N. H. Shubin, S. M. Gatesy, and K. Padian. 2001. A diminutive pterosaur from the Greenlandic Triassic. *Bulletin of the Museum of Comparative Zoology* 156:151–70.

Ji, S., and Q. Ji. 1997. Discovery of a new pterosaur in western Liaoning, China. *Acta Geologica Sinica* 71:115–21.

Jiang, S., and X. Wang. 2011. A new ctenochasmatid pterosaur from the Lower Cretaceous, western Liaoning, China. *Anais da Academia Brasileira de Ciências* 83:1243–49.

Jiang, S., et al. 2016. A new archaeopterodactyloid pterosaur from the Jiufotang Formation of western Liaoning, China, with a comparison of sterna in Pterodactylomorpha. *Journal of Vertebrate Paleontology* 36:e1212058.

Jiang, S., et al. 2020. The first pterosaur basihyal, shedding light on the evolution and function of pterosaur hyoid apparatuses. *PeerJ* 8:e8292.

Johansson, L. C., et al. 2018. Mechanical power curve measured in the wake of pied flycatchers indicates modulation of parasite power across flight speeds. *Journal of the Royal Society Interface* 15:20170814.

Kellner, A.W.A. 2013. A new unusual tapejarid from the Early Cretaceous Romualdo Formation, Araripe Basin, Brazil. *Earth and Environmental Science Transactions of the Royal Society of Edinburgh* 103:409–21.

———. 2015. Comments on Triassic pterosaurs with discussion about ontogeny and description of new taxa. *Anais da Academia Brasileira de Ciências* 87:669–89.

Kellner, A.W.A., et al. 2019a. First complete pterosaur from the Afro-Arabian continent: Insight into pterodactyloid diversity. *Scientific Reports* 9:17875.

Kellner, A.W.A., et al. 2019b. A new toothless pterosaur from southern Brazil with insights into the paleoecology of a Cretaceous desert. *Anais da Academia Brasileira de Ciências* 91:e20190768.

Kim, J. Y., et al. 2012. Enigmatic giant pterosaur tracks and associated ichnofauna from the Cretaceous of Korea: Implication for the bipedal locomotion of pterosaurs. *Ichnos* 19:50–65.

Larramendi, A., G. S. Paul, and S. Hu. 2021. A review and reappraisal of the specific gravities of present and part multicellular organisms, with an emphasis on tetrapods. *Anatomical Record.* https://doi.org/10.1002/ar.24574.

Larson, D. W., C. M. Brown, and D. C. Evans. 2016. Dental disparity and ecological stability in bird-like dinosaurs prior to the end Cretaceous mass extinction. *Current Biology* 26:1325–33.

Li, Y., Wang, X., and Jiang, S. 2021. A new pterosaur tracksite from the Lower Cretaceous of Wuerho, Junggar Basin, China: Inferring the first putative pterosaur trackmaker. *PeerJ* 9:e11361

Li, Z., Z. Zhou, and J. A. Clarke. 2018. Convergent evolution of a mobile bony tongue in flighted dinosaurs and pterosaurs. *PLoS ONE* 13:e0198078.

Lockley, M. G., and J. Wright. 2003. Pterosaur swim tracks and other ichnological evidence of behavior and ecology. *Geological Society, London, Special Publications* 217:297–313.

Longrich, N. R., D. M. Martill, and B. Andres. 2018. Late Maastrichtian pterosaurs from North Africa and mass extinctions of Pterosauria at the Cretaceous-Paleogene. *PLoS Biology* 16:e2001663.

Lü, J. 2003. A new pterosaur: *Beipiaopterus chenianus*, gen. et sp. nov. from western Liaoning Province, China. *Memoir of the Fukui Prefectural Dinosaur Museum* 2:153–60.

———. 2009. A new non-pterodactyloid pterosaur from Qinglong County, Hebei Province of China. *Acta Geologica Sinica* 83:189–99.

Lü, J., X. Fucha, and J. Chen. 2010. A new scaphognathine pterosaur from the Middle Jurassic of western Liaoning, China. *Acta Geoscientica Sinica* 31:263–66.

Lü, J., and Q. Ji. 2005. A new ornithocheirid from the Early Cretaceous of Liaoning Province, China. *Acta Geologica Sinica* 79:157–63.

Lü, J., M. Kundrát, and C. Shen. 2016. New material of the pterosaur *Gladocephaloideus* Lü et al., 2012 from the Early Cretaceous of Liaoning Province, China, with comments on its systematic position. *PLoS ONE* 11:e0154888.

Lü, J., and B. K. Zhang. 2005. New pterodactyloid pterosaur from the Yixian Formation of western Liaoning. *Geological Review* 51:458–62.

Lü, J., et al. 2006. New pterodactyloid pterosaur from the Lower Cretaceous Yixian Formation of western Liaoning. In *Papers from the 2005 Heyuan International Dinosaur Symposium*, 195–203. Beijing: Geological Publishing House.

Lü, J., et al. 2008. A new azhdarchoid pterosaur from the Lower Cretaceous of China and its implications for pterosaur phylogeny and evolution. *Naturwissenschaften* 95:891–97.

Lü, J., et al. 2012. Largest toothed pterosaur skull from the Early Cretaceous Yixian Formation of western Liaoning, China, with comments on the family Boreopteridae. *Acta Geologica Sinica* 86:287–93.

Lü, J., et al. 2015. A new rhamphorhynchid pterosaur from Jurassic deposits of Liaoning Province, China. *Zootaxa* 3911:119–29.

Lü, J., et al. 2017. Short note on a new anurognathid pterosaur with evidence of perching behavior from Jianchang of Liaoning Province, China. *Geological Society, London, Special Publications* 455:95–104.

Manafzadeh, A. R., and K. Padian. 2018. ROM mapping of ligamentous constraints on avian hip mobility: Implications for extinct ornithodirans. *Proceedings of the Royal Society B* 285:20180727.

Manzig, P. C., et al. 2014. Discovery of a rare pterosaur bone bed in a Cretaceous desert with insights on ontogeny and behavior of flying reptiles. *PLoS ONE* 9:e100005.

Martill, D. M., and S. Etches. 2013. A new monofenestratan pterosaur from the Kimmeridge Clay Formation of Dorset, England. *Acta Palaeontologica Polonica* 58:285–94.

Martill, D. M., et al. 1998. Discovery of the holotype of the giant pterosaur *Titanopteryx philadelphiae* Arambourg 1959, and the status of *Arambourgiania* and *Quetzalcoatlus*. *Neues Jahrbuch für Geologie und Paläontologie, Abhandlungen* 207:57–76.

Martill, D. M., et al. 2020. Evidence for tactile feeding in pterosaurs: A sensitive tip to the beak of *Lonchodraco giganteus* from the Upper Cretaceous of southern England. *Cretaceous Research* 117:104637.

Martín, D. Ezcurra, et al. 2020. Enigmatic dinosaur precursors bridge the gap to the origin of Pterosauria. *Nature* 588:445–49.

Martin-Silverstone, E., et al. 2016. A small azhdarchoid pterosaur from the latest Cretaceous, the age of flying giants. *Royal Society Open Science* 3:160333.

Martin-Silverstone, E., et al. 2017. Reassessment of *Dawndraco kanzai* Kellner, 2010 and reassignment of the type specimen to *Pteranodon sternbergi* Harksen, 1966. *Vertebrate Anatomy Morphology Palaeontology* 3:47–59.

Mazin, J.-M., J.-P. Billon-Bruyat, and K. Padian. 2009. First record of a pterosaur landing trackway. *Proceedings of the Royal Society B* 276:3881–86.

Mazin, J.-M., and J. Pouech. 2020. The first non-pterodactyloid pterosaurian trackways and the terrestrial ability of non-pterodactyloid pterosaurs. *Geobios* 58:39–53.

McGowen, M. R., et al. 2002. Description of *Montanazhdarcho minor*, an azhdarchid pterosaur from the Two Medicine Formation of Montana. *PaleoBios* 22:1–9.

Naish, D., and M. P. Witton. 2017. Neck biomechanics indicate that giant Transylvanian azhdarchid pterosaurs were short-necked arch predators. *PeerJ* 5:e2908.

Novas, F. E., et al. 2012. A new large pterosaur from the Late Cretaceous of Patagonia. *Journal of Vertebrate Paleontology* 32:1447–52.

O'Sullivan, M., and D. M. Martill. 2017. The taxonomy and systematics of *Parapsicephalus purdoni* from the Lower Jurassic Whitby Mudstone Formation, Whitby, U.K. *Historical Biology* 29:1009–18.

Padian, K. 2008. The Early Jurassic pterosaur *Dorygnathus banthensis* (Theodori, 1830) and the Early Jurassic pterosaur *Campylognathoides* (Strand, 1928). *Special Papers in Palaeontology* 80:1–107.

Palmer, C., and G. Dyke. 2012. Constraints on the wing morphology of pterosaurs. *Proceedings of the Royal Society B* 279:1218–24.

Paul, G. S. 1991. The many myths, some old, some new, of dinosaurology. *Modern Geology* 16:69–99.

———. 2002. *Dinosaurs of the Air*. Baltimore: Johns Hopkins University Press.

———. 2012. Evidence for avian-mammalian aerobic capacity and thermoregulation in Mesozoic dinosaurs. In *The Complete Dinosaur*, vol. 2, edited by M. Brett-Surman, J. Farlow, and T. Holtz, 819–71. Bloomington: Indiana University Press.

———. 2016. *The Princeton Field Guide to Dinosaurs*. 2nd ed. Princeton, NJ: Princeton University Press.

———. 2017a. Polar and K/Pg nonavian dinosaurs were low-metabolic rate reptiles vulnerable to cold-induced extinction, rather than more survivable tachyenergetic bird relatives: Comment on an obsolete hypothesis. *International Journal of Earth Sciences* 106:2991–98.

———. 2017b. Restoring maximum vertical browsing reach in sauropod dinosaurs. *The Anatomical Record* 300:1802–25.

Pêgas, R. V., F. R. Costa, and A.W.A. Kellner. 2021. Reconstruction of the adductor chamber and predicted bite force in pterodactyloids (Pterosauria). *Zoological Journal of the Linnean Society* 20:1–34. https://doi.org/10.1093/zoolinnean/zlaa163.

Pentland, A. H., and S. F. Poropat. 2018. Reappraisal of *Mythunga camara* Molnar & Thulborn, 2007 from the upper Albian Toolebuc Formation of Queensland, Australia. *Cretaceous Research* 93:151–69.

Pentland, A. H., et al. 2019. *Ferrodraco lentoni* gen. et sp. nov., a new ornithocheirid pterosaur from the Winton Formation of Queensland, Australia. *Scientific Reports* 9:13454.

Perry, S., et al. 2009. Implications of an avian-style respiratory system for gigantism in sauropod dinosaurs. *Journal of Experimental Biology* 311A:600–610.

Peters, D. 2009. A reinterpretation of pteroid articulation in pterosaurs. *Journal of Vertebrate Paleontology* 29:1327–30.

Pinheiro, F. L., and T. Rodrigues. 2017. *Anhanguera* taxonomy revisited: Is our understanding of Santana Group pterosaur diversity biased by poor biological and stratigraphic control? *PeerJ* 5:e3285.

Pinheiro, F. L., et al. 2011. New information on *Tupandactylus imperator*, with comments on the relationships of Tapejaridae. *Acta Palaeontologica Polonica* 56:567–80.

Prondvai, E., E. R. Bodor, and A. Ősi. 2014. Does morphology reflect osteohistology-based ontogeny? A case study of Late Cretaceous pterosaur jaw symphyses from Hungary reveals hidden taxonomic diversity. *Paleobiology* 40:288–321.

Qvarnström, M., et al. 2019. Filter feeding in Late Jurassic pterosaurs supported by coprolite contents. *PeerJ* 7:e7375.

Rampino, M. R. 2015. Disc matter in the galaxy and potential cycles of extraterrestrial impacts, mass extinctions and geological events. *Monthly Notices of the Royal Astronomical Society* 448:1816–20.

Rodrigues, T., and A. Kellner. 2013. Taxonomic review of the *Ornithocheirus* complex from the Cretaceous of England. *ZooKeys* 308:1–112.

Rogalla, S., et al. 2019. Hot wings: Thermal impacts of wing coloration on surface temperature during bird flight. *Journal of the Royal Society Interface* 16:20190032.

Schmitz, L., and R. Motani. 2011. Nocturnality in dinosaurs inferred from scleral ring and orbit morphology. *Science* 332:705–8.

Schoene, B., et al. 2019. U-Pb constraints on pulsed eruption of the Deccan Traps across the end-Cretaceous mass extinction. *Science* 363:862–66.

Smith, R. E., et al. 2020. Edentulous pterosaurs from the Cambridge Greensand of eastern England with a review of *Ornithostoma* Seeley, 1871. *Proceedings of the Geologists' Association* 132:110–26.

Stokes, I. A., and A. J. Lucas. 2021. Wave-slope soaring of the brown pelican. *Movement Ecology* 9:13.

Unwin, D. M. 1997. Pterosaur tracks and the terrestrial ability of pterosaurs. *Lethaia* 29:373–86.

REFERENCES

Unwin, D. M., and D. C. Deeming. 2019. Prenatal development in pterosaurs and its implications for their postnatal locomotory ability. *Proceedings of the Royal Society B* 286:20190409.

Venditti, C., et al. 2020. 150 million years of sustained increase in pterosaur flight efficiency. *Nature* 587:83–86.

Vremir, M.T.S., et al. 2013. A new azhdarchid pterosaur from the Late Cretaceous of the Transylvanian Basin, Romania: Implications for azhdarchid diversity and distribution. *PLoS ONE* 8:e54268.

Vullo, R., et al. 2012. A new crested pterosaur from the Early Cretaceous of Spain: The first European tapejarid. *PLoS ONE* 7:e38900.

Wang, X., and J. Lü. 2001. Discovery of a pterodactyloid pterosaur from the Yixian Formation of western Liaoning, China. *Chinese Science Bulletin* 45:447–54.

Wang, X., and Z. Zhou. 2003a. A new pterosaur (Pterodactyloidea, Tapejaridae) from the Early Cretaceous Jiufotang Formation of western Liaoning, China and its implications for biostratigraphy. *Chinese Science Bulletin* 48:16–23.

——. 2003b. Two new pterodactyloid pterosaurs from the Early Cretaceous Jiufotang Formation of western Liaoning, China. *Vertebrata PalAsiatica* 41:34–41.

Wang, X., et al. 2005. Pterosaur diversity and faunal turnover in Cretaceous terrestrial ecosystems in China. *Nature* 437:875–79.

Wang, X., et al. 2007. A new pterosaur from the Lower Cretaceous Yixian Formation of China. *Cretaceous Research* 28:2245–60.

Wang, X., et al. 2008. A primitive istiodactylid pterosaur from the Jiufotang Formation, northeastern China. *Zootaxa* 1813:1–18.

Wang, X., et al. 2009. An unusual long-tailed pterosaur with elongated neck from western Liaoning of China. *Anais da Academia Brasileira de Ciências* 81:793–812.

Wang, X., et al. 2012. New toothed flying reptile from Asia: Close similarities between early Cretaceous pterosaur faunas from China and Brazil. *Naturwissenschaften* 94:249–57.

Wang, X., et al. 2014. An Early Cretaceous pterosaur with an unusual mandibular crest from China and a potential novel feeding strategy. *Scientific Reports* 4:6329.

Wang, X., et al. 2015. Eggshell and histology provide insight on the life history of a pterosaur with two functional ovaries. *Anais da Academia Brasileira de Ciências* 87:1599–1609.

Wang, X., et al. 2017. New evidence from China for the nature of the pterosaur evolutionary transition. *Scientific Reports* 7:42763.

Wei, X. et al. 2021. *Sinomacrops bondei*, a new anurognathid pterosaur from the Jurassic of China and comments on the group. *PeerJ* 9:e11161.

Witton, M. P. 2013. *Pterosaurs*. Princeton, NJ: Princeton University Press.

Yang, Z., et al. 2018. Pterosaur integumentary structures with complex feather-like branching. *Nature Ecology & Evolution* 3:24–30.

Yang, Z., et al. 2020. Reply to: No protofeathers on pterosaurs. *Nature Ecology & Evolution* 4:1592–1593.

Zhou, C.-F., et al. 2017. Earliest filter-feeding pterosaur from the Jurassic of China and ecological evolution of Pterodactyloidea. *Royal Society Open Science* 4:160672.

Zhou, X., et al. 2019. *Nurhachius luei*, a new istiodactylid pterosaur from the Early Cretaceous Jiufotang Formation of Chaoyang City, Liaoning Province (China) and comments on the Istiodactylidae. *PeerJ* 7:e7688.

Zhou, X. et al. 2021. A new darwinopteran pterosaur reveals arborealism and an opposed thumb. *Current Biology*. doi:10.1016.j.cub.2021.03.030.

INDEX

This index covers pterosaur groups, genera and species, as well as pterosaur-bearing formation described in the main directory starting on page 97.

PTEROSAUR TAXA

Formations

When a formation is cited more than once on a page, the number of times is indicated in parentheses.